Metaverse, Generative AI, and Brain–Computer Interfaces

This compelling exploration delves into the heart of cutting-edge technologies, specifically the metaverse, brain–computer interfaces (BCIs), and generative artificial intelligence (AI). The authors unravel their transformative applications in shaping the future and contemplate how the next generation of virtual environments will manifest, as well as how the digital landscape can be affected.

This book encompasses various examples of the metaverse, BCIs, and generative AI across diverse fields, including education, engineering, and medical science. In the dynamic realm of the metaverse, where virtual realities converge, this book showcases how these innovations can transform the way we learn, interact, and build. Through BCIs, users can directly interface with technology using neural signals, opening up new avenues for communication and control. Additionally, generative AI algorithms empower us to create immersive virtual environments and generate content with unprecedented creativity.

Whether you're a student, educator, or industry professional, this book is your ticket to understanding and harnessing the power of the future. It provides a pathway for those who believe in the limitless possibilities that the metaverse, BCIs, and generative AI can bring to our daily experiences.

Ali Darejeh is a faculty member and researcher at UNSW specializing in AR/VR, brain–computer interfaces, serious games, and educational technology. His work integrates emerging technologies with educational psychology. He also brings extensive industry experience as a UX researcher and gamification designer across healthcare, oil and gas, and finance sectors.

Samad Sepasgozar is a global leader, ranked among the top 5 in Digital Twin and recognized as a world leader in Architecture (Computer Science and Engineering), with numerous scholarly publications. A leading educator, he pioneers innovative teaching tools and sustainable solutions, engaging more than 50,000 global attendees with transformative digital technology insights.

Advanced Digital Technologies for the Built Environment
Series Editor: Samad Sepasgozar

Digital Twin Adoption and BIM-GIS Implementation
Edited by Samad Sepasgozar and Sara Shirowzhan

Metaverse, Generative AI, and Brain–Computer Interfaces
Shaping the Future of the Digital Landscape
Edited by Ali Darejeh and Samad Sepasgozar

Metaverse, Generative AI, and Brain–Computer Interfaces

Shaping the Future of the Digital Landscape

Edited by Ali Darejeh and Samad Sepasgozar

Routledge
Taylor & Francis Group
LONDON AND NEW YORK

Designed cover image: Shutterstock

First published 2026
by Routledge
4 Park Square, Milton Park, Abingdon, Oxon OX14 4RN

and by Routledge
605 Third Avenue, New York, NY 10158

Routledge is an imprint of the Taylor & Francis Group, an informa business

British Library Cataloguing-in-Publication Data
A catalogue record for this book is available from the British Library

ISBN: 9781032896786 (hbk)
ISBN: 9781032903538 (pbk)
ISBN: 9781003557555 (ebk)

DOI: 10.1201/9781003557555

Typeset in Times New Roman
by codeMantra

Contents

Contributors

Angel Ayala, Universidade de Pernambuco, Brazil

Mahdi Bamdad, University of New South Wales, Sydney, Australia

Nemat Bhullar, University of New South Wales, Sydney, Australia

Guy Peter Chilcott, University of New South Wales, Sydney, Australia

Francisco Cruz, University of New South Wales, Sydney, Australia

Ali Darejeh, University of New South Wales, Sydney, Australia

Rovielyn Espiritu, University of New South Wales, Sydney, Australia

Arifah Anwar Fadzil, University of New South Wales, Sydney, Australia

Rachid Hamadi, University of New South Wales, Sydney, Australia

Lauren Huynh, University of New South Wales, Sydney, Australia

Xiaohan Ji, University of New South Wales, Sydney, Australia

Eliza Lee, University of New South Wales, Sydney, Australia

Jihyun Lee, University of New South Wales, Sydney, Australia

Wendy Lee, University of New South Wales, Sydney, Australia

David Liusun, University of New South Wales, Sydney, Australia

Stephen Lukins, University of New South Wales, Sydney, Australia

Sara Mashayekh, University of New South Wales, Sydney, Australia

Arkapravo Nandi, University of New South Wales, Sydney, Australia

Dylan Ngo, University of New South Wales, Sydney, Australia

Ebrahim Oromiehie, University of New South Wales, Sydney, Australia

Colby Rehn, University of New South Wales, Sydney, Australia

Rubin Roy, University of New South Wales, Sydney, Australia

Neelam Samachar, University of New South Wales, Sydney, Australia

Samad M. E. Sepasgozar, University of New South Wales, Sydney, Australia

Michael Shi, University of New South Wales, Sydney, Australia

Dyllanson So, University of New South Wales, Sydney, Australia

Kanit Srihakorth, University of New South Wales, Sydney, Australia

Sudaksh Tiwari, University of New South Wales, Sydney, Australia

Sawat Wongsaroj, University of New South Wales, Sydney, Australia

Tianxing Xu, University of New South Wales, Sydney, Australia

Jingbang Zhou, University of New South Wales, Sydney, Australia

Foreword

It is with immense honor and great pleasure that I write this foreword for *The Metaverse and Emerging Technologies: Transforming Learning and Beyond*, a groundbreaking volume edited by Dr. Ali Darejeh and Professor Samad M.E. Sepasgozar. Having had the privilege of personally experiencing the innovative virtual laboratories developed at the University of New South Wales (UNSW), under the expert guidance of Dr. Darejeh, I can attest to the transformative power of metaverse technologies, immersive environments, and generative artificial intelligence (AI) showcased in this book. This firsthand encounter has deepened my appreciation for the pioneering research conducted over three years by a dedicated team of 33 authors, including UNSW students and academics, culminating in this remarkable publication by Taylor & Francis.

As a researcher in cognitive load theory and instructional design, I have long been fascinated by how emerging technologies can optimize learning by aligning with human cognitive architecture. The convergence of the metaverse, generative AI, and brain–computer interfaces (BCIs), as meticulously explored in this book, offers unprecedented opportunities to reimagine education. These technologies move us beyond traditional, often passive, learning environments toward dynamic, immersive, and personalized experiences that engage learners in ways previously unimaginable. My experience in UNSW's virtual laboratories—where students interact with sophisticated equipment and AI-driven tutors in real time—vividly demonstrates how these innovations enhance cognitive engagement, accessibility, and equity in education, inspiring the comprehensive insights presented in this volume.

What sets this book apart is its rigorous blend of theory, empirical research, and practical application. Drawing on UNSW's custom-built metaverse platform, the 13 chapters provide a rich tapestry of case studies, frameworks, and forward-looking recommendations that span disciplines as diverse as engineering, medicine, computer science, filmmaking, and architecture. Each chapter is grounded in a comprehensive literature review, ensuring that the insights are both evidence-based and contextually relevant. The editors have skillfully woven these contributions into a cohesive narrative that not only highlights the successes of immersive technologies but also candidly addresses challenges such as privacy, digital equity, and interoperability—issues that are critical to the responsible adoption of these tools.

For educators, researchers, developers, and policymakers, this book serves as both an inspiration and a practical guide. It proposes robust frameworks for designing inclusive and pedagogically sound metaverse-based learning environments, emphasizing ethical considerations and scalability. The final chapter's strategic roadmap is particularly compelling, offering a vision for how these technologies can be harnessed to create future-ready learning ecosystems that empower learners across the globe.

I am particularly struck by this book's focus on real-world impact. By showcasing how UNSW's Metaverse platform has enabled students to collaborate in virtual laboratories, engage with costly equipment in safe settings, and develop critical 21st-century skills, it underscores the potential of these technologies to democratize education and bridge geographical and socioeconomic divides. This aligns closely with my own belief that technology, when thoughtfully applied, can reduce cognitive barriers and foster deeper, more meaningful learning experiences.

In closing, I congratulate Professors Darejeh and Sepasgozar, along with their team of contributors, for producing a work that is as visionary as it is pragmatic. *The Metaverse and Emerging Technologies: Transforming Learning and Beyond* is a landmark contribution to the field of educational technology, offering a blueprint for how we can shape the future of learning in an increasingly digital world. I have no doubt that it will inspire and guide those who seek to harness the power of immersive technologies to transform education for generations to come.

Fred Paas, PhD
Professor of Educational Psychology
Erasmus University Rotterdam, The Netherlands

1 The Metaverse and Emerging Technologies

Transforming Education and Beyond

Ali Darejeh and Samad M. E. Sepasgozar

1 Introduction

The convergence of transformative technologies such as the metaverse, generative artificial intelligence (GenAI), and brain–computer interfaces (BCIs) is reshaping how humans interact with digital systems, particularly in the context of education and professional development. The metaverse introduces immersive, collaborative virtual environments that transcend the limitations of traditional learning spaces (Jovanović & Milosavljević, 2022; Prakash et al., 2023). Generative AI offers capabilities for real-time adaptive instruction, intelligent virtual assistants, and automated content creation—making learning more personalized and responsive (Lv, 2023; Banh & Strobel, 2023). BCIs, meanwhile, open up new frontiers by enabling direct brain-to-interface communication, allowing users to engage with digital content in ways never before possible (Gu et al., 2021).

In recent years, the metaverse has emerged as a global trend across academia, industry, and government, capturing widespread interest as a platform for digital innovation (Kastrenakes, 2021; Meta, 2025). Educational institutions around the world are increasingly investing in metaverse initiatives to modernize their curricula and expand access to high-impact learning experiences (Ommundsen & Jumbe, 2023; European Space Agency, 2022). In the field of education, the metaverse represents a novel paradigm shift—moving from passive content consumption to active, experiential learning. Its immersive nature allows for the creation of highly interactive, contextualized, and engaging learning environments that closely mimic real-world scenarios (Cacchione, 2024; Swami, 2024). This is particularly valuable in domains requiring hands-on practice, such as medicine, engineering, and vocational training, where traditional methods often fall short (Alharbi et al., 2020; Singh et al., 2020; Soliman et al., 2021). Furthermore, the metaverse offers scalable and accessible solutions for remote and underserved communities, breaking down geographical and socioeconomic barriers to education (Hatami et al., 2024). As a result, it has the potential to democratize learning, promote deeper understanding through experiential simulations, and support the development of critical 21st-century skills (Prakash et al., 2023; JPMorgan, 2022).

DOI: 10.1201/9781003557555-1

When combined with generative AI and BCIs, the educational potential of the metaverse expands even further. These technologies do not exist in silos; rather, they are increasingly converging to form a more intelligent, inclusive, and immersive learning ecosystem. GenAI enables real-time personalized feedback, dynamic content generation, and intelligent tutoring systems that adapt to individual learner needs (Lv, 2023; Sargazi Moghadam et al., 2024; Zhai et al., 2021). BCIs enable more intimate and responsive interactions with digital environments by leveraging neurotechnology to establish direct communication between the brain and computer systems (Darejeh & Mashayekh, 2023; Gu et al., 2021). Together, these innovations redefine the landscape of higher education and professional training, offering unprecedented opportunities for enhancing engagement, accessibility, and outcomes through experiential, simulation-based learning environments.

This book explores these convergences in depth, presenting a range of use cases and theoretical insights that underscore the practical utility of emerging technologies in higher education. Each chapter contributes to a comprehensive understanding of how these tools can be used to improve learning outcomes, foster interdisciplinary collaboration, and bridge the gap between virtual and real-world applications. The chapters that follow highlight real-world applications across fields such as medical science, engineering, and computer science offering comparative analyses and cross-disciplinary perspectives. These examples provide readers with actionable strategies and best practices while also inspiring the development of new, creative modules for immersive learning.

The central aim of this book is to explore various applications of the metaverse, GenAI, and BCI—individually and in combination—and to propose frameworks for developing inclusive and effective metaverse-based learning environments. As the convergence of immersive technologies, GenAI, and BCI reshapes educational possibilities, there is a pressing need for structured design approaches that ensure these innovations are used effectively. These frameworks are essential not only for guiding the creation of accessible, engaging, and pedagogically sound digital environments but also for supporting broader systemic adoption. They serve as critical resources for developers, who need design standards; educators, who seek to implement immersive teaching strategies; researchers, who explore cognitive and experiential learning; and policymakers, who must ensure equity, scalability, and ethical integrity. By outlining how GenAI and BCIs can be meaningfully integrated into metaverse platforms, the proposed frameworks enable the development of scalable, adaptable, and learner-centered solutions across diverse educational and professional domains.

The remainder of this chapter delves deeper into the evolution and technological underpinnings of the metaverse. It outlines key enablers—including virtual reality (VR), augmented reality (AR), artificial intelligence (AI), and BCIs—and analyzes how these technologies are shaping immersive digital experiences. Real-world examples are discussed to illustrate both the promise and the challenges of this transformation. Finally, this chapter provides an overview of the book's structure, summarizing the content and goals of each subsequent chapter and setting the stage for a deeper exploration into the metaverse's role in reshaping learning, accessibility, and professional development.

2 Understanding the Metaverse: Evolution and Its Transformative Role in Society

Originally a concept from science fiction, the metaverse has evolved into a transformative digital environment with broad applications across multiple industries. Often referred to as the next evolution of the internet (Cacchione, 2024), the metaverse extends beyond conventional online interactions by offering immersive, three-dimensional experiences where users can navigate, collaborate, and engage in real time. Unlike traditional two-dimensional interfaces, it functions as a vast, interconnected digital realm that integrates VR, AI, and the internet to create an interactive and dynamic ecosystem (Jovanović & Milosavljević, 2022).

VR technology serves as the primary means of accessing the metaverse, immersing users in digital environments that closely mimic real-world settings. Devices such as head-mounted displays (HMDs), motion tracking systems, and haptic feedback tools enhance the sense of presence, making interactions within these spaces more engaging and realistic (Jensen & Konradsen, 2018). Within the metaverse, individuals use customized avatars to represent themselves in virtual settings, participating in activities such as gaming, networking, professional meetings, and education. This digital transformation is redefining how people interact with online environments, enabling new ways of socializing, working, and learning.

The metaverse's impact spans multiple sectors, including entertainment, tourism, business, healthcare, and education. In the entertainment industry, virtual concerts, interactive experiences, and cultural exhibitions are redefining how people interact with media and culture. (Swami, 2024). Similarly, the tourism sector is being transformed by metaverse-based experiences that provide both hedonic (fun, entertainment) and utilitarian (practical) benefits. Shamim, Gupta, and Shin (2024) predict that immersive tourism will reshape user behaviors, increasing positive feedback, engagement with 3D environments, and shared experiences.

Beyond leisure, businesses utilize the metaverse for virtual offices, remote collaboration, and networking events, thereby reducing geographical constraints and enhancing productivity. In healthcare and engineering, metaverse applications enhance training by providing simulated environments, enabling professionals to refine their skills in a risk-free setting. Likewise, in professional training, the fields of emergency response, construction, risk analysis, and technical specialization use scenario-based simulations to prepare for real-world challenges (Huang et al., 2024; Wang et al., 2023). Higher education is another domain undergoing significant transformation. The metaverse fosters interactive learning environments, eliminating geographical barriers while enabling students to participate in virtual classrooms, conduct lab experiments, and engage in experiential learning (Darejeh et al., 2024a; Darejeh, 2023; Prakash et al., 2023).

By merging digital and physical realities, the metaverse is redefining human interaction, commerce, education, and healthcare. Although there are some barriers to widespread adoption of the metaverse, its use continues to expand in certain fields (Gupta et al., 2024). This may enhance its capacity to foster innovation, collaboration, and accessibility, shaping the future of various industries and ultimately transforming how society engages with technology at a fundamental level.

3 Key Technologies Powering the Metaverse

The metaverse is more than just a virtual space or an extension of VR. It represents the potential for a revolutionary shift in how users interact with digital objects in immersive environments, offering a seamless blend of virtual, augmented, and real-world experiences, as well as enhanced communication and collaboration among users. By integrating technologies such as VR, AR, blockchain, AI, cloud computing, and the Internet of Things (IoT), the metaverse enables immersive interactions where users can engage with digital assets, control avatars, and collaborate in ways that feel natural and tangible (Shi et al., 2023). Since these technologies and their integration are crucial to the functionality and practical utility of the metaverse, it is essential to clearly define them and outline their usage. They are discussed as follows:

Virtual Reality –A fully immersive digital environment that users can enter through VR headsets. It creates a three-dimensional virtual world where users can explore, interact with objects, and engage in simulated experiences. VR supports applications ranging from education and training to entertainment and virtual collaboration, providing sensory-rich environments that enhance presence and engagement. As one of the foundational technologies of the metaverse, VR enables users to inhabit and navigate virtual spaces with a high degree of realism and interactivity (Darejeh et al., 2024b; Kavanagh et al., 2017).

Augmented Reality –A technology that overlays digital elements onto the real-world environment, enhancing users' perception of and interaction with their physical surroundings. AR allows users to see and engage with virtual objects in real time without leaving the real world, enabling seamless integration between digital and physical contexts. This capability supports a wide range of metaverse applications, including interactive education, remote assistance, digital workspaces, location-based gaming, and immersive retail experiences. AR is particularly valuable for enhancing real-world tasks with contextual information and interactive digital layers (Radu & Schneider, 2019).

Mixed Reality (MR) –A fusion of VR and AR where virtual and real-world elements coexist and interact in real time. MR enables digital content to anchor to and respond dynamically to the physical environment, creating highly interactive and context-aware experiences. This allows users to manipulate both physical and virtual objects simultaneously, making MR ideal for immersive simulations, advanced training, industrial design, and collaborative workspaces that blend digital tools with real-world interaction (Speicher, Hall, & Nebeling, 2019).

Digital Twins –Highly detailed virtual replicas of real-world objects, processes, or environments. In the metaverse, digital twins are leveraged for urban planning, industrial simulations, smart city management, and real-time monitoring of physical assets. By continuously synchronizing with real-world data through IoTs and sensor networks, they enable predictive analysis, performance optimization, and data-driven decision-making. Digital twins also support collaborative planning and remote diagnostics, making them vital tools for enhancing operational efficiency and strategic foresight in both public and private sectors (Mihai et al., 2022). Unlike mixed reality, which emphasizes user interaction within blended environments, digital twins focus on accurately mirroring and analyzing physical systems for insight and optimization.

Internet of Things –A network of interconnected smart devices that collect, exchange, and process real-world data. In the metaverse, IoT bridges the physical and digital realms by enabling seamless, real-time data integration, supporting applications such as mixed reality and digital twins. For example, IoT sensors can feed live environmental data—such as temperature, movement, or energy usage—into digital twins or VR environments, making metaverse experiences more dynamic, context-aware, and interactive. This real-time responsiveness enhances simulations, decision-making, and user engagement in areas such as smart cities, healthcare, and remote industrial monitoring (Greengard, 2021).

Artificial Intelligence –AI is the backbone of automation and intelligent systems within the metaverse. It enhances user experience by powering virtual assistants, real-time data processing, intelligent non-player characters (NPCs), personalized content recommendations, and predictive analytics, making interactions more seamless and adaptive (Darejeh et al., 2024c; Zhai et al., 2021).

Generative Artificial Intelligence –A specialized subset of AI that creates new content, including images, videos, 3D models, music, and text, by learning from existing data patterns (Saparamadu et al., 2025). In the metaverse, GenAI plays a crucial role in dynamically generating immersive landscapes, designing virtual objects, crafting AI-driven assistants, and personalizing digital experiences. It enables procedural content generation, facilitating the automated design of diverse environments, avatars, and interactive elements, while also enhancing learning outcomes through the use of intelligent virtual tutors (Darejeh, Marcus, & Sweller, 2022). Beyond education, it also drives the development of responsive virtual assistants, fostering more natural and realistic user interactions (Banh & Strobel, 2023; Lv, 2023).

AI Image Synthesis–GenAI-powered tools that transform text descriptions or sketches into realistic or stylized images, enabling intuitive and dynamic content generation. Widely used in the metaverse, these tools support building construction, avatar customization, virtual product visualization, and digital art creation. They allow users to generate and modify their surroundings, clothing, and digital personas in real time simply by typing or speaking their desired changes, making virtual experiences more immersive, interactive, and personalized (Lv, 2023; Wang et al., 2020).

Brain–Computer Interfaces –Advanced systems that create a direct communication pathway between the human brain and digital devices. In the metaverse, BCIs allow users to control avatars, interact with virtual objects, type, and navigate immersive environments using only their brain signals, making digital interactions more intuitive and accessible. This technology holds immense potential for accessibility, enabling individuals with motor disabilities to participate in digital learning environments without relying on traditional input devices. From hands-free control in VR-based simulations to mind-controlled interfaces in educational platforms,

BCIs are paving the way for a future where human cognition seamlessly integrates with digital interactions (Gu et al., 2021).

Blockchain & Decentralization – Blockchain technology ensures security, transparency, and digital ownership within the metaverse by leveraging decentralized ledgers that record every transaction in a tamper-resistant and verifiable manner. This transparency not only prevents fraud and unauthorized alterations but also reinforces users' trust in virtual environments. As a result, users can confidently create, own, and trade digital assets, smart contracts, NFTs, virtual land, and digital currencies within immersive virtual economies (Mourtzis, Angelopoulos, & Panopoulos, 2023).

Cloud Computing & Edge Computing –The metaverse relies on cloud-based infrastructure to support large-scale data processing, storage, rendering, and real-time interactions across distributed environments. Edge computing complements this by processing data closer to the user, significantly reducing latency and bandwidth demands. This synergy enables seamless, scalable, and lag-free metaverse experiences, particularly in applications requiring immediate feedback, such as VR training, multiplayer gaming, and remote collaboration (Andriulo et al., 2024).

5G & High-Speed Connectivity –The success of the metaverse depends on ultra-fast, stable, and low-latency internet connections. 5G technology supports real-time streaming of immersive content, enabling smooth, high-fidelity interactions within virtual environments. Its high bandwidth and low latency make it ideal for data-intensive applications such as multiplayer VR, AR navigation, and remote collaboration, ensuring users experience minimal lag and uninterrupted connectivity in complex metaverse scenarios (Jamshidi et al., 2023).

These technologies collectively shape the metaverse, making it an intelligent, immersive, and scalable digital ecosystem.

4 The Growing Role of the Metaverse: Investments, Innovations, and Global Impact

The metaverse is rapidly evolving, driven by significant investments and technological advancements across various sectors. These efforts are not only reshaping entertainment and gaming but also fostering innovation in humanitarian work, environmental conservation, and workforce training.

4.1 Metaverse Investments by Leading Companies

The increasing investment in metaverse applications signals a potential for the widespread adoption of immersive platforms in the near future. Major technology companies such as Meta, Epic Games, and NVIDIA are at the forefront of shaping the metaverse, investing heavily in virtual environments, simulation tools, and user-generated experiences.

In October 2021, Meta Platforms, Inc. announced a $10 billion investment in its Reality Labs division, which includes Horizon Worlds and other VR and

AR initiatives (Kastrenakes, 2021). Horizon Worlds is Meta's social VR platform, enabling users to create, explore, and interact within immersive digital environments that support gaming, socializing, and collaboration. In October 2023, Meta further expanded its investment by launching a $50 million Creator Fund to incentivize content creators, aiming to enhance user engagement and platform development (Meta, 2025). These substantial investments reflect Meta's long-term vision of developing socially immersive digital experiences within the metaverse.

Similarly, in April 2022, Epic Games, the company behind Unreal Engine, one of the most powerful gaming engines, secured $2 billion in funding to expand its metaverse initiatives, receiving $1 billion each from Sony and Kirkbi, the parent company of The Lego Group (Webster, 2022). This investment reinforced Epic's commitment to creating expansive, interactive virtual environments that integrate gaming, social experiences, and creative collaboration.

NVIDIA has also demonstrated a strong commitment to metaverse-related technologies, particularly through its Omniverse platform—a real-time, multi-GPU simulation and collaboration tool designed for creating, operating, and simulating metaverse applications. Omniverse enables 3D designers, developers, and enterprises to build and connect virtual worlds using advanced AI, physics-based simulations, and ray-traced graphics. While specific investment figures remain undisclosed, NVIDIA's financial performance suggests a substantial allocation of resources to this area. In 2024, NVIDIA's professional visualization segment, which includes Omniverse, reported revenues of $486 million, reflecting a 17% year-over-year increase (Chauhan, 2024; NVIDIA, 2024). This growth highlights the increasing investment in and adoption of Omniverse and related metaverse technologies, signaling a long-term commitment to developing advanced immersive experiences.

These strategic investments from Meta, Epic Games, and NVIDIA show the metaverse's growing economic potential and the commitment of industry leaders to shaping virtual experiences, social interaction, and digital economies in the years to come.

4.2 The Metaverse as a Tool for Global Change

Beyond corporate investments, the metaverse is increasingly being leveraged to address global challenges. The World Economic Forum's Global Collaboration Village, for instance, demonstrates how immersive technologies can enhance humanitarian training, conservation efforts, and workforce upskilling (Ommundsen & Jumbe, 2023). One of its featured applications, Drop in the Ocean, immerses audiences in the plastic pollution crisis, fostering empathy and promoting collective action. Similarly, humanitarian organizations such as the International Federation of Red Cross and Red Crescent Societies (IFRC) use VR and AR to improve disaster response training, providing local communities with cost-effective and scalable learning solutions (Ommundsen & Jumbe, 2023).

These strategic investments and real-world applications highlight the metaverse's growing significance. From corporate funding in gaming and virtual collaboration to global initiatives addressing social and environmental challenges, the metaverse is evolving into a multifaceted digital ecosystem that is shaping the future of entertainment, education, business, and humanitarian efforts.

5 The Metaverse in Higher Education: Empowering Learning and Upskilling

As a pioneering educational tool, the metaverse has the potential to revolutionize traditional teaching methods across disciplines, especially in higher education—the focus of this book—by making learning more interactive, engaging, and collaborative. It enables the recreation of historical events, immersive geographical explorations, virtual geology and mining simulations, scientific experiments, medical training, and engineering design.

5.1 History and Geography Education

The metaverse serves as a powerful tool in history education by reconstructing historical sites and events, allowing students to immerse themselves in different eras and contexts. A growing number of researchers have developed applications that leverage metaverse technologies to enhance historical learning. For example, Gaafar (2021) created a virtual environment for architectural history courses that simulates heritage buildings. Agustini et al. (2023) developed an interactive platform for exploring prehistoric objects, while Son et al. (2023) designed a metaverse museum centered on the Heungdeoksa Temple from the Koryeo Dynasty, integrating the historical data related to metal type production. In a similar vein, Chehab and Nakhal (2023) facilitated virtual tours of historic Beirut City Center, and Cabero-Almenara, Llorente-Cejudo, and Martinez-Roig (2022) created a VR model of the Church of the Annunciation in Seville for use in art history education.

In geography education, the metaverse enables students to embark on virtual field trips across various landscapes, ecosystems, and cultural landmarks (Prisille & Ellerbrake, 2020). Learners can explore locations such as the Amazon rainforest, observe natural phenomena like volcanic eruptions (Tibaldi et al., 2020), or visit historically significant sites around the world. Several notable studies have explored the integration of immersive technologies in geographic education. Habibah et al. (2023) developed a VR environment centered on eco-spatial and estuary conservation, while Tsou and Mejia (2023) analyzed metaverse applications in digital cartography. Similarly, Hagge (2021) and Shakirova, Said, and Konyushenko (2020) leveraged Google Earth's VR platform to enhance geography instruction. In addition, Shen et al. (2022) and Prisille and Ellerbrake (2020) investigated the impact of 360-degree VR videos on immersive learning experiences in geographic education.

5.2 Science and Medical Education

In the fields of chemistry, physics, and biology, the metaverse extends learning beyond traditional laboratories by offering immersive virtual experiments and simulations.

In chemistry education, for instance, Reeves, Crippen, and McCray (2021) developed a VR-based lab for conducting fundamental experiments; Kumar et al. (2021) and Ferrell et al. (2019) developed a VR system for learning biochemical engineering; and Cortes Rodriguez et al. (2023) introduced a 3D molecular exploration tool to facilitate spatial understanding of chemical structures. Similarly, Edwards et al. (2019) created the VR Multisensory Classroom, allowing students to construct hydrocarbon molecules using hand gestures. Complementing these efforts, Fung et al. (2019) simulated international field trips to support environmental chemistry education.

Among various applications in physics education, Tsivitanidou, Georgiou, and Ioannou (2021) designed a VR system to teach the Special Theory of Relativity. Daineko et al. (2020) developed a virtual laboratory for conducting physics-related experiments, while Pirker et al. (2019) evaluated learning outcomes using *Maroon VR*, an immersive physics lab environment. Additional noteworthy contributions include a water cycle simulation developed and tested by Bogusevschi, Muntean, and Muntean (2020), and a nuclear physics educational module created by Šiđanin et al. (2020).

In the context of biology education, VR technologies have been employed to support anatomical and molecular understanding. Choi and Kim (2020) developed a VR application for visualizing the human digestive system, while Bennett and Saunders (2019) created an immersive tour of human cell structures. Thompson et al. (2020) designed a tool for teaching cell characteristics, and Paxinou et al. (2022, 2020) evaluated the *OnLabs* platform for training students in the use of laboratory equipment. Expanding the scope further, Hemme et al. (2023) introduced VR/AR tools addressing biomedical topics such as molecular visualization, drug mechanisms, and nanotechnology.

Medical education has also seen significant advancements through realistic simulations enabled by the metaverse (Darejeh et al., 2024a). For instance, Singh et al. (2020) compared the effectiveness of VR and traditional video-based methods in teaching medical communication skills, while Lopez et al. (2021) evaluated learning outcomes using a VR simulator focused on cardiology and neurology. Alharbi et al. (2020) developed a VR system for teaching anatomy, Chen et al. (2020) created a platform tailored to nursing education, and Behmadi et al. (2022) introduced a VR tool designed for emergency triage training.

5.3 Engineering Education

The metaverse empowers engineering students to design and test solutions within immersive virtual environments that replicate real-world challenges. These simulations enable learners to visualize structural behavior, evaluate engineering performance, and refine designs prior to physical implementation—fostering both cost-effective and innovative learning experiences (Soliman et al., 2021).

Several applications have demonstrated the potential of VR in various engineering disciplines. For instance, Salah et al. (2019) introduced a platform focused on Industry 4.0 and reconfigurable manufacturing systems. Yang et al. (2021) created a VR learning environment for urban railway engineering, and Vergara et al. (2019) designed tools to support materials engineering education. In other areas, Singh et al. (2021) developed a VR learning system for electrical engineering students, Hernández-Chávez et al. (2021) created a VR application for Automotive Systems Engineering that allows manipulation of 3D engine components, and Pérez et al. (2019) simulated a manufacturing robot, enabling users to interact with it through virtual controllers.

5.4 Building Design and Construction

The metaverse is increasingly being adopted in architecture, engineering, and construction (AEC) education to simulate complex environments and enhance professional training. For instance, Wu et al. (2019) developed an avatar-based VR communication platform that integrates Building Information Modelling (BIM) to facilitate collaborative design and coordination. Similarly, Sepasgozar (2020) introduced a series of immersive learning modules—including a VR piling simulator and a virtual tunnel boring machine—that allow users to visualize core construction processes and explore underground infrastructure.

These modules allow students to engage with scenarios that are typically inaccessible in real-world settings—such as subterranean construction projects (Figure 1.1). This immersive approach fosters experiential learning and supports deeper conceptual understanding, moving beyond the limitations of traditional textbooks and lectures. The tools also support multi-user interaction, enabling learners to collaborate through avatars in shared virtual environments. This feature is particularly beneficial for remote, hands-on training—especially during disruptions like the COVID-19 pandemic. When integrated into learning management systems such as Moodle, these VR environments have demonstrated strong potential for engaging digitally native students and bridging the gap between theoretical instruction and real-world practice.

Figure 1.1 A group of students using a virtual underground module.

6 The Road Ahead: Opportunities and Challenges

The economic potential of the metaverse is substantial, with major financial institutions offering optimistic projections. J. P. Morgan estimated in 2022 that the metaverse could become a $1 trillion market opportunity (JPMorgan, 2022). McKinsey & Company projected in June 2022 that it could generate up to $5 trillion in value by 2030, while Citigroup forecasted in 2023 that it could represent an opportunity of $8 to $13 trillion by the same year (Denton, 2023). These figures highlight the massive financial impact the metaverse could have across various industries as it continues to develop and integrate into global markets.

6.1 Barriers to Adoption

Despite its promising potential, several barriers hinder the metaverse's widespread adoption. Privacy concerns remain a critical issue, as immersive environments require extensive data collection, including biometric information, raising significant data protection risks. Security threats, including hacking, data breaches, and identity theft, pose significant risks due to the interconnected nature of the metaverse (Gupta et al., 2024; Zhao et al., 2023). Additionally, cyber safety for children is a growing concern, as platforms like Meta's Horizon Worlds become more accessible to younger audiences, increasing their exposure to inappropriate content and online threats.

Another major challenge is digital equity, as many underserved communities lack access to high-speed internet and advanced computing devices, thereby limiting their ability to participate in the metaverse (Hatami et al., 2024). Also, interoperability across platforms remains a challenge, requiring companies to make substantial investments in software and hardware to align with metaverse technologies. These technical and financial barriers could slow down its widespread adoption. However, advances in technology infrastructure, along with supporting tools, protective measures, and evolving metaverse ecosystems, including all the associated tools, collectively enable these barriers to be addressed over time (Al-Sharafi et al., 2024; Hwang & Seo, 2025).

6.2 The Need for Strategic Investments and Regulations

To fully harness the metaverse's potential, strategic planning and targeted investments are essential. Governments and corporations must invest in infrastructure development to support the high data demands of metaverse applications. Additionally, policy formulation is crucial to regulate data privacy, security, and digital equity, ensuring responsible and inclusive digital interactions (Gray & Tang, 2025).

As more industries adopt immersive technologies, there is an increasing need for educational initiatives to train individuals on navigating and interacting with metaverse environments effectively. By addressing these challenges through strategic investments, regulations, and innovative solutions, the metaverse can evolve into a secure, inclusive, and profitable digital ecosystem. If developed responsibly, it has the potential to reshape industries, drive economic growth, and transform digital experiences, enabling people to interact, learn, and conduct business more effectively in virtual spaces.

7 Our Successful Education-Focused Metaverse Experience

As discussed in the early part of this chapter, the metaverse has demonstrated its transformative potential in higher education by offering immersive virtual spaces where students can engage in hands-on learning experiences without real-world risks. Universities often struggle to provide all students with access to expensive, high-tech engineering and medical lab equipment. The metaverse and VR present a cost-efficient alternative by replicating lecture halls and laboratory settings in an immersive digital environment. Research highlights the potential of VR to reduce educational costs (Cacchione, 2024) while enhancing student motivation (Girijakumari Sreekantan Nair & Chandrasekaran, 2024). Its immersive qualities also contribute to improved student engagement in STEM education (Shamim, Gupta, & Shin, 2024), allowing for safe, simulated laboratory practice (Swami, 2024).

Various institutions have already begun integrating metaverse technologies into their academic programs. For example, the National University of Singapore (NUS) employs the My Virtual Lab platform to prepare students for STEM fields (Ommundsen & Jumbe, 2023). Similarly, the European Space Agency has developed collaborative virtual learning environments to facilitate the study of advanced technologies and space exploration (European Space Agency, 2022).

Although numerous studies and projects have explored the use of metaverse platforms in education, to the best of our knowledge, no existing system fully replicates an entire university campus—including lecture halls and laboratory spaces with interactive equipment—while simultaneously supporting real-time student collaboration in a virtual lab environment. There are limited publications exhibiting diverse examples and use cases of virtual environments that can potentially be extended to be used as a metaverse. Consequently, there is less evidence of publications demonstrating how various metaverse applications can be utilized for training and learning purposes.

Some companies and institutions have introduced VR-based lab environments, but these implementations often lack a fully immersive and comprehensive simulation of the entire laboratory space. In particular, they do not adequately replicate complex lab equipment with multiplayer capabilities, limiting opportunities for students to collaborate in a realistic, shared virtual setting. These systems are typically designed for single-task activities primarily aimed at K-12 education, making them less suitable for higher education (VRLab Academy, 2025; ENGAGE XR, 2025).

To address this gap, we at the University of New South Wales (UNSW) have developed our own metaverse platform, including a virtual campus and different labs for higher education, and conducted various research experiments to assess its effectiveness across different fields of engineering, medical science, and art. These studies serve as the foundation for several chapters in this book. Our metaverse recreates the university environment as a digital campus (Figures 1.2 and 1.3), with distinct buildings representing academic disciplines such as computer science, mechanical engineering, and medical science. Within this virtual environment, students engage through avatars, attending lectures, collaborating in lab activities, and participating in discussions. Each academic department is represented by a specific building, allowing students to navigate to their respective labs and facilities, fostering a dynamic and interactive learning experience.

Figure 1.2 The metaverse virtual campus environment.

Figure 1.3 The metaverse virtual campus environment and user's avatar.

This metaverse platform has been developed using Unreal Engine 5 for both Windows and Android-based headsets, including the HTC VIVE Pro and Meta Quest. The virtual equipment was modeled using Blender and then imported into Unreal Engine for integration. To foster collaboration and enhance the learning experience, the system supports a multiplayer environment where multiple users can join simultaneously, interact through their avatars, and communicate via voice chat. This functionality is powered by the Photon plugin, which has been seamlessly incorporated to enable real-time communication and synchronization among users, regardless of their physical location.

A defining aspect of this project is its highly realistic simulation of laboratory environments, which not only grants students access to expensive and specialized equipment but also integrates an intelligent agent that serves as a virtual tutor. Powered by GenAI, this agent guides students through experiments, providing step-by-step instructions and answering their questions in real time. Within this immersive digital space, students can engage with a wide range of laboratory instruments, including costly and potentially hazardous equipment, in a safe and controlled setting.

The simulated tools are fully interactive, enabling students to manipulate, assemble, and conduct hands-on experiments that closely replicate real-world procedures. These are not static models but functional virtual instruments designed for authentic experimentation. The platform successfully recreates essential lab equipment from various disciplines, such as a manufacturing robot from the School of Mechanical Engineering, a Programmable Logic Controller (PLC) and oscilloscope from the School of Electrical Engineering, and a Polymerase Chain Reaction (PCR) machine from the School of Medical Science. By combining advanced simulations with AI-driven guidance, this system offers students a unique and engaging learning experience, enhancing both technical skills and conceptual understanding. An example of one of these simulated labs is shown in Figure 1.4.

Figure 1.4 A simulated lab environment.

Beyond the laboratory environments and the university campus, this metaverse also features Virtual Lecture Theaters designed for conducting online classes, where students can participate in live sessions or access recorded lectures (see Figure 1.5). These virtual theaters offer an immersive alternative to traditional telecommunication platforms like Zoom or Microsoft Teams by fostering a shared presence that enhances engagement and interactivity. Unlike conventional video conferencing tools, this environment enables students to see themselves, their peers, and the lecturer's avatar within a shared virtual space, fostering a more natural and collaborative learning experience.

Figure 1.5 The lecture theatre.

8 Overview and Structure of the Book

This book is divided into 13 chapters, each beginning with a comprehensive literature review that establishes foundational knowledge and contextual background for the topic under discussion. Following the review, the chapters present empirical studies conducted by our research team using UNSW's custom-built metaverse platform, as outlined in the previous section. Each study investigates a specific component of this platform through experiments designed to evaluate its effectiveness in enhancing student learning and interaction. All participants in these studies were recruited in accordance with a human research ethics protocol approved by the University of New South Wales and its institutional review board.

The primary purpose of these studies is to highlight the diverse applications of the metaverse and VR in various fields, rather than to focus on research methodology or results. Although the limited number of participants in some studies may affect the generalizability of the findings, the breadth of these studies provides valuable insights into design possibilities across a wide range of disciplines—from medical science to engineering education and robotics, and from museum simulation to filmmaking education within the metaverse.

Each chapter also includes forward-looking recommendations informed by these empirical studies, literature reviews, and the authors' insights. These findings are further analyzed and reviewed by the book's editors, culminating in a final chapter that presents a roadmap and framework for the future of the metaverse in education, industry, and beyond. This roadmap serves as a practical guide for stakeholders, including educators, policymakers, and industry leaders, who seek to leverage immersive technologies for innovation and growth.

Please note that throughout this book, we primarily refer to VR-based environments rather than the metaverse. However, it is essential to clarify that all VR applications discussed in different chapters originate from the custom metaverse platform developed at UNSW. Therefore, when we refer to a VR-based application, such as a VR-based medical science lab, it denotes explicitly a medical science VR lab developed as part of the UNSW metaverse.

It should also be noted that this book is the product of more than two years of original research across diverse domains of the metaverse and VR. While the chapters reflect the authors' own scholarly work, some non-native English-speaking contributors made use of ChatGPT-4 solely for grammar checking to ensure clarity of expression. No AI tools were involved in generating or shaping the substantive content.

The summary of each chapter that will be presented in this book is as follows:

Chapter 2 explores the integration of metaverse technologies—including VR and digital twins—into mechanical engineering education and practice. It begins by examining how immersive and interactive virtual environments can enhance laboratory experiences and simulation workflows in mechanical engineering. Subsequently, this chapter presents two empirical studies that investigate the practical applications of metaverse technologies in both mechanical engineering laboratories and industrial training contexts. The first study evaluates a VR-based composite manufacturing laboratory, demonstrating reduced cognitive load and improved knowledge retention compared to physical labs. The second compares VR and video-based training for industrial robotics, highlighting VR's ability to lower mental demand while increasing engagement. Finally, this chapter discusses the advantages and disadvantages of developing VR-based environments in mechanical engineering, providing a balanced perspective on their impact and feasibility.

Chapter 3 explores the integration of the metaverse into electrical engineering education, emphasizing its transformative impact on traditional pedagogical approaches. It begins with an overview of how VR-based environments and metaverse technologies are leveraged to create immersive and interactive virtual laboratories, particularly in electrical engineering. To evaluate the effectiveness of VR-based learning, this chapter presents two empirical studies. The first study compares VR simulations for circuit construction with physical laboratories, examining their impact on learning outcomes, student engagement, and usability. The second assesses VR-based training against conventional circuit simulation tools, such as TinkerCAD, for learning oscilloscope functionality. Finally, this chapter discusses the broader implications of these findings for curriculum design and the future of immersive learning technologies.

Chapter 4 explores the transformative impact of the metaverse on medical science education, particularly in laboratory techniques and equipment training. This chapter begins with a comprehensive literature review on the applications of VR in medical laboratories, surgical training, and patient treatment, emphasizing its role in enhancing engagement, skill acquisition, and knowledge retention. A detailed empirical study featuring two experiments on a VR-based platform for Polymerase Chain Reaction (PCR) training is presented to evaluate the effectiveness of virtual laboratories in terms of cognitive load and learning outcomes. This chapter concludes by discussing the findings and offering insights into how the effectiveness of virtual medical science laboratories can be further enhanced.

Chapter 5 explores the application of the metaverse in enhancing technical skills in computer science education. It begins with a comprehensive literature review on the use of VR for learning programming, offering insights into how immersive and gamified environments can transform programming instruction. This is followed by an empirical study on a VR-based escape room platform designed to teach programming. The study investigates whether the gamified VR environment improves students' perceptions of learning performance and motivation in both fundamental programming and object-oriented programming (OOP). This chapter concludes by emphasizing the potential of VR-based learning environments to make programming education more immersive, interactive, and engaging.

Chapter 6 explores the integration of the metaverse in filmmaking education, examining how VR and GenAI enhance traditional cinematography training. This chapter begins with a comprehensive literature review on the applications of the metaverse and VR in filmmaking education. It then presents an empirical study investigating the impact of metaverse applications, powered by a GenAI feedback system, in facilitating student learning of cinematographic elements—such as camera positioning and lighting balance—across different film genres. This chapter concludes with discussions on ethical considerations and future directions, emphasizing the transformative potential of metaverse technologies in reshaping filmmaking education and training methodologies.

Chapter 7 explores the role of the metaverse in developing virtual museums, enabling immersive, collaborative, and accessible cultural heritage experiences. It begins by examining the evolution and technological foundations of virtual museums, followed by a systematic literature review and an analysis of existing virtual museum applications. Additionally, an empirical study presents our newly developed virtual museum system within the metaverse, designed to address the limitations of existing applications through interactive storytelling, dynamic artifact exploration, and social engagement. This chapter concludes with a discussion on future directions, ethical considerations, and best practices for designing virtual museums that effectively preserve and present cultural heritage in the metaverse.

Chapter 8 explores the integration of robotics and the metaverse, focusing on enhancing human-robot interaction through virtual collaboration and simulation. It provides an overview of automation advancements, from early robotic control systems to modern AI-driven and IoT-enabled robotics, highlighting digital twins as a bridge between virtual simulations and real-world control. A review of existing research examines robotic path planning, associated algorithms and programming models, and metaverse-based manipulation, with an emphasis on efficiency, cost reduction, and remote collaboration. Through an empirical study, this chapter proposes a framework for real-time robotic control in the metaverse, incorporating digital twins, reinforcement learning models, and anthropomorphic robot designs to enhance control accuracy and user experience. Finally, this chapter outlines future directions for metaverse-based robotic interaction, focusing on optimizing user experience, improving simulation-to-reality transfer, and addressing ethical considerations in AI-driven robotics.

Chapter 9 explores privacy preservation and awareness in metaverse platforms, addressing challenges, ethical considerations, and best practices to ensure data security and user trust. It begins with a review of related work, key privacy considerations, and strategies for enhancing privacy protection in metaverse-based interactions. This chapter identifies critical factors such as data collection, regulatory compliance (GDPR, FERPA, COPPA), user consent, transparency, and data minimization. It further examines privacy-preservation techniques, including encryption, anonymization, access control, and privacy-by-design approaches. Additionally, it discusses challenges such as managing large data volumes, ensuring interoperability, mitigating cybersecurity threats, and balancing innovation with ethical concerns. Finally, this chapter explores future directions, emphasizing emerging technologies, regulatory developments, and best practices for safeguarding privacy while fostering innovation in metaverse environments.

Chapter 10 explores the transformative potential of GenAI in education and the metaverse, highlighting its applications, challenges, and future implications. It begins by introducing the fundamental principles and capabilities of GenAI, providing a foundation for understanding its role in reshaping digital and physical environments. This chapter then examines how GenAI is revolutionizing education within the metaverse through intelligent agents, virtual avatars, and immersive learning systems. To illustrate its impact, four empirical studies are presented, demonstrating how GenAI provides career guidance, predicts university admission outcomes, and supports the acquisition of technical skills in VR environments. This chapter also addresses key challenges, including concerns related to data privacy and algorithmic biases. Finally, it underscores the importance of ethical guidelines, policy development, and future research to ensure GenAI's responsible and equitable adoption.

Chapter 11 explores the profound impact of GenAI in revolutionizing visualization within architecture and construction, both in the metaverse and the real world. It highlights how this field can derive significant benefits from GenAI, mainly through image synthesis technologies. This chapter begins with an introduction to Generative AI Image Synthesis (GAI-IS) tools, which are powered by advanced deep learning techniques, including Generative Adversarial Networks and Diffusion Models. It then examines the diverse applications and tools of GAI-IS in construction and metaverse design, demonstrating how these technologies enhance visualization processes through various examples, improve early-stage architectural modelling, and foster innovative design solutions. Finally, this chapter concludes by speculating on the future of GAI-IS in construction and metaverse design, envisioning potential developments and emerging trends that may shape the industry's landscape.

Chapter 12 explores the transformative potential of BCI as an advanced interaction modality, highlighting its impact on individuals with motor impairments, creative professionals, and gamers seeking enhanced interactivity. It begins with an introduction to BCI and its integration into virtual environments, outlining the foundational principles behind this innovative technology. A comprehensive literature review provides insights into the advantages and challenges of utilizing BCIs, including signal reliability, cognitive load, and ethical concerns related to data privacy. To illustrate real-world applications, three empirical studies are presented. The first explores a BCI-powered communication tool designed to assist users with motor impairments, demonstrating its potential in accessibility. The second study showcases a BCI-based painting application, enabling users to create digital artwork using neural commands. The third study focuses on gaming, analyzing how BCI technology enhances immersion, gameplay dynamics, and user interaction within gaming environments. This chapter concludes with actionable insights for refining BCI systems, emphasizing the importance of interdisciplinary collaboration and user-centric design to unlock their full capabilities in the metaverse and beyond.

Chapter 13 presents a strategic roadmap for designing and implementing intelligent, immersive learning environments in the metaverse. Building on the findings of previous chapters, it introduces two comprehensive frameworks: one focused on the core design principles necessary for effective metaverse-based education, and the other on the integration of GenAI and BCI. These frameworks guide the development of adaptive, collaborative, and user-centred learning platforms, while also addressing critical ethical considerations such as privacy, accessibility, and data transparency. This chapter also presents forward-looking perspectives on how BCI and GenAI technologies could transform future educational experiences. It explores conceptual ideas such as metaverse-based content that adapts dynamically to learners' neural data, intelligent tutors capable of adjusting instruction in real time based on cognitive signals, and AI-driven mechanisms that promote equitable collaboration by monitoring neural engagement and managing group dynamics. Rather than focusing solely on technological innovation, this chapter emphasizes the importance of pedagogical purpose and interdisciplinary collaboration in shaping meaningful learning environments. Serving as both a synthesis of prior insights and a forward-looking guide, this final chapter offers practical tools and visionary direction for educators, developers, and institutions seeking to leverage immersive technologies responsibly and effectively.

References

Agustini, K., Putrama, I. M., Wahyuni, D. S., & Mertayasa, I. N. E. (2023). Applying gamification technique and virtual reality for prehistoric learning toward the metaverse. *International Journal of Information and Education Technology*, *13*(2), 247–256.

Alharbi, Y., Al-Mansour, M., Al-Saffar, R., Garman, A., & Alraddadi, A. (2020). Three-dimensional virtual reality as an innovative teaching and learning tool for human anatomy courses in medical education: A mixed methods study. *Cureus*, *12*(2): e7085. https://doi.org/10.7759/cureus.7085

Al-Sharafi, M. A., Al-Emran, M., Al-Qaysi, N., Iranmanesh, M., & Ibrahim, N. (2024). Drivers and barriers affecting metaverse adoption: A systematic review, theoretical framework, and avenues for future research. *International Journal of Human–Computer Interaction*, *40*(22), 7043–7064.

Andriulo, F. C., Fiore, M., Mongiello, M., Traversa, E., & Zizzo, V. (2024, September). Edge computing and cloud computing for internet of things: A review. *Informatics*, *11*(4), Article 71. https://doi.org/10.3390/informatics11040071

Banh, L., & Strobel, G. (2023). Generative artificial intelligence. *Electronic Markets*, *33*(1), 63.

Behmadi, S., Asadi, F., Okhovati, M., & Sarabi, R. E. (2022). Virtual reality-based medical education versus lecture-based method in teaching start triage lessons in emergency medical students: Virtual reality in medical education. *Journal of Advances in Medical Education & Professionalism*, *10*(1), 48.

Bennett, J. A., & Saunders, C. P. (2019). A virtual tour of the cell: Impact of virtual reality on student learning and engagement in the STEM classroom. *Journal of Microbiology & Biology Education*, *20*(2), 10–1128.

Bogusevschi, D., Muntean, C., & Muntean, G. M. (2020). Teaching and learning physics using 3D virtual learning environment: A case study of combined virtual reality and virtual laboratory in secondary school. *Journal of Computers in Mathematics and Science Teaching*, *39*(1), 5–18.

Cabero-Almenara, J., Llorente-Cejudo, C., & Martinez-Roig, R. (2022). The use of mixed, augmented and virtual reality in history of art teaching: A case study. *Applied System Innovation*, *5*(3), 44.

Cacchione, A. (2024). Beyond the hype and the crisis: A possible metaverse for education. In V. Geroimenko (Ed.), *Augmented and virtual reality in the metaverse* (pp. 25–40). Springer Nature Switzerland. https://doi.org/10.1007/978-3-031-57746-8_2

Chauhan, H. (2024, June 6). *Prediction: This could be Nvidia stock's next massive-growth opportunity*. Nasdaq.

Chehab, A., & Nakhal, B. (2023). Exploring virtual reality as an approach to resurrect destroyed historical buildings-an approach to revive the destroyed "Egg Building" through Vr. *Architecture And Planning Journal (Apj)*, *28*(3), 17.

Chen, F. Q., Leng, Y. F., Ge, J. F., Wang, D. W., Li, C., Chen, B., & Sun, Z. L. (2020). Effectiveness of virtual reality in nursing education: Meta-analysis. *Journal of Medical Internet Research*, *22*(9), e18290.

Choi, S., & Kim, H. B. (2020). Application and effects of VR-based biology class reflecting characteristics of virtual reality. *Journal of the Korean Association for Science Education*, *40*(2), 203–216.

Cortes Rodriguez, F. J., Frattini, G., Teixeira Pinto Meireles, F., Terrien, D. A., Cruz-Leon, S., Dal Peraro, M., ... & Abriata, L. A. (2023). MolecularWebXR: Multiuser discussions about chemistry and biology in immersive and inclusive VR. *bioRxiv*. Preprint. https://doi.org/10.1101/2023.11.01.564623

Daineko, Y., Ipalakova, M., Tsoy, D., Bolatov, Z., Baurzhan, Z., & Yelgondy, Y. (2020). Augmented and virtual reality for physics: Experience of Kazakhstan secondary educational institutions. *Computer Applications in Engineering Education*, *28*(5), 1220–1231.

Darejeh, A. (2023). Empowering education through eerp: A customizable educational vr escape room platform. In *2023 IEEE International Symposium on Mixed and Augmented Reality Adjunct (ISMAR-Adjunct)* (pp. 764–766). IEEE.

Darejeh, A., & Mashayekh, S. (2023). Mind-object control as a new form of interaction in metaverse for users with motor disability. *Journal of Biomedical Physics & Engineering*, *13*(1), 257.

Darejeh, A., Huynh, L., Ji, X., Marcus, N., & Polly, P. (2024a). Assessing the Efficacy of Virtual Reality-Based Laboratories Versus Real-Life Settings for PCR Procedures, within Medical Science Laboratories. In *2024 IEEE International Symposium on Mixed and Augmented Reality Adjunct (ISMAR-Adjunct)* (pp. 489–492). IEEE.

Darejeh, A., Marcus, N., & Sweller, J. (2022). Increasing learner interactions with E-learning systems can either decrease or increase cognitive load depending on the nature of the interaction. *L'Année psychologique*, *122*(3), 405–437.

Darejeh, A., Moghadam, T. S., Delaramifar, M., & Mashayekh, S. (2024c). A framework for AI-powered decision making in developing adaptive e-learning systems to impact learners' emotional responses. In *2024 11th International and the 17th National Conference on E-Learning and E-Teaching (ICeLeT)* (pp. 1–6). IEEE.

Darejeh, A., Mondal, M., Marcus, N., & Vassar, A. (2024b). Exploring the cognitive load effects of diverse virtual reality interaction methods in interactive educational platforms. In *2024 IEEE International Symposium on Mixed and Augmented Reality Adjunct (ISMAR-Adjunct)* (pp. 435–438). IEEE.

Denton, J. (2023). *Metaverse may be worth $13 trillion, Citi says. What's behind the bullish take on Web3?* Barron's. Retrieved from https://www.barrons.com/articles/stock-market-movers-a8bcae99

Edwards, B. I., Bielawski, K. S., Prada, R., & Cheok, A. D. (2019). Haptic virtual reality and immersive learning for enhanced organic chemistry instruction. *Virtual Reality*, *23*, 363–373.

ENGAGE XR Holdings. (2025). *K12 Virtual Reality Education*. ENGAGE. Retrieved [12/2/2025], https://engagevr.io/k12-virtual-reality-education/

European Space Agency. (2022, November 14). *Inside ESA's COP27 metaverse*. European Space Agency. Retrieved from https://www.esa.int/ESA_Multimedia/Images/2022/11/Inside_ESA_s_COP27_metaverse

Ferrell, J. B., Campbell, J. P., McCarthy, D. R., McKay, K. T., Hensinger, M., Srinivasan, R., ...& Schneebeli, S. T. (2019). Chemical exploration with virtual reality in organic teaching laboratories. *Journal of Chemical Education*, *96*(9), 1961–1966.

Fung, F. M., Choo, W. Y., Ardisara, A., Zimmermann, C. D., Watts, S., Koscielniak, T., ...& Dumke, R. (2019). Applying a virtual reality platform in environmental chemistry education to conduct a field trip to an overseas site. *Journal of Chemical Education*, *96*(2), 382–386. https://doi.org/10.1021/acs.jchemed.8b00728

Gaafar, A. A. (2021). Metaverse in architectural heritage documentation & education. *Advances in Ecological and Environmental Research*, *6*(10), 66–86.

Girijakumari Sreekantan Nair, S., & Chandrasekaran, B. (2024). MetaHealth: Unlocking metaverse technologies in digital healthcare. *Expert Review of Medical Devices, 21*(8), 727–739.

Gray, J. E., & Tang, W. (2025). The Chinese metaverse: An analysis of China's policy agenda for extended reality (XR). *Policy & Internet*, *17*(1), e418.

Greengard, S. (2021). *The internet of things*. MIT press.

Gu, X., Cao, Z., Jolfaei, A., Xu, P., Wu, D., Jung, T. P., & Lin, C. T. (2021). EEG-based brain-computer interfaces (BCIs): A survey of recent studies on signal sensing technologies and computational intelligence approaches and their applications. *IEEE/ACM Transactions on Computational Biology and Bioinformatics*, *18*(5), 1645–1666.

Gupta, R., Rathore, B., Biswas, B., Jaiswal, M., & Singh, R. K. (2024). Are we ready for metaverse adoption in the service industry? Theoretically exploring the barriers to successful adoption. *Journal of Retailing and Consumer Services*, *79*, 103882.

Habibah, K., Putra, A. K., Nilsson, S., & Vielhaber, C. (2023). Digital estuaries: Exploring the pedagogical benefits of virtual reality media in geography and spatial analysis. *Jambura Geo Education Journal*, *4*(2), 187–198.

Hagge, P. (2021). Student perceptions of semester-long in-class virtual reality: Effectively using "Google Earth VR" in a higher education classroom. *Journal of Geography in Higher Education*, *45*(3), 342–360.

Hatami, M., Qu, Q., Chen, Y., Kholidy, H., Blasch, E., & Ardiles-Cruz, E. (2024). A survey of the real-time metaverse: Challenges and opportunities. *Future Internet*, *16*(10), 379.

Hemme, C. L., Carley, R., Norton, A., Ghumman, M., Nguyen, H., Ivone, R., …& Cho, B. (2023). Developing virtual and augmented reality applications for science, technology, engineering and math education. *BioTechniques*, *75*(1), 11–20. https://doi.org/10.2144/btn-2023-0029

Hernández-Chávez, M., Cortés-Caballero, J. M., Pérez-Martínez, Á. A., Hernández-Quintanar, L. F., Roa-Tort, K., Rivera-Fernández, J. D., & Fabila-Bustos, D. A. (2021). Development of virtual reality automotive lab for training in engineering students. *Sustainability*, *13*(17), 9776.

Huang, H., Yin, J., Lv, F., Lin, Y., & Zou, J. (2024). A study on the impact of open source metaverse immersive teaching method on emergency skills training for medical undergraduate students. *BMC Medical Education*, *24*(1), 859.

Hwang, I., & Seo, R. (2025). Organizational support for digital transformation in the metaverse: A contingent pathway from user experience to digitalization resistance. *R&D Management*, *55*(2), 420–437.

Jamshidi, M., Yahya, S. I., Nouri, L., Hashemi-Dezaki, H., Rezaei, A., & Chaudhary, M. A. (2023). A high-efficiency diplexer for sustainable 5G-enabled IOT in metaverse transportation system and smart grids. *Symmetry*, *15*(4), 821.

Jensen, L., & Konradsen, F. (2018). A review of the use of virtual reality head-mounted displays in education and training. *Education and Information Technologies*, *23*, 1515–1529.

Jovanović, A., & Milosavljević, A. (2022). VoRtex Metaverse platform for gamified collaborative learning. *Electronics*, *11*(3), 317.

JPMorgan. (2022). *Opportunities in the metaverse: How businesses can explore the metaverse*. JPMorgan. Retrieved from https://www.jpmorgan.com

Kastrenakes, J. (2021). *Facebook is spending at least $10 billion this year on its metaverse division*. The Verge. Archived from the original on February 15, 2022. Retrieved February 14, 2022.

Kavanagh, S., Luxton-Reilly, A., Wuensche, B., & Plimmer, B. (2017). A systematic review of virtual reality in education. *Themes in Science and Technology Education*, *10*(2), 85–119.

Kumar, V. V., Carberry, D., Beenfeldt, C., Andersson, M. P., Mansouri, S. S., & Gallucci, F. (2021). Virtual reality in chemical and biochemical engineering education and training. *Education for Chemical Engineers*, *36*, 143–153.

Lopez, M., Arriaga, J. G. C., Álvarez, J. P. N., González, R. T., Elizondo-Leal, J. A., Valdez-García, J. E., & Carrión, B. (2021). Virtual reality vs traditional education: Is there any advantage in human neuroanatomy teaching? *Computers & Electrical Engineering*, *93*, 107282.

Lv, Z. (2023). Generative artificial intelligence in the metaverse era. *Cognitive Robotics*, *3*, 208–217.

Meta. (2025). *Empowering Meta Horizon creators with a new $50M fund and advanced tools*. Meta Horizon Blog. Retrieved from https://developers.meta.com/horizon/blog/empowering-horizon-creators-with-a-new-50m-fund-and-advanced-tools-for-mobile-and-mr-worlds/

Mihai, S., Yaqoob, M., Hung, D. V., Davis, W., Towakel, P., Raza, M., …& Nguyen, H. X. (2022). Digital twins: A survey on enabling technologies, challenges, trends and future prospects. *IEEE Communications Surveys & Tutorials*, *24*(4), 2255–2291.

Mourtzis, D., Angelopoulos, J., & Panopoulos, N. (2023). Blockchain integration in the era of industrial metaverse. *Applied Sciences*, *13*(3), 1353.

NVIDIA. (2024, November 20). *NVIDIA announces financial results for third quarter fiscal 2025*. NVIDIA Newsroom. Retrieved from https://nvidianews.nvidia.com/news/nvidia-announces-financial-results-for-third-quarter-fiscal-2025

Ommundsen, K., & Jumbe, I. V. (2023, July 10). *Global Collaboration Village: This is how the metaverse could impact the real world of work, training and education*. World Economic Forum. https://www.weforum.org/stories/2023/07/global-collaboration-village-metaverse-real-world-impact/

Paxinou, E., Georgiou, M., Kakkos, V., Kalles, D., & Galani, L. (2022). Achieving educational goals in microscopy education by adopting virtual reality labs on top of face-to-face tutorials. *Research in Science & Technological Education*, *40*(3), 320–339.

Paxinou, E., Panagiotakopoulos, C. T., Karatrantou, A., Kalles, D., & Sgourou, A. (2020). Implementation and evaluation of a three-dimensional virtual reality biology lab versus conventional didactic practices in lab experimenting with the photonic microscope. *Biochemistry and Molecular Biology Education*, *48*(1), 21–27.

Pérez, L., Diez, E., Usamentiaga, R., & García, D. F. (2019). Industrial robot control and operator training using virtual reality interfaces. *Computers in Industry*, *109*, 114–120.

Pirker, J., Holly, M., Lesjak, I., Kopf, J., & Gütl, C. (2019). Maroonvr—an interactive and immersive virtual reality physics laboratory. In P. Díaz, A. Ioannou, K. K. Bhagat, & J. M. Spector (Eds.), *Learning in a Digital World: Perspective on Interactive Technologies for Formal and Informal Education* (pp. 213–238). Springer Singapore. https://doi.org/10.1007/978-981-13-8265-9_11

Prakash, A., Haque, A., Islam, F., & Sonal, D. (2023). Exploring the potential of metaverse for higher education: Opportunities, challenges, and implications. *Metaverse Basic and Applied Research*, 2, 13.

Prisille, C., & Ellerbrake, M. (2020). Virtual reality (VR) and geography education: Potentials of 360°'experiences' in secondary schools. In D. Edler, C. Jenal, & O. Kühne (Eds.), *Modern Approaches to the Visualization of Landscapes* (pp. 321–332). Springer. https://doi.org/10.1007/978-3-658-30956-5_18

Radu, I., & Schneider, B. (2019, May). What can we learn from augmented reality (AR)? Benefits and drawbacks of AR for inquiry-based learning of physics. In *Proceedings of the 2019 CHI Conference on Human Factors in Computing Systems* (pp. 1–12). ACM. https://doi.org/10.1145/3290605.3300774

Reeves, S. M., Crippen, K. J., & McCray, E. D. (2021). The varied experience of undergraduate students learning chemistry in virtual reality laboratories. *Computers & Education, 175*, 104320.

Salah, B., Abidi, M. H., Mian, S. H., Krid, M., Alkhalefah, H., & Abdo, A. (2019). Virtual reality-based engineering education to enhance manufacturing sustainability in industry 4.0. *Sustainability, 11*(5), 1477.

Saparamadu, P. V. I. N., Sepasgozar, S., Guruge, R. N. D., Jayasena, H. S., Darejeh, A., Ebrahimzadeh, S. M., & Eranga, B. A. I. (2025). Optimising contract interpretations with large language models: A comparative evaluation of a vector database-powered Chatbot vs. ChatGPT. *Buildings, 15*(7), 1144.

Sargazi Moghadam, T., Darejeh, A., Delaramifar, M., & Mashayekh, S. (2024). Toward an artificial intelligence-based decision framework for developing adaptive e-learning systems to impact learners' emotions. *Interactive Learning Environments, 32*(7), 3665–3685.

Sepasgozar, S. M. (2020). Digital twin and web-based virtual gaming technologies for online education: A case of construction management and engineering. *Applied Sciences, 10*(13), 4678.

Shakirova, N., Said, N., & Konyushenko, S. (2020). The use of virtual reality in geo-education. *International Journal of Emerging Technologies in Learning (iJET), 15*(20), 59–70.

Shamim, N., Gupta, S., & Shin, M. M. (2024). Evaluating user engagement via metaverse environment through immersive experience for travel and tourism websites. *International Journal of Contemporary Hospitality Management*, 37(4), 1132–1174. https://doi.org/10.1108/IJCHM-10-2023-1590

Shen, Y., Wang, Z., Li, M., Yuan, J., & Gu, Y. (2022, February). An empirical study of geography learning on students' emotions and motivation in immersive virtual reality. *Frontiers in Education* (Vol. 7, 831619). https://doi.org/10.3389/feduc.2022.831619

Shi, F., Ning, H., Zhang, X., Li, R., Tian, Q., Zhang, S., …& Daneshmand, M. (2023). A new technology perspective of the Metaverse: Its essence, framework and challenges. *Digital Communications and Networks*, 10(6), 1653–1665. https://doi.org/10.1016/j.dcan.2023.02.017

Šiđanin, P., Plavšić, J., Arsenić, I., & Krmar, M. (2020). Virtual reality (VR) simulation of a nuclear physics laboratory exercise. *European Journal of Physics, 41*(6), 065802.

Singh, A., Ferry, D., Ramakrishnan, A., & Balasubramanian, S. (2020). Using virtual reality in biomedical engineering education. *Journal of Biomechanical Engineering, 142*(11), 111013.

Singh, G., Mantri, A., Sharma, O., & Kaur, R. (2021). Virtual reality learning environment for enhancing electronics engineering laboratory experience. *Computer Applications in Engineering Education, 29*(1), 229–243.

Soliman, M., Pesyridis, A., Dalaymani-Zad, D., Gronfula, M., & Kourmpetis, M. (2021). The application of virtual reality in engineering education. *Applied Sciences, 11*(6), 2879.

Son, J. M., Woo, S. H., Han, J. H., & Kim, J. J. (2023). Development and application of social studies content based on augmented virtual Metaverse: Jikji Simche Yojeol. *Journal of Digital Contents Society, 24*(6), 1201–1207.

Speicher, M., Hall, B. D., & Nebeling, M. (2019, May). What is mixed reality?. In *Proceedings of the 2019 CHI Conference on Human Factors in Computing Systems* (pp. 1–15). ACM. https://doi.org/10.1145/3290605.3300767

Swami, P. (2024). Metaverse: Transforming the user experience in the gaming and entertainment industry. In J. Kumar, M. Arora, & G. E. Bayram (Eds.), *Research, Innovation, and Industry Impacts of the Metaverse* (pp. 115–128). IGI Global. https://doi.org/10.4018/979-8-3693-2607-7.ch007

Thompson, M. M., Wang, A., Uz Bilgin, C., Anteneh, M., Roy, D., Tan, P., ...& Klopfer, E. (2020). Influence of virtual reality on high school students' conceptions of cells. *Journal of Universal Computer Science,* 26(8), 929–946. https://doi.org/10.3897/jucs.2020.050

Tibaldi, A., Bonali, F. L., Vitello, F., Delage, E., Nomikou, P., Antoniou, V., ...& Whitworth, M. (2020). Real world–based immersive Virtual Reality for research, teaching and communication in volcanology. *Bulletin of Volcanology*, *82*, 1–12.

Tsivitanidou, O. E., Georgiou, Y., & Ioannou, A. (2021). A Learning experience in inquiry-based physics with immersive virtual reality: Student perceptions and an interaction effect between conceptual gains and attitudinal profiles. *Journal of Science Education and Technology*, *30*(6), 841–861.

Tsou, M. H., & Mejia, C. (2023). Beyond mapping: Extend the role of cartographers to user interface designers in the Metaverse using virtual reality, augmented reality, and mixed reality. *Cartography and Geographic Information Science*, 51(5), 659–673. https://doi.org/10.1080/15230406.2023.2264748

Vergara, D., Extremera, J., Rubio, M. P., & Dávila, L. P. (2019). Meaningful learning through virtual reality learning environments: A case study in materials engineering. *Applied Sciences*, *9*(21), 4625.

VRLab Academy. (2025). *VRLab Academy*. Retrieved [12/2/2025], https://www.vrlabacademy.com/en

Wang, H., Ning, H., Lin, Y., Wang, W., Dhelim, S., Farha, F., ...& Daneshmand, M. (2023). A survey on the metaverse: The state-of-the-art, technologies, applications, and challenges. *IEEE Internet of Things Journal*, *10*(16), 14671–14688.

Wang, L., Chen, W., Yang, W., Bi, F., & Yu, F. R. (2020). A state-of-the-art review on image synthesis with generative adversarial networks. *Ieee Access*, *8*, 63514–63537.

Webster, Andrew (2022). *Epic announces $2 billion in funding for its metaverse efforts". The Verge*. Archived from the original on April 11, 2022. Retrieved April 11, 2022.

Wu, T. H., Wu, F., Liang, C. J., Li, Y. F., Tseng, C. M., & Kang, S. C. (2019). A virtual reality tool for training in global engineering collaboration. *Universal Access in the Information Society*, *18*, 243–255.

Yang, J., Liu, F., Wang, J., Kou, Z., Zhu, A., & Yao, D. (2021). Effect of virtual reality technology on the teaching of urban railway vehicle engineering. *Computer Applications in Engineering Education*, *29*(5), 1163–1175.

Zhai, X., Chu, X., Chai, C. S., Jong, M. S. Y., Istenic, A., Spector, M., ...& Li, Y. (2021). A review of Artificial Intelligence (AI) in education from 2010 to 2020. *Complexity*, *2021*(1), 8812542.

Zhao, R., Zhang, Y., Zhu, Y., Lan, R., & Hua, Z. (2023). Metaverse: Security and privacy concerns. *Journal of Metaverse*, *3*(2), 93–99.

2 Metaverse and Virtual Reality Integration in Mechanical Engineering

Applications in Virtual Laboratories and Training

Ali Darejeh, Guy Peter Chilcott, Eliza Lee, Ebrahim Oromiehie, Jingbang Zhou, and Samad M. E. Sepasgozar

1 Introduction

The rapid advancement of digital technologies has ushered in a new era of innovation across various fields, with the metaverse emerging as a transformative force in education, research, and industry. Defined as an interconnected, immersive, and interactive virtual environment, the metaverse offers unprecedented opportunities to revolutionize traditional engineering practices, particularly in mechanical engineering (Alghamdi, 2024). Through the integration of cutting-edge technologies such as virtual reality (VR), artificial intelligence (AI), and digital twins, the metaverse is poised to redefine how engineers simulate laboratory procedures and train for complex industrial tasks (Mystakidis, 2022; Sidhu et al., 2024a, 2024b).

This transformation is particularly timely, as the increasing complexity of engineering processes—coupled with growing demands for sustainability and operational efficiency—has highlighted the limitations of conventional training methods. Immersive learning environments, such as VR and metaverse-based simulations, effectively address these limitations by eliminating the physical constraints of traditional laboratories. They provide on-demand access, scalable instruction, and risk-free environments for experimentation. These features contribute to higher levels of student engagement, improved knowledge retention, and enhanced practical skills, making such technologies especially valuable in mechanical engineering education and industrial training (Ma et al., 2019; Kapilan, Vidhya, & Gao, 2021).

In this context, traditional mechanical engineering training—particularly laboratory sessions in areas like composite manufacturing and robotic manipulation—often presents significant barriers, including high costs, safety risks, limited accessibility, and logistical challenges. VR-based simulation environments offer targeted solutions to these issues by replicating complex physical processes within fully immersive, interactive, and hazard-free virtual settings. These simulations enable learners to engage with realistic scenarios through step-by-step instructional

DOI: 10.1201/9781003557555-2

systems and receive immediate feedback, thereby reducing cognitive overload while enhancing procedural understanding and safety (Falcão & Soares, 2013; Georgiou, Dimitropoulos, & Manitsaris, 2007). For instance, students can perform the hand-lay-up process for composites or operate industrial robot arms virtually, gaining essential hands-on experience without the need for physical materials or equipment.

This chapter explores the multifaceted applications of the metaverse in mechanical engineering by emphasizing laboratory simulation. It begins with a literature review, examining the historical evolution of VR in mechanical engineering and its implementation in manufacturing laboratories, robotics, and design environments (Ellis, 1991; Fisher, 1999; Syed et al., 2019). It then delves into contemporary advancements, highlighting how digital twins and extended reality (XR) technologies are reshaping both educational and industrial practices (Seth, Vance, & Oliver, 2001; Yahyaeian, 2023).

To illustrate these advancements, this chapter presents two empirical studies. The first investigates a VR-based laboratory for teaching composite manufacturing, demonstrating its impact on cognitive load and learning outcomes. The second compares VR and video-based learning for industrial robotics, showcasing how immersive simulations improve engagement and reduce mental effort. Through these studies, this chapter offers insights into how VR and the metaverse can bridge the gap between conceptual knowledge and practical experience in mechanical engineering.

2 Literature Review

This section reviews the integration and application of VR, XR, and metaverse technologies in engineering and mechanical engineering education, emphasizing their roles in virtual laboratory simulations, collaborative training, educational enhancement, and practical industry applications. It covers historical developments, contemporary advancements, and key applications.

2.1 Historical Development of VR in Engineering Education

VR technologies have evolved significantly from their emergence in the 1980s, primarily through computer-aided design (CAD) systems, to their current role in Industry 4.0 and education (Xu, Xu, & Li, 2018). Early VR applications included astronaut training in aerospace engineering (Ellis, 1991) and automotive industry simulations for design visualization, ergonomic assessments, and manufacturing process improvements (Fisher, 1999). Tools such as CATIA and Pro/ENGINEER were crucial in facilitating early VR adoption, allowing engineers and students to visualize and manipulate complex mechanical systems interactively (Seth, Vance, & Oliver, 2001).

As computing power and graphics rendering capabilities improved, the 2000s saw the emergence of VR platforms tailored to academic environments. Institutions began exploring VR not only for professional development but also as a pedagogical medium. These efforts laid the groundwork for what would become an increasingly immersive and learner-centered approach to engineering education.

2.2 VR, XR, and Virtual Laboratories in Mechanical Engineering

VR-based virtual laboratories have significantly transformed traditional training methods by offering accessible, interactive, and immersive simulations. These technologies proved especially valuable during the COVID-19 pandemic, enabling educational continuity despite widespread physical restrictions (Srinivasa et al., 2020). A growing body of research highlights the effectiveness of VR-based laboratories in enhancing student engagement, knowledge retention, and the development of cognitive and technical skills in mechanical engineering education. For instance, Srinivasa et al. (2020) reported that students trained with VR in material testing scored higher in assessments and expressed greater confidence. Ma et al. (2019) found that a VR module for manufacturing system design improved conceptual understanding and student satisfaction, while Kapilan, Vidhya, and Gao (2021) emphasized the benefits of interactive visualization for learning fluid mechanics. Similarly, Pfotenhauer and Gagnon (2021) developed a VR system for teaching thermodynamics, which was shown to improve knowledge retention and boost student engagement. Also, Hernández-Chávez et al. (2021) demonstrated that an automotive VR laboratory-enhanced design visualization, manufacturing simulation, and ergonomic assessment.

Building upon these advances, metaverse platforms are extending the potential of VR by enabling collaborative, distributed learning environments. Prasetya et al. (2023a, 2023b) demonstrated that virtual laboratories within metaverse platforms, particularly in CNC training, significantly improved both cognitive and practical skills. Chuang and Smith (2024) also showcased how cross-platform metaverse laboratories support immersive, remote training while fostering teamwork through collaborative interaction.

Complementing these educational innovations, recent advancements in nanomaterial-based flexible sensors have enhanced the realism and interactivity of VR and metaverse environments. As noted by Wang et al. (2023), these sensors—due to their flexibility, biocompatibility, and high sensitivity—enable precise user input and data collection, significantly enriching the learning experience. By facilitating more responsive and tactile virtual interactions, these technologies hold promise for improving training outcomes and providing a more authentic simulation of hands-on mechanical engineering tasks.

2.3 Integration of Metaverse, Digital Twins, and Industry 4.0

Digital twins and metaverse integration offer powerful tools for enhancing engineering education and training. Digital twins provide real-time, interactive virtual replicas of physical systems, supporting safe experimentation and system monitoring (Bordegoni & Ferrise, 2023; Khan & Mekid, 2024; Tadeja et al., 2020).

Nedeva, Dineva, and Atanasov (2025) emphasized that digital twins are integral to metaverse environments, enabling interactive and experiential learning that bridges the gap between physical and virtual worlds. This integration significantly enhances collaboration and the overall quality of education in engineering disciplines.

In addition to core engineering education, the metaverse has shown strong potential in engineering management and logistics. Ooi et al. (2024) demonstrated how metaverse-based platforms can improve decision-making processes, streamline supply chain logistics, and enhance knowledge management—factors critical for boosting organizational performance and competitiveness in engineering industries.

Advancements in VR infrastructure further support this trend. Xu et al. (2025) introduced wireless, multi-user interactive VR environments powered by mobile edge computing (MEC), which enable real-time collaborative training and significantly increase both interactivity and immersion in engineering education. Complementing this, Saravanan et al. (2025) developed the Battery Manufacturing Metaverse (BMM), a VR-based platform for geographically distributed collaborative learning, particularly in lithium-ion battery production. This underscores the role of VR and the metaverse in facilitating global, interdisciplinary engineering education.

To address platform compatibility, Jaimini, Zhang, and Brikis (2022) proposed the Industrial Metaverse Knowledge Graph (iMetaverseKG), a framework designed to overcome interoperability challenges among disparate engineering tools. Their solution enables seamless data exchange across virtual environments, providing a scalable infrastructure crucial for integrated digital twin modelling and collaborative engineering design—an essential feature for educational institutions adopting diverse toolsets.

The application of XR technologies in product design and demonstration has also gained traction. Chu et al. (2023, 2024) introduced a three-phase industrial metaverse framework that utilizes VR, AR, and MR for product visualization. Their case study on industrial cooler systems showcased how features like virtual exhibition halls, AR catalogs, and MR assessments can enrich technical understanding and support informed decision-making among students and industry stakeholders alike. Moreover, Alpala et al. (2022) illustrated how multi-user VR environments in smart factories can bolster collaboration and communication during training. These environments enable realistic 3D interactions and real-time synchronization between digital and physical systems, creating effective learning and operational scenarios.

This collaborative dimension is echoed by Bellalouna and Puljiz (2023), who applied metaverse-driven VR systems in engineering design reviews. Their study demonstrated that immersive environments enhance the review process and foster remote collaboration among dispersed teams. Such applications emphasize the importance of equipping engineering students with both technical and digital collaboration skills that align with contemporary industry practices. Similarly, Hosseinia et al. (2024) showed how immersive interaction within the metaverse enhances collaboration, monitoring, and process management in digital factories—demonstrating the profound educational and operational benefits of these innovations.

Extending beyond human-to-human collaboration, metaverse technologies are also transforming human-robot interaction in both educational and industrial contexts. Xie et al. (2024) introduced an XR-based system that integrates blockchain and digital twins to support real-time human-robot collaboration, system optimization, and remote control. This convergence of technologies reflects the metaverse's growing role in enabling more dynamic, responsive, and interconnected training and production systems.

The reviewed literature consistently highlights immersive technologies—such as VR, XR, and the metaverse—as transformative tools in engineering education. As educational institutions strive to modernize learning environments in line with the demands of Industry 4.0 and beyond, the integration of these technologies offers a compelling path forward. To further explore their application in mechanical engineering laboratory training, the following empirical studies examine the use of VR in composite manufacturing.

3 Empirical Studies

The following studies explore the impact of VR in engineering education, focusing on its effectiveness in reducing cognitive load, increasing engagement, and improving learning outcomes. The first study investigates a VR laboratory designed for hand lay-up composite manufacturing, while the second compares VR-based and video-based learning methods for training with industrial robot arms.

All participants across the studies were recruited in accordance with a human research ethics protocol approved by the University of New South Wales (UNSW) and sanctioned by its institutional review board.

3.1 *Empirical Study One: VR Laboratory for Hand-Lay-Up Composite Manufacturing*

This study investigates the cognitive load effects of a VR laboratory compared to a physical laboratory for teaching hand-lay-up composite part manufacturing. The research evaluates whether VR can affect cognitive load and enhance learning outcomes, engagement, and motivation. The study also examines user experiences, comparing mental demand, task completion rates, and participant feedback between VR and traditional laboratory environments. By applying Cognitive Load Theory (CLT), the research seeks to determine the feasibility of replacing physical mechanical engineering laboratory experiences with VR simulations.

3.1.1 Application Development for Study One

A VR simulation was developed using Unreal Engine 5 (UE5) for HTC VIVE PRO 2 headsets to replicate the mechanical engineering laboratory for composite manufacturing. As detailed in Chapter 1 of this book, this application was created as part of a broader metaverse platform initiative at the UNSW. The system included interactive elements, guided learning instructions, and simulated laboratory equipment to ensure an accurate replication of the real-world manufacturing process. This digital environment eliminated hazards, material costs, and equipment constraints, making it accessible to a broader range of students. In this application, users are required to use the VR controller to navigate to a designated area of the laboratory, cut the necessary materials, and transport them to the workspace. They then place the materials on a table, apply resin, and cover them. Finally, they connect a pump and a pipe to initiate the composite manufacturing process. Figure 2.1 presents the virtual mechanical engineering laboratory environment, while Figure 2.2 illustrates the virtual glass fiber mesh along with other materials involved in the composite manufacturing process.

Figure 2.1 The simulated mechanical engineering laboratory environment.

Figure 2.2 The virtual glass fiber, mesh, and other materials involved in the composite manufacturing.

Figure 2.3 depicts the cutting and mixing of materials to create the composite, and Figure 2.4 demonstrates the final stage of assembling all necessary components. The entire process is interactive and performed by users.

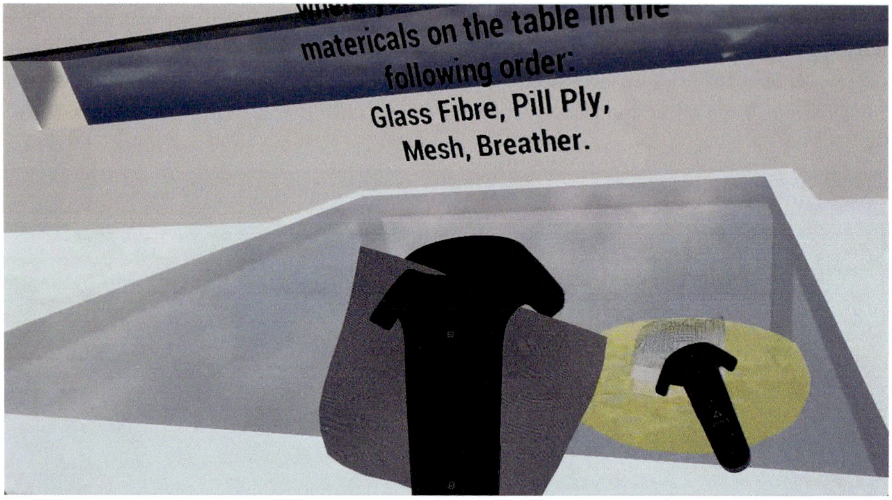

Figure 2.3 Cutting and mixing of materials to create the composite.

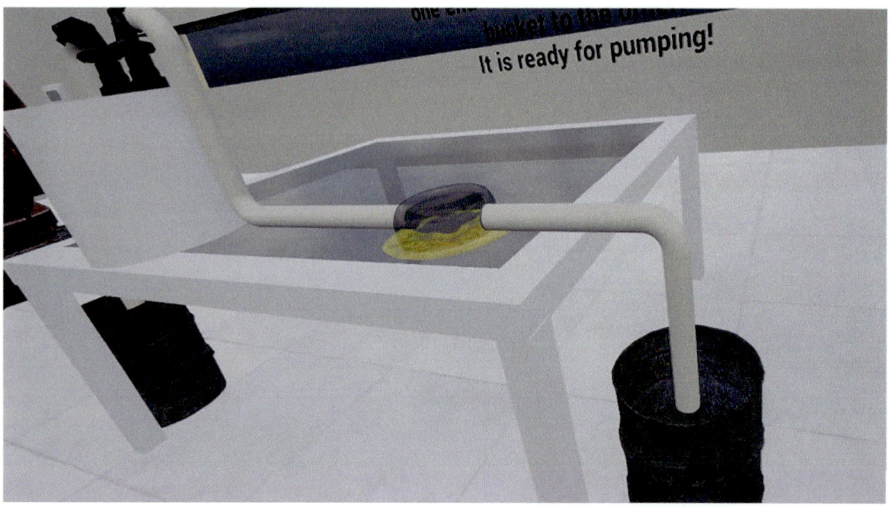

Figure 2.4 The final stage of assembling all necessary components to make the composite.

3.1.2 Methodology for Conducting Study One

The study recruited 17 participants aged 18–25 from the mechanical engineering school at UNSW, all of whom had no prior experience with the hand lay-up composite manufacturing process. The experiment was divided into three phases: pre-experiment, task engagement (in either the VR or physical laboratory), and post-experiment. In the pre-experiment phase, participants completed a questionnaire to assess their baseline knowledge of composite manufacturing to ensure they met the requirements for inclusion in the study. They also signed a consent form acknowledging their voluntary participation.

During the experimental phase, participants were randomly divided into two groups: 50% completed the physical laboratory task first, followed by the VR version, while the other 50% followed the reverse order. In the physical laboratory, participants performed the composite manufacturing procedure using real materials under the guidance of a tutor. In contrast, the VR laboratory simulated the same process within an immersive environment, incorporating a step-by-step instructional system to guide learners through the task. While participants completed each task, the researcher recorded the time taken and the number of steps completed to evaluate performance across both conditions.

After completing each version of the task—physical and VR—participants filled out the NASA-TLX Likert scale questionnaire (scored out of 15) to assess perceived cognitive load. The within-subject design ensured that each participant completed the questionnaire twice—once for each condition—enabling direct comparison. The NASA-TLX measures six dimensions of cognitive workload: mental demand, physical demand, temporal demand, performance, effort, and frustration.

Finally, in the post-experiment phase, participants took part in semi-structured interviews to share their feedback and reflections on both learning experiences.

3.1.3 Results from Study One

The study identified significant differences in task performance, cognitive load, and user experience between the VR and physical laboratory environments. Participants took significantly longer to complete the task in VR ($M = 12.46$, $SD = 2.31$) than in the physical laboratory ($M = 10.00$, $SD = 1.24$), $t(16) = 6.001$, $p < .001$. Although participants retained more procedural steps in VR ($M = 4.59$, $SD = 2.14$) than in the physical setting ($M = 3.94$, $SD = 0.97$), this difference was not statistically significant, $t(16) = 0.070$, $p = .945$, suggesting comparable effectiveness in retention.

However, cognitive load assessments using the NASA-TLX questionnaire indicated a clear advantage for VR. Mental demand was significantly lower in the VR environment ($M = 11.23$, $SD = 3.35$) than in the physical laboratory ($M = 13.82$, $SD = 1.67$), $t(16) = -4.278$, $p = .0006$. Similarly, physical demand was markedly reduced in VR ($M = 6.06$, $SD = 2.46$) compared to the physical laboratory ($M = 15.00$, $SD = 1.44$), $t(16) = -9.746$, $p < .0001$. Participants also reported significantly less effort in VR ($M = 3.41$, $SD = 1.85$) versus the physical environment ($M = 10.29$, $SD = 2.31$), $t(16) = -8.416$, $p < .0001$.

Despite these advantages, frustration levels were notably higher in VR ($M = 5.06$, $SD = 2.25$) than in the physical laboratory ($M = 1.71$, $SD = 1.83$), $t(16) = 4.858$, $p = .0002$. This increase in frustration is likely attributed to navigation challenges and motion-related discomfort inherent in immersive virtual environments.

Post-experiment interviews further revealed that all participants preferred the VR laboratory for future learning experiences. The gamified nature of VR was particularly engaging, with 75% of participants favoring it over traditional methods. However, some users found the VR controls slightly unintuitive, suggesting that further refinement in interface design could improve usability. While VR removed physical hazards and time constraints, some participants struggled with motion sickness, highlighting a limitation that must be addressed before widespread adoption.

3.1.4 Insights from Study One

The findings suggest that VR laboratories provide a viable and effective alternative to traditional laboratory settings, particularly for engineering education. The significant reduction in mental demand, physical effort, and task difficulty highlights VR as a promising tool for improving learning efficiency and accessibility, supporting the view that immersive, interactive environments enhance cognitive processing and learner engagement (Darejeh et al., 2024a, 2024b). These findings align with CLT, which suggests that reducing extraneous cognitive load can improve knowledge acquisition and retention (Darejeh, Marcus, & Sweller, 2022). In the VR setting, the elimination of physical hazards and the presence of a controlled environment contributed to an enhanced learning experience.

Despite these advantages, the study also highlights important limitations. Frustration levels were higher in VR, suggesting that user interface design and system usability must be optimized to ensure a seamless experience. Participants experienced motion sickness and navigation difficulties, which can negatively impact engagement and effectiveness. Addressing these issues through better ergonomic design, improved motion tracking, and adaptive control mechanisms will be essential for refining VR-based education.

From an educational perspective, VR laboratories hold great potential for institutions that lack physical resources for large-scale practical training. By eliminating equipment costs, safety risks, and material constraints, VR can democratize access to high-quality training experiences. Moreover, as VR technology advances, it could be used not only as a supplement but also as a complete replacement for traditional laboratory training in specific domains. However, careful design considerations are necessary to mitigate the challenges of adaptation, motion sickness, and frustration.

This research contributes to the broader discussion on technology-enhanced learning, demonstrating that VR can lower cognitive load while maintaining high engagement and motivation. While current limitations must be addressed, the overall results indicate that VR is not just a feasible alternative but a potentially superior medium for laboratory-based education. Further research should explore long-term retention rates, usability improvements, and broader applications of VR learning in other engineering disciplines and educational contexts.

3.2 Empirical Study Two: Industrial Robot Arms in VR Versus Video Tutorial

This study investigates the cognitive load effects of VR-based and video-based learning methods for manipulating industrial robot arms in manufacturing engineering laboratories. With the increasing adoption of VR for education and training, it is crucial to assess whether VR-based learning provides cognitive benefits over traditional instructional methods such as video-based learning, which is widely used for technical education. The research aims to determine whether VR learning reduces cognitive load, enhances learning efficiency, and improves user engagement and satisfaction compared to traditional video-based instruction.

3.2.1 Application Development for Study Two

The experiment utilized a VR-based simulation of an industrial robot arm and a video-based learning method to compare cognitive load and learning outcomes. The VR system, developed in UE5, allowed users to interactively manipulate a virtual robot arm using VR controllers, offering an immersive, real-time learning experience. As detailed in Chapter 1 of this book, this application was created as part of a broader metaverse platform initiative at UNSW. The video-based learning method provided a pre-recorded instructional video demonstrating the same robotic tasks, ensuring that the content remained identical between both learning conditions. Figure 2.5 shows the laboratory environment and the manufacturing robot. Figure 2.6 presents the robot's control panel, where users can press different buttons to move the robotic arm. Figure 2.7 illustrates how the robot operates and fabricates the cylinder-shaped part based on the adjustments made through the control panel.

Figure 2.5 The laboratory environment and the manufacturing robot.

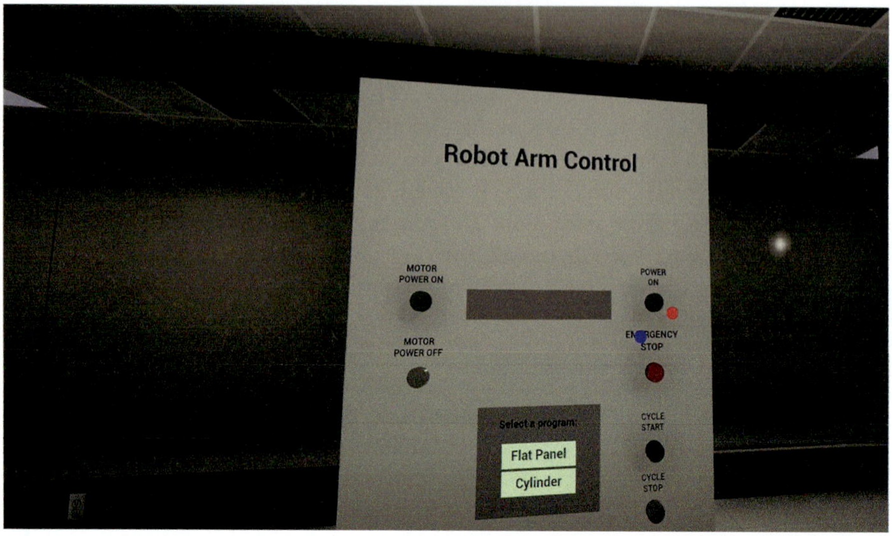

Figure 2.6 The robot's control panel.

Figure 2.7 The robot operates and fabricates the cylinder-shaped part.

3.2.2 Methodology for Conducting Study Two

A total of 12 participants were recruited, all of whom were tertiary-level engineering students at UNSW with prior exposure to manufacturing robotics. To allow for a within-subject comparison, each participant completed the experiment using both learning conditions—VR and video-based learning—on separate occasions. Fifty percent of the participants completed the VR version first, followed by the video tutorial, while the remaining fifty percent followed the reverse order.

The experiment followed a structured three-phase approach. In the pre-experiment phase, participants completed a pre-test questionnaire to assess their technical background, prior experience with VR and robotic systems, and learning preferences. This phase ensured that participants had comparable baseline knowledge before engaging with the instructional material.

In the learning phase, participants engaged in two different learning conditions. In the VR condition, participants wore VR headsets and used controllers to manipulate the robotic arm in a virtual training environment, following an interactive guided tutorial. In the video-based condition, participants watched a two-minute instructional video explaining the same robotic manipulation process but without any direct interactivity.

After completing the learning tasks through either video or VR, participants completed the NASA-TLX Likert scale questionnaire (scored out of 15) to assess perceived cognitive load. The within-subject design ensured that each participant completed the questionnaire twice—once after each learning condition—allowing for a direct comparison of cognitive load across modalities. Finally, in the post-experiment phase, participants took part in semi-structured interviews to share their feedback and reflections on both learning experiences.

3.2.3 Results from Study Two

Cognitive load assessments through the NASA-TLX questionnaire (nine-point scale) revealed that mental demand in the VR condition *($M = 4.11$, $SD = 2.33$)* was significantly lower than in the video condition *($M = 7.18$, $SD = 5.51$)*, $t(11) = -3.591$, $p = 0.0042$. However, this cognitive benefit came at the cost of increased physical demand, with VR participants reporting significantly higher physical effort *($M = 6.51$, $SD = 7.13$)* compared to video learners *($M = 1.05$, $SD = 2.20$)*, $t(11) = 5.625$, $p = 0.0002$.

In terms of temporal demand, VR users reported less time pressure *($M = 2.25$, $SD = 4.06$)* compared to video-based learners *($M = 5.42$, $SD = 4.70$)*, though this difference was not statistically significant, $t(11) = -1.450$, $p = 0.175$. This trend may reflect the self-paced nature of VR environments.

Regarding performance, ratings were comparable between conditions, with VR *($M = 2.70$, $SD = 4.33$)* slightly lower than video *($M = 3.79$, $SD = 3.05$)*, $t(11) = -0.373$, $p = 0.716$. Similarly, the effort required to complete tasks was slightly lower in VR *($M = 4.46$, $SD = 3.30$)* than in the video condition *($M = 5.43$, $SD = 4.79$)*, but this difference also did not reach statistical significance, $t(11) = 1.363$, $p = 0.200$.

One noted drawback of VR was the level of frustration reported. VR participants had higher frustration scores *($M = 5.80$, $SD = 6.01$)* compared to video learners *($M = 4.54$, $SD = 5.65$)*, though this difference was not statistically significant, $t(11) = -0.140$, $p = 0.891$. This increase in frustration may be attributed to navigation difficulties or motion-related discomfort in the immersive environment.

The post-questionnaire responses reinforced these findings, with most participants favoring VR for future training due to its immersive and interactive nature. They reported high satisfaction and viewed VR as an effective learning tool, though some found the initial adaptation challenging. Positive feedback highlighted the realism of the virtual environment, while suggestions included enhancing environmental details and providing clearer task instructions, such as an in-environment checklist. Participants appreciated VR's physical interactivity and cost-effectiveness but noted challenges like motion sickness, system glitches, and high immersion causing occasional discomfort.

3.2.4 Insights from Study Two

The results of this study provide valuable insights into the strengths and limitations of VR-based learning in industrial robotics education. VR exhibited clear advantages in reducing mental demand and improving temporal efficiency, suggesting that interactive learning methods may enhance cognitive engagement. The finding that VR increased physical demand aligns with previous research, which suggests that prolonged VR exposure can lead to discomfort and fatigue. This result indicates that VR-based learning environments should incorporate features that reduce physical strain, such as shorter learning sessions, optimized VR hardware, or alternative interaction techniques.

Another important consideration is the role of engagement and motivation in learning. Participants overwhelmingly preferred VR learning, despite the higher physical demand, suggesting that engagement plays a critical role in perceived learning effectiveness. The immersive nature of VR likely contributed to a higher sense of presence, which has been linked to improved learning retention and skill acquisition in prior studies. Qualitative feedback further reinforced these findings, with participants praising the realism of the VR environment, particularly its physics and movement systems. However, some suggested additional environmental details, such as fencing, to enhance the authenticity of the simulated laboratory experience.

Despite these benefits, motion sickness and user fatigue remain significant challenges. Several participants reported mild discomfort, which may affect long-term usability. Additionally, a few participants noted that high immersion levels sometimes led to a sense of disconnection from their surroundings, raising concerns about unintended anxiety. Others highlighted usability issues, such as overly sensitive controllers and the need for clearer task instructions, suggesting that an in-environment checklist could improve navigation and task completion.

Given that most participants favored VR for future learning and emphasized its cost-effectiveness in reducing risks and material expenses, addressing current usability concerns could further support its widespread adoption. Future research should focus on ergonomic improvements, such as optimizing frame rates, enhancing head tracking accuracy, and refining interaction mechanisms, to mitigate discomfort and improve the overall user experience.

4 Discussion

The integration of VR and metaverse technologies in mechanical engineering education marks a transformative shift from conventional pedagogies to immersive, learner-centered environments. This chapter presented two empirical studies—composite manufacturing and industrial robotics training—to explore the practical implications of simulation-based learning environments. The findings from these studies, combined with insights from current literature, suggest that VR and the metaverse are not merely supplementary tools but hold the potential to redefine how technical knowledge and hands-on skills are taught, practiced, and retained.

The composite manufacturing simulation demonstrated VR's capacity to reduce both mental and physical cognitive load while maintaining instructional clarity and procedural fidelity. Participants in the VR environment reported significantly lower levels of mental demand, effort, and physical strain, a result that aligns well with the principles of CLT (Sweller, 1988; Darejeh, Marcus, & Sweller, 2022). By minimizing extraneous load through structured guidance and interactivity, VR enables learners to allocate cognitive resources to meaningful learning processes. These findings reinforce similar results from other domains in mechanical engineering education, including material testing (Srinivasa et al., 2020) and fluid mechanics (Kapilan, Vidhya, & Gao, 2021), where VR use led to enhanced knowledge retention and improved learner confidence.

The robotics study further emphasizes VR's advantages over traditional learning formats. Participants reported lower mental and temporal demands in the VR condition compared to the video-based tutorial, while also expressing a clear preference for the immersive modality. These results confirm the growing body of research showing that VR's embodied learning environments improve spatial reasoning and procedural memory (Ma et al., 2019; Syed et al., 2019), both of which are critical in mastering mechanical systems. While physical demand and user frustration were slightly higher in VR—primarily due to the constraints of current hardware and interface design—participants still favored the immersive method for future learning. This supports the assertions of Halabi (2020) and Darejeh (2023) that immersive VR enhances learner motivation and satisfaction, even when moderate usability challenges are present.

Beyond the individual outcomes of the studies, their broader significance lies in how they reflect and contribute to the evolving role of the metaverse in mechanical engineering education. The composite manufacturing and industrial robotics simulations were not merely stand-alone applications but early examples of how immersive environments can serve as foundational components within interconnected, persistent digital ecosystems. As highlighted by Mystakidis (2022), the metaverse supports continuous, collaborative, and adaptive learning—qualities that are increasingly relevant as educational settings seek to mirror real-world engineering workflows. In this context, the VR composite laboratory, with its structured instructional flow and safety-enhanced interactivity, serves as a prototype for scalable, networked laboratories that could be accessed remotely and collaboratively by multiple users. Similarly, the robotics study—where learners manipulated robot arms through VR simulations—demonstrates the feasibility of integrating hands-on training into broader metaverse environments, such as Siemens Xcelerator or NVIDIA Omniverse (Chu et al., 2023), where real-time co-design and testing are core functionalities. These systems, if deployed in educational contexts, could prepare students for global, team-based design and manufacturing tasks (Bellalouna & Puljiz, 2023; Hosseinia et al., 2024).

The concept of digital twins is particularly relevant when reflecting on the simulation design in both studies. In the composite manufacturing VR laboratory, students engaged with a high-fidelity virtual replica of a physical laboratory setup, simulating every step from material handling to resin application and vacuum setup. While this system stopped short of full real-time synchronization with physical machinery, it nevertheless provided an experiential learning model consistent with the foundational principles of digital twins—namely, real-time interaction with system behavior in a simulated environment (Khan & Mekid, 2024). Likewise, the industrial robot arm simulation mirrored not just the structure but also the functional logic of a programmable robotic system, enabling learners to practice sequential control tasks. These simulations could be extended into full digital twin implementations by integrating real-time data or failure modeling, as envisioned by Khalaj et al. (2023) and explored in design courses using platforms like Saravanan et al.'s, (2025) BMM. Such developments would support systems-level thinking—critical for modern engineers navigating the complexities of evolving mechanical systems (Hawthorne, 2023; Yahyaeian, 2023).

The potential for AI-driven personalization within metaverse environments is closely reflected in the feedback from the robotics simulation study. Participants reported difficulties with navigation and task comprehension, highlighting the need for dynamic support mechanisms such as real-time prompts or contextual guidance—features that align well with the capabilities of AI-powered virtual tutors and emotion-aware systems (Sargazi Moghadam et al., 2024). For example, a virtual assistant capable of recognizing user hesitation or repeated mistakes could provide tailored feedback, thereby reducing frustration and enhancing learning efficiency. This concept is consistent with emerging AI frameworks proposed by Darejeh et al. (2024c), in which learner performance and emotional states inform adaptive instructional delivery. Additionally, the integration of large language models (LLMs) as interactive agents in virtual laboratories (Saparamadu et al., 2025) presents promising opportunities for hands-free tutoring, intelligent support, and contextualized guidance—further advancing learner autonomy and instructional responsiveness. Although not yet implemented in the studies, such AI-driven personalization represents a logical and impactful progression for future VR-based learning systems in the metaverse.

Despite the benefits observed for VR learning in both studies, several challenges remain. A notable concern across both studies was user discomfort, primarily related to motion sickness and unfamiliar control interfaces. These issues have been widely acknowledged in VR literature and are tied to the limitations of current hardware ergonomics and interaction fidelity (Ghazali et al., 2024; Kamińska et al., 2021). Furthermore, the lack of haptic feedback continues to restrict VR's ability to fully simulate tasks involving tactile or force-based interaction, such as torque application or surface testing. While progress is being made in integrating flexible sensors and advanced haptic devices (Wang et al., 2023), these systems have yet to reach mainstream deployment in educational settings.

Another constraint concerns platform interoperability and infrastructure requirements. Effective implementation of metaverse-based education relies on seamless integration between various systems—including CAD tools, digital twin platforms, and learning management systems (Jaimini, Zhang, & Brikis, 2022). Without standardized protocols, cross-platform compatibility remains a barrier to scalability. Additionally, while larger institutions may possess the financial and technical capacity to adopt these systems, smaller universities or organizations may face cost-related challenges (Ghazali et al., 2024; Ooi et al., 2024). These issues highlight the need for open-source solutions and institutional collaborations to make immersive education more equitable.

Finally, it is important to acknowledge that the limited sample size in the studies presented in this chapter constrains the generalizability of results from cognitive load assessments and user feedback. Ensuring statistical robustness and wider applicability will require future studies to engage larger, more heterogeneous participant groups reflective of the broader robotics and composite manufacturing training population.

5 Conclusion

This chapter has demonstrated how integrating VR and metaverse technologies into mechanical engineering education can transform traditional training methods by offering immersive, scalable, and cognitively efficient learning environments. The two studies—composite manufacturing and industrial robotics—highlight VR's effectiveness in reducing mental and physical demand, increasing engagement, and enhancing procedural understanding. Beyond their immediate educational impact, both studies illustrate how these technologies can serve as foundational elements of future metaverse-based ecosystems, incorporating digital twins, AI-driven personalization, and collaborative remote learning. Despite current limitations such as motion sickness and interface usability, ongoing advancements in haptics, AI agents, and interoperability promise to address these challenges. As institutions increasingly adopt immersive technologies, VR is poised to become not just an alternative but a central component of mechanical engineering education—bridging the gap between theory and practice in innovative, engaging ways.

References

Alghamdi, T. A. (2024). Harnessing metaverse technologies: A paradigm shift in engineering education and professional development. *International Journal of Computer Science and Information Security, 22*(5), 1–12.

Alpala, L. O., Quiroga-Parra, D. J., Torres, J. C., & Peluffo-Ordóñez, D. H. (2022). Smart factory using virtual reality and online multi-user: Towards a metaverse for experimental frameworks. *Applied Sciences, 12*(12), 6258. https://doi.org/10.3390/app12126258

Bellalouna, F., & Puljiz, D. (2023). Use case for the application of the industrial metaverse approach for engineering design review. *Procedia CIRP, 119*, 638–643. https://doi.org/10.1016/j.procir.2023.03.116

Bordegoni, M., & Ferrise, F. (2023). Exploring the intersection of metaverse, digital twins, and artificial intelligence in training and maintenance. *Journal of Computing and Information Science in Engineering, 23*(6), 060806. https://doi.org/10.1115/1.4062455

Chu, C. H., Baroroh, D. K., Pan, J. K., & Chen, S. M. (2023). An exemplary case of industrial metaverse: Engineering product demonstration using extended reality technologies. *International Journal of Precision Engineering and Manufacturing, 1*, 243–250. https://doi.org/10.57062/ijpem-st.2023.0038

Chu, C. H., Bernstein, W., Zhang, Y. W., Krishnamurthy, V. R., & Ma, J. (Eds.). (2024). Extended reality in design and manufacturing. *Journal of Computing and Information Science in Engineering, 24*(3), 030301.

Chuang, T. J., & Smith, S. (2024). A multi-user cross-platform hands-on virtual lab within the metaverse – The case of machining training. *Virtual Reality, 28*(1), 62. https://doi.org/10.1007/s10055-024-00974-5

Darejeh, A. (2023). Empowering education through eerp: A customizable educational vr escape room platform. In *2023 IEEE International Symposium on Mixed and Augmented Reality Adjunct (ISMAR-Adjunct)* (pp. 764–766). IEEE.

Darejeh, A., Marcus, N., & Sweller, J. (2022). Increasing learner interactions with E-learning systems can either decrease or increase cognitive load depending on the nature of the interaction. *L'Année psychologique, 122*(3), 405–437.

Darejeh, A., Huynh, L., Ji, X., Marcus, N., & Polly, P. (2024a). Assessing the efficacy of virtual reality-based laboratories versus real-life settings for PCR procedures, within medical science laboratories. In *2024 IEEE International Symposium on Mixed and Augmented Reality Adjunct (ISMAR-Adjunct)* (pp. 489–492). IEEE.

Darejeh, A., Mondal, M., Marcus, N., & Vassar, A. (2024b). Exploring the cognitive load effects of diverse virtual reality interaction methods in interactive educational platforms. In *2024 IEEE International Symposium on Mixed and Augmented Reality Adjunct (ISMAR-Adjunct)* (pp. 435–438). IEEE.

Darejeh, A., Moghadam, T. S., Delaramifar, M., & Mashayekh, S. (2024c). A framework for AI-powered decision making in developing adaptive e-learning systems to impact learners' emotional responses. In *2024 11th International and the 17th National Conference on E-Learning and E-Teaching (ICeLeT)* (pp. 1–6). IEEE.

Ellis, S. R. (1991). Nature and origin of virtual environments: A bibliographical essay. *Computing Systems in Engineering, 2*(4), 321–347.

Falcão, C. S., & Soares, M. M. (2013). Application of virtual reality technologies in consumer product usability. In *Design, User Experience, and Usability. Web, Mobile, and Product Design: Second International Conference, DUXU 2013, Held as Part of HCI International 2013, Las Vegas, NV, USA, July 21–26, 2013, Proceedings, Part IV 2* (pp. 342–351). Springer Berlin Heidelberg.

Fisher, M. (1999). The use of virtual reality in automotive design: Lessons from Ford's VR center. *Automotive Engineer, 24*(3), 17–20.

Georgiou, J., Dimitropoulos, K., & Manitsaris, A. (2007). A virtual reality laboratory for distance education in chemistry. *International Journal of Social Sciences, 2*, 306–313. https://doi.org/10.5281/zenodo.1082545

Ghazali, A. K., Ab. Aziz, N. A., Ab. Aziz, K., & Tse Kian, N. (2024). The usage of virtual reality in engineering education. *Cogent Education, 11*(1), 2319441. https://doi.org/10.1080/2331186X.2024.2319441

Halabi, O. (2020). Immersive virtual reality to enforce teaching in engineering education. *Multimedia Tools and Applications, 79*(4), 2987–3004. https://doi.org/10.1007/s11042-019-08214-8

Hawthorne, W. (2023). *Using immersive technology to improve mechanical design: Use cases, a review of current technology, and an experiment in requirement elicitation in virtual reality. Master's Thesis,* Clemson University.

Hernández-Chávez, M., Cortés-Caballero, J. M., Pérez-Martínez, Á. A., Hernández-Quintanar, L. F., Roa-Tort, K., Rivera-Fernández, J. D., & Fabila-Bustos, D. A. (2021). Development of virtual reality automotive lab for training in engineering students. *Sustainability, 13*(17), 9776. https://doi.org/10.3390/su13179776

Hosseinia, S., Abbasi, A., Magalhaes, L. G., Fonseca, J. C., da Costa, N. M. C., Moreira, A. H. J., & Borges, J. (2024). Immersive interaction in digital factory: Metaverse in manufacturing. *Procedia Computer Science, 232*, 2310–2320. https://doi.org/10.1016/j.procs.2024.02.050

Jaimini, U., Zhang, T., & Brikis, G. O. (2022). iMetaverseKG: Industrial metaverse knowledge graph to promote interoperability in design and engineering applications. *IEEE Internet Computing, 26*(6), 59–69. https://doi.org/10.1109/MIC.2022.3212085

Kamińska, D., Zwoliński, G., Wiak, S., Petkovska, L., Cvetkovski, G., Barba, P. D., … & Anbarjafari, G. (2021). Virtual reality-based training: Case study in mechatronics. *Technology, Knowledge and Learning, 26*, 1043–1059.

Kapilan, N., Vidhya, P., & Gao, X.-Z. (2021). Virtual laboratory: A boon to the mechanical engineering education during COVID-19 pandemic. *Higher Education for the Future, 8*(1), 31–46. https://doi.org/10.1177/2347631120970757

Khalaj, O., Jamshidi, M., Hassas, P., Mašek, B., Štadler, C., & Svoboda, J. (2023). Digital twinning of a magnetic forging holder to enhance productivity for Industry 4.0 and metaverse. *Processes, 11*(6), 1703. https://doi.org/10.3390/pr11061703

Khan, Z. H., & Mekid, S. (2024). Designing remote labs in outcome-based education using digital twin and metaverse. *Journal of Engineering Education and Practice*. IEEE. https://teaching.ieee.org/designing-remote-labs-in-outcome-based-education-using-digital-twin-and-metaverse/

Ma, J., Jaradat, R., Ashour, O., Hamilton, M., Jones, P., & Dayarathna, V. L. (2019). Efficacy investigation of virtual reality teaching module in manufacturing system design course. *Journal of Mechanical Design, 141*(1), 012002. https://doi.org/10.1115/1.4041428

Mystakidis, S. (2022). Metaverse. *Encyclopedia, 2*(1), 486–497.

Nedeva, V., Dineva, S., & Atanasov, S. (2025). Engineering education meets the metaverse. *Acta Infologica*. Advance online publication. https://doi.org/10.26650/acin.1612225

Ooi, K. B., Tan, G. W. H., Al-Emran, M., Al-Sharafi, M. A., Arpaci, I., Zaidan, A. A., …& Iranmanesh, M. (2024). The metaverse in engineering management: Overview, opportunities, challenges, and future research agenda. *IEEE Transactions on Engineering Management, 71*(1), 13882. https://doi.org/10.1109/TEM.2023.3307562

Pfotenhauer, J. M., & Gagnon, D. J. (2021, July). ThermoVR: A virtual laboratory to enhance learning in undergraduate thermodynamics. In *2021 ASEE Virtual Annual Conference Content Access.*

Prasetya, F., Fajri, B. R., Wulansari, R. E., Primawati, P., & Fortuna, A. (2023a). Virtual reality adventures as an effort to improve the quality of welding technology learning during a pandemic. *International Journal of Online and Biomedical (iJOE), 19*(2), 4–22. https://doi.org/10.3991/ijoe.v19i02.35447

Prasetya, F., Fortuna, A., Samala, A. D., Fajri, B. R., Efendi, F., & Nyamapfene, A. (2023b). Effectiveness of distance learning computer numerical control based on virtual laboratory using a metaverse platform to improve students' cognitive ability and practice skills. *International Journal of Interactive Mobile Technologies, 17*(24), 4–21. https://doi.org/10.3991/ijim.v17i24.45019

Saparamadu, P. V. I. N., Sepasgozar, S., Guruge, R. N. D., Jayasena, H. S., Darejeh, A., Ebrahimzadeh, S. M., & Eranga, B. A. I. (2025). Optimising contract interpretations with large language models: A comparative evaluation of a vector database-powered Chatbot vs. ChatGPT. *Buildings*, *15*(7), 1144.

Saravanan, S., Vijay, U., Tran, S., Mathew, M. M., Apostolova, D. Y., Gandarias, I., Leclere, A., Lelong, R., & Franco, A. A. (2025). Metaverse for battery manufacturing: Connecting students from different geographical locations to solve battery manufacturing problems in the virtual reality space. *Batteries & Supercaps*. Advance online publication. https://doi.org/10.1002/batt.202500098

Sargazi Moghadam, T., Darejeh, A., Delaramifar, M., & Mashayekh, S. (2024). Toward an artificial intelligence-based decision framework for developing adaptive e-learning systems to impact learners' emotions. *Interactive Learning Environments*, *32*(7), 3665–3685.

Seth, A., Vance, J. M., & Oliver, J. H. (2001). Virtual reality for assembly methods prototyping: A review. *Virtual Reality*, *5*(3), 161–174.

Sidhu, M. S., Lee, C. K., Hassan, Z., Mousakhani, S., Sidhu, K. K., & Anuar, A. (2024a). Preliminary study on the perception of engineering students on metaverse. In *2024 IEEE 6th Symposium on Computers & Informatics (ISCI)* (pp. 12–17). IEEE.

Sidhu, M. S., Mousakhani, S., Lee, C. K., & Sidhu, K. K. (2024b). Educational impact of metaverse learning environment for engineering mechanics dynamics. *Computer Applications in Engineering Education, 32*(3), e22772. https://doi.org/10.1002/cae.22772

Srinivasa, A. R., Jha, R., Ozkan, T., & Wang, Z. (2020). Virtual reality and its role in improving student knowledge, self-efficacy, and attitude in the materials testing laboratory. *International Journal of Mechanical Engineering Education, 49*(4), 382–409. https://doi.org/10.1177/0306419019898824

Sweller, J. (1988). Cognitive load during problem solving: Effects on learning. *Cognitive Science*, *12*(2), 257–285.

Syed, Z. A., Trabookis, Z., Bertrand, J., Chalil Madathil, K., Hartley, R. S., Frady, K. K., …& Gramopadhye, A. K. (2019). Evaluation of virtual reality based learning materials as a supplement to the undergraduate mechanical engineering laboratory experience. *International Journal of Engineering Education*, *35*(3), 1–11.

Tadeja, S. K., Seshadri, P., & Kristensson, P. O. (2020). AeroVR: An immersive visualisation system for aerospace design and digital twinning in virtual reality. *The Aeronautical Journal*, *124*(1280), 1615–1635.

Wang, J., Suo, J., Song, Z., Li, W. J., & Wang, Z. (2023). Nanomaterial-based flexible sensors for metaverse and virtual reality applications. *International Journal of Extreme Manufacturing, 5*, 032013. https://doi.org/10.1088/2631-7990/acded1

Xie, J., Liu, Y., Wang, X., Fang, S., & Liu, S. (2024). A new XR-based human-robot collaboration assembly system based on industrial metaverse. *Journal of Manufacturing Systems, 74*, 949–964. https://doi.org/10.1016/j.jmsy.2024.05.001

Xu, C., Chen, Z., Tao, M., & Zhang, W. (2025). Wireless multi-user interactive virtual reality in metaverse with edge-device collaborative computing. *IEEE Transactions on Wireless Communications*. Advance online publication. https://doi.org/10.1109/TWC.2025.3551938

Xu, L. D., Xu, E. L., & Li, L. (2018). Industry 4.0: state of the art and future trends. *International Journal of Production Research*, *56*(8), 2941–2962.

Yahyaeian, A. A. (2023). Enhancing mechanical engineering education through a virtual instructor in an Ai-driven virtual reality fatigue test lab (Doctoral dissertation).

3 Metaverse and Virtual Reality Integration in Electrical Engineering

Applications in Virtual Laboratories and Training

Arifah Anwar Fadzil, Ali Darejeh, Rovielyn Espiritu, Sudaksh Tiwari, Neelam Samachar, Wendy Lee, and Samad M. E. Sepasgozar

1 Introduction

The integration of virtual reality (VR) and the metaverse into engineering education represents a groundbreaking shift in how students engage with complex technical concepts. Traditional engineering education relies heavily on a combination of theoretical instruction and physical laboratory work. However, access to cutting-edge laboratory equipment is often constrained by financial limitations, logistical barriers, and safety concerns. These challenges necessitate the exploration of innovative, technology-driven solutions to enhance the learning experience (Lin et al., 2024).

The metaverse extends the capabilities of VR by creating persistent, collaborative virtual spaces where students and instructors can interact regardless of geographical constraints. In an engineering education context, the metaverse allows students to collaborate on design projects, troubleshoot complex systems, and engage in interactive problem-solving activities (Sidhu et al., 2024a). This represents a significant evolution from static, text-based learning to dynamic, experiential education (Lin et al., 2024). Additionally, the metaverse plays a crucial role in expanding access to engineering education. Traditional laboratory-based learning is often limited by physical infrastructure, requiring students to be present in designated laboratory spaces. The metaverse eliminates these constraints by allowing students to access virtual laboratories from anywhere in the world. This is particularly beneficial for remote learners, international students, and individuals who may face barriers to attending in-person classes. Virtual laboratories provide a scalable solution to the increasing demand for engineering education while maintaining high-quality learning experiences (Darejeh et al., 2022a; Zhang et al., 2024).

Building on these capabilities, a growing body of research has explored how VR—an integral component of the metaverse—is being applied in specific engineering disciplines such as electrical engineering. These studies highlight VR's potential to enhance conceptual understanding and practical skill development. For

DOI: 10.1201/9781003557555-3

example, VR-based electrical engineering simulations enable students to interact with virtual circuit boards, analyze signal behaviors, and practice troubleshooting techniques without the risk of damaging real-world components (Asham, Bakr, & Emadi, 2023; Shi et al., 2011). Furthermore, VR environments allow students to experiment with advanced electrical engineering equipment that may be cost-prohibitive or otherwise inaccessible in traditional university laboratories (Liang & Xiaoming, 2013; Salutina, Platunina, & Frank, 2024).

Overall, VR offers a promising alternative to traditional engineering education by providing immersive, interactive environments that closely mimic real-world settings (Darejeh et al., 2024a). By leveraging VR, students can explore engineering concepts in a simulated space, manipulate digital models of laboratory equipment, and conduct experiments with unprecedented flexibility. Unlike conventional simulation software, VR introduces depth perception, spatial interaction, and real-time feedback, allowing for a more comprehensive understanding of engineering principles (Wallace et al., 2010).

One of the most significant advantages of VR and the metaverse in engineering education is their ability to bridge the gap between theory and practice. Many students struggle with abstract engineering concepts when learning solely from textbooks and lectures. VR enables students to visualize and manipulate complex systems, leading to a deeper understanding of theoretical frameworks. Studies have demonstrated that VR-based learning improves knowledge retention and engagement compared to traditional methods, making it a valuable tool in engineering education (Valdez et al., 2015).

In addition to improving conceptual clarity, immersive technologies contribute meaningfully to student motivation and engagement (Darejeh et al., 2024a, 2024b). Traditional instructional methods often rely on passive learning, where students absorb information without interaction. VR transforms this into an active learning process, allowing students to experiment with virtual prototypes, simulate engineering failures, and iterate on designs in real time. This hands-on, exploratory approach fosters curiosity and encourages deeper engagement with engineering concepts (Orlosky et al., 2017).

Moreover, VR has the potential to address cognitive load issues associated with learning complex technical subjects (Darejeh, 2023). By breaking down intricate engineering concepts into interactive simulations, VR helps students process information more effectively. Immersive learning environments provide contextual cues that aid in reducing the mental effort required to grasp difficult topics. Furthermore, VR's multimodal approach—incorporating visual, auditory, and haptic feedback—caters to diverse learning styles, ensuring a more inclusive educational experience (Paas et al., 2003; Van Merrienboer & Ayres, 2005).

This chapter explores the rise of metaverse and immersive technologies and their transformative impact on engineering education, with a particular focus on their application in electrical engineering laboratories. It begins with a comprehensive literature review on the use of VR and the metaverse in engineering education, followed by a detailed presentation of our empirical study on simulating electrical engineering laboratories in VR. By examining how these technologies enhance learning, this chapter aims to assess their potential for shaping the future of education and their suitability for integration into current curricula.

2 Literature Review

The following literature review explores the role of VR and the metaverse in engineering education, focusing on their application in electrical engineering. The section begins with an overview of how VR enhances learning through simulation, addressing gaps in traditional engineering education tools. It then examines the specific application of VR and metaverse-based learning environments in electrical engineering education, highlighting their role in virtual laboratories, practical skill development, and collaborative problem-solving. Finally, the challenges of these technologies are discussed, assessing their potential to transform engineering education.

2.1 Application of VR and Metaverse in Engineering Education

VR has revolutionized engineering education by offering interactive simulations of real-world environments, enabling immersive and engaging learning experiences (Burdea, 2003; Sidhu et al., 2024a). Through VR simulations, students can interact with digital models of laboratory equipment, manipulate virtual prototypes, and conduct experiments that closely replicate hands-on scenarios. This immersive, active learning approach fosters curiosity, enhances motivation, and facilitates a deeper understanding of engineering principles (Asham, Bakr, & Emadi, 2023; Orlosky et al., 2017).

The use of VR in engineering education has evolved significantly since its early adoption in flight simulators, such as the Visually Coupled Airborne Systems Simulator (VCASS), which demonstrated the effectiveness of simulation-based training in developing practical skills and reducing instructional costs (Mazuryk & Gervautz, 1995). Today, VR is widely applied across multiple engineering subfields, including mechanical engineering, civil engineering, computer-aided design, and electrical engineering. It allows students to simulate machinery operations, test virtual components, and visualize complex systems—experiences that are often constrained in physical laboratories due to budget, infrastructure, or safety limitations (Soliman et al., 2021).

VR also addresses the limitations of traditional engineering simulation tools, such as MATLAB, TinkerCAD, and Autodesk Maya, which lack depth perception, real-time interactivity, and full-body engagement (Asham, Bakr, & Emadi, 2023). By offering full immersion and real-time interaction, VR bridges the gap between theoretical knowledge and practical application (Valdez et al., 2015), allowing students to gain firsthand experience with engineering systems in a safe, controlled environment.

The increasing affordability of VR headsets and the development of user-friendly interactive software have further accelerated its integration into engineering curricula. Unlike physical laboratories, which require significant infrastructure and pose potential safety risks, VR-based simulations provide scalable and cost-effective alternatives, enabling students to experiment freely without fear of damaging equipment or incurring additional costs (Shi et al., 2011). Moreover, the incorporation of multimodal feedback—visual, auditory, and haptic—enhances knowledge retention and supports diverse learning preferences (Paas et al., 2003; Van Merrienboer & Ayres, 2005).

In addition, advanced VR features such as body-tracking sensors allow instructors to monitor student progress and performance through data analytics. These insights help educators identify knowledge gaps and tailor instruction accordingly—particularly in simulations that require procedural accuracy and hands-on experience (Caserman et al., 2019). By embedding such technologies into engineering education, institutions can offer more flexible, responsive, and inclusive learning environments.

Extending beyond the standalone capabilities of VR, metaverse-based platforms introduce persistent, collaborative, and globally accessible virtual spaces. These environments eliminate geographical barriers, enabling students to participate in virtual laboratories, engage in group projects, and interact with instructors from anywhere in the world. Such platforms significantly enhance accessibility and inclusivity, supporting broader participation in engineering education (Alghamdi, 2024; Zhang et al., 2024). The metaverse also introduces opportunities for interactive lectures and collaborative learning. Students can participate in multiplayer educational games that reinforce engineering concepts while improving engagement and teamwork (Franco et al., 2023). As metaverse-based learning environments continue to evolve, they offer new ways for students to interact with engineering concepts beyond traditional classroom settings.

2.2 Application of VR and Metaverse in Electrical Engineering Education

Electrical engineering education is highly reliant on laboratory-based experimentation. VR facilitates the development of fully functional virtual laboratories where students can simulate circuit design, test components, and troubleshoot electrical systems. Real-time visualization of electrical grids, power systems, and smart circuits enhances students' ability to comprehend and apply theoretical knowledge in practical scenarios (Jamil et al., 2022). Studies have demonstrated that VR-based electrical engineering simulations improve conceptual understanding and practical skills by allowing students to interact with virtual circuit boards and analyze signal behaviors in a risk-free environment (Shi et al., 2011). Virtual laboratories offer students the ability to perform electrical engineering experiments remotely, making them particularly beneficial for online and distance learning programs (Salutina, Platunina, & Frank, 2024).

One example is the Virtual Electrical Lab (VE Lab), which divides learning into theoretical instruction and interactive simulation. Research findings indicate that students exposed to VE Lab environments demonstrate improved retention and comprehension of direct-current (DC) electrical circuits (Albarracin-Acero et al., 2024). Also, research by Ghazali et al. (2024) indicates that VR-based laboratories in electromagnetism allowed students to interact with electric fields and forces in a controlled virtual environment, leading to improved comprehension of electromagnetic wave propagation. VE Labs can replicate a wide range of scenarios, from basic circuit construction to advanced power grid management, allowing students to develop technical proficiency in a controlled environment. The ability to simulate errors and failures also enhances troubleshooting skills, preparing students for real-world engineering challenges.

VR's ability to replicate real-world applications extends beyond theoretical instruction. Studies have shown that VR enhances students' understanding of technical concepts, such as oscilloscope usage and circuit debugging (Hasanah et al., 2017). Simulated experiments provide hands-on practice without the risk of equipment damage, allowing students to build confidence and reduce cognitive load when handling real-world electrical devices (Singh et al., 2021). The interactive nature of VR allows students to engage with complex electrical components, visualize electromagnetic fields, and test circuit behavior under various conditions, all of which contribute to a more profound learning experience.

The metaverse offers an advanced framework for collaborative learning in electrical engineering by combining immersive simulations with real-time interactions. Students can engage in complex problem-solving exercises—such as circuit diagnostics, system optimization, and control logic testing—within realistic virtual environments. They can also network with peers in shared virtual laboratories and visualize intricate electrical systems in three dimensions, enhancing both spatial understanding and technical competence (Lin et al., 2024). The collaborative potential of the metaverse enables students to work on group projects such as designing smart grid systems, developing automated control panels, or simulating power distribution networks in real time. These experiences promote essential teamwork and communication skills that are critical for success in engineering careers (Zhang et al., 2024).

2.3 Challenges of Using VR and Metaverse in Engineering Education

Implementing VR and metaverse-based education involves substantial investment in hardware, software, and faculty training. Institutions must also ensure system compatibility and the seamless operation of VR applications to maintain usability and learning continuity (Alnagrat et al., 2022). However, poorly designed VR environments may lead to cognitive overload, reducing the effectiveness of the learning experience. Additionally, issues such as motion sickness and physical discomfort are common with prolonged VR headset use, which can hinder student engagement and limit session duration (Kirschner, 2002).

Integrating VR into existing engineering curricula presents further challenges, as it often requires extensive modifications to course content, learning outcomes, and assessment strategies. Aligning traditional course objectives with immersive, simulation-based learning experiences remains a complex task for educators (Liegmann et al., 2022). Moreover, while VR provides highly immersive environments, it cannot fully replicate the tactile and sensory experiences of physical laboratories. As such, access to real-world laboratory settings may still be necessary for comprehensive hands-on training in certain disciplines (Long et al., 2005). Finally, the adoption of metaverse learning environments introduces important considerations around data privacy, intellectual property protection, and the broader ethical implications of fully virtualized education (Huggett, 2020).

To further evaluate the impact of VR and metaverse-based environments on electrical engineering education, the next section presents an empirical study on the development of an electrical engineering VR laboratory. This study assesses the effectiveness of VR laboratories on students' learning performance, providing insights into the potential benefits and limitations of immersive learning technologies.

3 Empirical Study

This study presents two experiments designed to evaluate the effectiveness of VR in electrical engineering education. The first experiment investigates the use of VR as a substitute for traditional physical laboratories, focusing on its impact on students' understanding and practical skills in assembling different components on circuits. The second experiment examines the role of VR in improving learning outcomes related to digital oscilloscopes, comparing VR-based training to traditional circuit simulation tools like TinkerCAD. By analyzing engagement, motivation, and learning performance, these experiments provide insights into the feasibility of integrating VR into electrical engineering curricula.

All participants in both experiments were recruited in accordance with a human research ethics protocol approved by the University of New South Wales (UNSW) and sanctioned by its institutional review board.

3.1 Experiment One: VR Simulation in Circuit Assembly Education

In this experiment, we investigated the effectiveness of VR simulations in enhancing students' understanding and practical skills related to circuit assembly. Specifically, the study focused on students' ability to assemble components such as resistors, LED lights, and connecting wires within a virtual environment. The research evaluated whether VR could enhance students' understanding of electrical laboratory equipment and improve their hands-on proficiency, compared to performing the same procedures in a physical electrical engineering laboratory.

3.1.1 Application Development

A VR-based electrical engineering laboratory was developed using Unreal Engine 5 (UE5) and deployed on HTC VIVE Pro 2 headsets. As outlined in Chapter 1 of this book, this application was created as part of a broader metaverse platform initiative at the UNSW. The system featured a landing page (Figure 3.1) where users could select a topic related to circuits to explore. After selecting the desired topic, users were presented with learning instructions and interactive challenges related to that topic (Figure 3.2). They could then perform real-time circuit manipulation and assemble their own circuits based on the provided instructions (Figure 3.3). To test their circuits, users had access to an interactive oscilloscope (Figure 3.4).

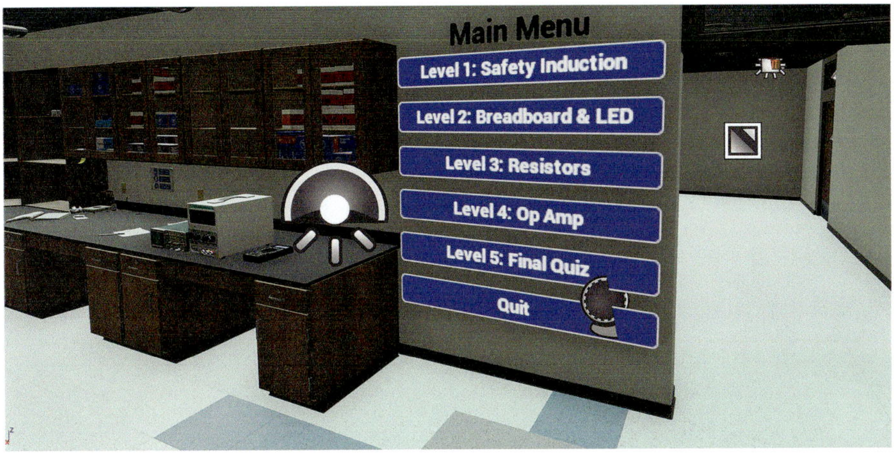

Figure 3.1 Our VR-based electrical engineering laboratory landing screen.

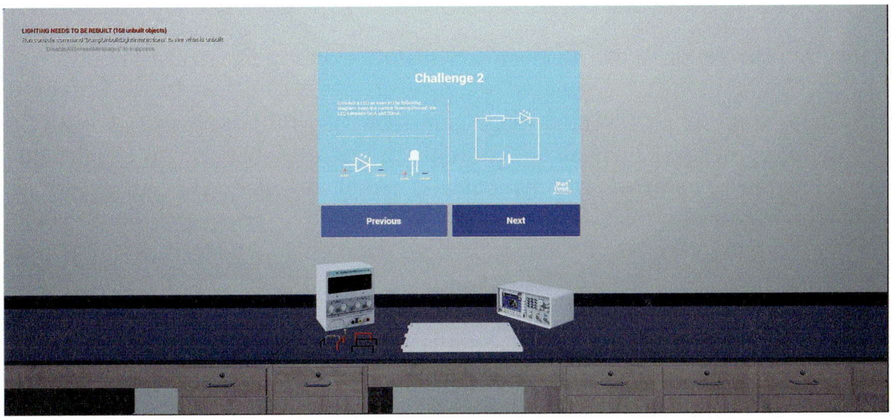

Figure 3.2 Guided learning instructions for making a circuit.

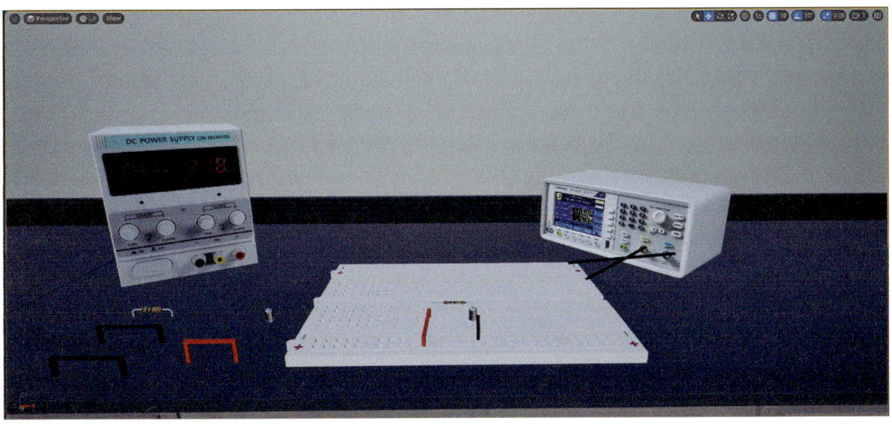

Figure 3.3 Users can make their own circuit by assembling wires, resistors, and LEDs.

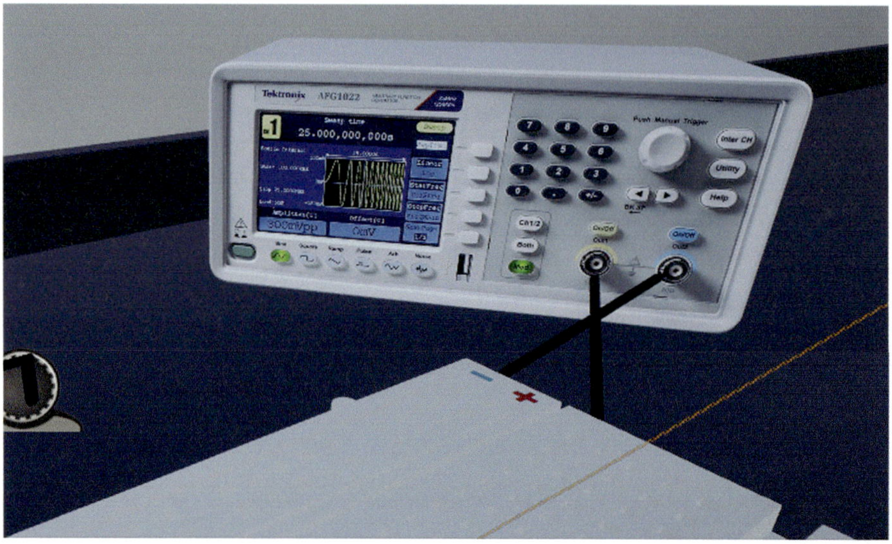

Figure 3.4 The VR interactive oscilloscope visualizes electrical signals dynamically.

An oscilloscope is an essential diagnostic tool used to display and analyze the waveform of electronic signals in real time. In this VR laboratory, it enabled students to adjust parameters and dynamically visualize voltage changes and signal behavior in their electrical circuits.

The designed VR environment replicates a traditional laboratory, enabling users to wire breadboards, connect circuit components, and troubleshoot issues. It provides real-time feedback and interactive learning experiences to enhance student engagement. Additionally, the VR simulation allows learners to practice circuit construction without physical risks or material costs, making it a viable alternative for institutions with budget constraints.

3.1.2 Methodology for Conducting Experiment One

This experiment involved ten participants enrolled in an electrical engineering course at UNSW, who were recruited to evaluate the comparative effectiveness of VR simulations versus traditional physical laboratories. To control for order effects, participants were randomly assigned to two groups. The first group completed the VR simulation before performing the same task in the physical laboratory, while the second group followed the reverse sequence. Tasks in both environments were equivalent in scope and complexity to ensure consistency across conditions.

To evaluate learning performance, quantitative data were collected during both sessions to assess participants' task execution. Key performance metrics included the number of errors made during circuit assembly (e.g., incorrect component placement), the number of hints requested (i.e., assistance prompts provided by the system or instructor), and the total time taken to complete each task.

3.1.3 Results from Experiment One

The mean completion time was M = 152.49, SD = 27.45 seconds for the VR experiment, compared to M = 126.69, SD = 22.80 seconds in the physical laboratory. A paired-samples t-test indicated that this difference was statistically significant, t(9) = 2.69, p = 0.025, suggesting that VR tasks required additional cognitive effort.

The average number of participant errors was M = 2.83, SD = 0.51 in the VR environment and M = 3.25, SD = 0.59 in the physical laboratory. Contrary to expectations, this difference was statistically significant, t(9) = −2.36, p = 0.042, indicating fewer errors in the VR condition.

The number of hints requested was also slightly higher in VR (M = 3.67, SD = 0.66) than in the physical laboratory (M = 3.15, SD = 0.57). However, this difference was not statistically significant, t(9) = 1.37, p = 0.205.

3.1.4 Insights from Experiment One

The results of Experiment 1 suggest that VR simulations can serve as an effective alternative to traditional physical laboratories in electrical engineering education. While tasks took slightly longer to complete in VR—likely due to participants' unfamiliarity with the technology—students made significantly fewer errors in the VR environment compared to the physical laboratory. This indicates that the guided and interactive nature of the VR system may support better procedural accuracy. Although more hints were requested in VR, the difference was not statistically significant. Overall, the findings highlight the potential of VR to enhance hands-on learning experiences while offering a scalable and accessible solution. Minor usability limitations, such as control sensitivity, should be addressed to further improve the learning experience.

3.2 Experiment Two: VR Simulation in Digital Oscilloscope Education

This experiment evaluates whether VR simulations can serve as a superior alternative to desktop simulation applications, such as TinkerCAD, particularly in teaching digital oscilloscopes. TinkerCAD is a widely adopted online 3D modelling and circuit simulation platform that enables students to design and test electrical circuits in a virtual environment. It provides an accessible and interactive means of learning circuit design without requiring physical components, making it a practical and widely used resource in engineering education.

3.2.1 Methodology for Conducting Experiment Two

This study involved 12 undergraduate students enrolled in an electrical engineering course at UNSW. Participants were randomly assigned to two groups of six. One group used TinkerCAD, a desktop-based simulation tool, to design a circuit and test its voltage using a digital oscilloscope. The second group performed the same task using the VR application developed in Experiment 1. Each participant engaged in a 30-minute guided learning session within their assigned environment.

Following the session, all participants completed a short closed-book exam consisting of structured questions designed to assess learning performance and conceptual understanding. In addition, a feedback survey was administered to evaluate user experience, focusing on usability, engagement, and perceived learning effectiveness. Collected data were analyzed to compare outcomes between the two groups and determine whether the VR simulation offered educational advantages over the traditional desktop tool.

3.2.2 Results from Experiment Two

A one-way ANOVA revealed a statistically significant difference in task completion times (measured in minutes), with the VR group completing tasks faster ($M = 2.52$, $SD = 0.68$) compared to the TinkerCAD group ($M = 3.53$, $SD = 0.65$), $F(1,10) = 7.44$, $p = 0.021$. While the VR group also showed higher perceived helpfulness ($M = 2.99$, $SD = 0.83$) than the TinkerCAD group ($M = 2.83$, $SD = 0.75$), this difference was not statistically significant, $F(1,10) = 2.60$, $p = 0.138$. Similarly, confidence levels were slightly higher in the VR group ($M = 3.83$, $SD = 0.75$) compared to the TinkerCAD group ($M = 3.67$, $SD = 0.82$), but this difference was also not statistically significant, $F(1,10) = 1.48$, $p = 0.251$. Knowledge scores followed the same trend, with the VR group scoring slightly higher ($M = 3.33$, $SD = 0.79$) than the TinkerCAD group ($M = 3.17$, $SD = 0.75$), though this difference was not significant, $F(1,10) = 1.35$, $p = 0.272$.

Based on the qualitative data, the study found notable differences in knowledge retention, engagement, and perceived ease of learning between the VR and non-VR groups. 73% of students in the VR group reported that the immersive environment helped them better understand oscilloscope functionality. 67% of participants indicated that the VR interface significantly enhanced their spatial awareness and conceptual understanding of circuit behaviors.

3.2.3 Insights from Experiment Two

The findings suggest that VR laboratories serve as a viable and effective alternative to traditional simulation applications, such as TinkerCAD, for electrical engineering education. The statistically significant difference in task completion times supports the argument that VR-based learning environments enable more efficient interaction and engagement. Although helpfulness, confidence, and knowledge levels did not show statistically significant differences, the observed trends suggest that VR may still offer slight advantages in improving user perception and knowledge acquisition.

Despite these advantages, certain challenges remain. Participants highlighted issues related to navigation and occasional frustration with VR controls. These usability concerns indicate the need for continued refinement of the VR interface to optimize the learning experience.

4 Discussion

The integration of VR and metaverse-based environments into electrical engineering education represents a paradigm shift in how technical concepts are taught and learned. This chapter's findings, drawn from two experimental studies, underscore VR's potential to enhance learner engagement, improve conceptual understanding, and expand access to high-quality technical training. At the same time, they reveal key challenges and point to promising directions for future development.

Both experiments demonstrated VR's ability to bridge the gap between theoretical knowledge and practical application. In Experiment 1, the learning outcomes achieved in VR-based laboratories were comparable to those from traditional physical laboratories, echoing prior research on VR's effectiveness in delivering immersive, realistic simulations (Asham, Bakr, & Emadi, 2023; Wallace et al., 2010). Importantly, VR offers long-term cost benefits by reducing reliance on physical infrastructure and consumable materials—an important consideration for resource-limited institutions. In addition, its remote accessibility helps overcome geographic and logistical barriers to education. These benefits are further amplified through the metaverse, where persistent, collaborative virtual spaces allow students to interact regardless of physical location—an especially valuable feature for international and remote learners (Lin et al., 2024; Zhang et al., 2024).

Experiment 2 reinforced these findings by demonstrating that students using VR were better able to engage with and comprehend complex circuitry compared to those using traditional tools like TinkerCAD. The spatial immersion provided by VR enhanced contextual understanding, which is often a limitation in screen-based simulations. These findings are supported by Shi et al. (2011) and Valdez et al. (2015), who highlight the importance of hands-on, low-risk environments in facilitating learning of abstract engineering concepts. However, it is also important to note that confidence levels and knowledge scores did not significantly differ between the VR and non-VR groups. This suggests that while VR provides added value in engagement and interaction, traditional simulation tools remain valuable for their accessibility and ease of use (Hasanah et al., 2017; Singh et al., 2021).

Learner motivation and engagement emerged as pivotal factors. Students consistently reported that VR enhanced their sense of immersion and involvement. The ability to experiment with virtual prototypes, simulate failures, and receive immediate feedback contributed to an active learning environment that sustained interest and deepened cognitive processing (Orlosky et al., 2017). These benefits align with Cognitive Load Theory (CLT), which helps explain how immersive systems can reduce extraneous load by contextualizing information and enabling multimodal interaction (Haryana et al., 2022). Yet, the same immersive features can also increase intrinsic or extraneous cognitive load if not carefully designed. For instance, in Experiment 1, participants required more time to complete VR tasks due to interface unfamiliarity. Similarly, some students in Experiment 2 experienced navigation difficulties and minor frustrations, consistent with concerns raised by Kirschner (2002) and Liegmann et al. (2022) about cognitive overload in poorly optimized simulated environments.

These usability issues reflect broader limitations of VR systems in education. While immersive, VR cannot fully replicate the tactile feedback of real-world laboratory equipment, and it may induce motion sickness or discomfort with prolonged use. To mitigate these challenges, a hybrid approach that combines physical laboratories with VR simulations is recommended, as it allows students to benefit from both immersive practice and tangible, sensory-rich experimentation (Long et al., 2005). Enhancing usability through intuitive interface design, minimizing cognitive friction, and aligning VR activities with learning objectives are crucial for maximizing educational value.

The future of immersive learning in electrical engineering lies in deeper integration with intelligent and adaptive technologies. One particularly promising direction is the incorporation of AI-powered instructors within VR environments. These systems could deliver real-time guidance, personalized feedback, and dynamically tailored instructional pathways that respond to learners' progress and preferences (Darejeh et al., 2024c; Buentello-Montoya et al., 2021). AI-enhanced VR platforms have the potential to replicate many benefits of face-to-face instruction, narrowing the experiential gap between virtual and physical learning environments.

Emerging technologies such as advanced haptic feedback systems and brain–computer interfaces (BCIs) may further enrich the learning experience. BCIs, for example, offer intuitive, hands-free control over VR systems, expanding accessibility for students with physical disabilities (Saparamadu et al., 2025; Sargazi Moghadam et al., 2024; Darejeh & Mashayekh, 2023). As VR becomes more integrated with these technologies, the potential for inclusive and personalized learning will continue to grow.

To validate and refine these innovations, future research must prioritize longitudinal studies that assess long-term knowledge retention, practical skill acquisition, and sustained learner engagement. While short-term experiments offer valuable insights, only extended evaluation can reveal the true impact of immersive systems on educational outcomes and guide their effective, scalable implementation in engineering curricula.

5 Conclusions

This chapter has demonstrated that VR and metaverse-based learning offer significant advantages over traditional methods by providing immersive, interactive, and cost-effective alternatives to physical laboratories. Through two controlled experiments, we established that VR-based simulations can effectively enhance student engagement, motivation, and conceptual understanding, making them a viable complement or substitute for conventional laboratory-based instruction. Additionally, VR offers a powerful means to bridge the accessibility gap, allowing remote and international students to engage in practical learning without requiring physical presence.

Beyond VR alone, the broader principles of the metaverse open new possibilities for electrical engineering education by supporting discipline convergence and collaborative learning. Students from diverse engineering domains can work together in shared virtual environments to design smart systems, simulate IoT networks, or explore sustainable energy solutions. As these platforms evolve, they can incorporate advanced features such as blockchain-based credentialing, AI-driven learning analytics, and digital twin integration—transforming metaverse environments into holistic ecosystems for both learning and career development.

Despite the promise of these technologies, challenges remain in their full integration. Issues like navigation complexity, lack of tactile feedback, and cognitive overload must be addressed through improved interface design, ergonomic enhancements, and adaptive learning systems. Future research should also investigate long-term knowledge retention, the effectiveness of AI-driven instructional agents, and the scalability of these platforms across various engineering disciplines.

Ultimately, the future of electrical engineering education lies not just in digitizing existing content but in reimagining the entire educational ecosystem. Immersive, metaverse-based platforms offer fertile ground for innovation, enabling more inclusive, collaborative, and future-ready learning experiences. As institutions continue to explore and refine these tools, they are not only enhancing the quality of engineering education but also actively shaping the next generation of engineers equipped for a connected and rapidly evolving world.

References

Albarracin-Acero, D. A., Romero-Toledo, F. A., Saavedra-Bautista, C. E., & Ariza-Echeverri, E. A. (2024). Virtual reality in the classroom: Transforming the teaching of electrical circuits in the digital age. *Future Internet*, *16*(8), 279–279.

Alghamdi, T. A. (2024). Harnessing metaverse technologies: A paradigm shift in engineering education and professional development. *International Journal of Computer Science and Information Security (IJCSIS)*, *22*(5),131–143.

Alnagrat, A. J. A., Ismail, R. C., & Idrus, S. Z. S. (2022). The opportunities and challenges in virtual reality for virtual laboratories. *Innovative Teaching and Learning Journal*, *6*(2), 73–89.

Asham, Y., Bakr, M. H., & Emadi, A. (2023). Applications of augmented and virtual reality in electrical engineering education: A review. *IEEE Access*, *11*, 134717–134738.

Buentello-Montoya, D. A., Lomel´ı-Plascencia, M., & Medina-Herrera, L. (2021). The role of reality enhancing technologies in teaching and learning of mathematics. *Computers & Electrical Engineering*, *94*, 107287.

Burdea, G. (2003). *Virtual reality technology*. John Wiley & Sons.

Caserman, P., Garcia-Agundez, A., Konrad, R., Göbel, S., & Steinmetz, R. (2019). Real-time body tracking in virtual reality using a Vive tracker. *Virtual Reality*, *23*, 155–168.

Darejeh, A. (2023). Empowering education through eerp: A customizable educational vr escape room platform. In *2023 IEEE International Symposium on Mixed and Augmented Reality Adjunct (ISMAR-Adjunct)* (pp. 764–766). IEEE.

Darejeh, A., Huynh, L., Ji, X., Marcus, N., & Polly, P. (2024a). Assessing the efficacy of virtual reality-based laboratories versus real-life settings for PCR procedures, within medical science laboratories. In *2024 IEEE International Symposium on Mixed and Augmented Reality Adjunct (ISMAR-Adjunct)* (pp. 489–492). IEEE.

Darejeh, A., Marcus, N., & Sweller, J. (2022). Increasing learner interactions with E-learning systems can either decrease or increase cognitive load depending on the nature of the interaction. *L'Année psychologique*, *122*(3), 405–437.

Darejeh, A., & Mashayekh, S. (2023). Mind-object control as a new form of interaction in metaverse for users with motor disability. *Journal of Biomedical Physics & Engineering, 13*(1), 257.

Darejeh, A., Moghadam, T. S., Delaramifar, M., & Mashayekh, S. (2024c). A framework for AI-powered decision making in developing adaptive e-learning systems to impact learners' emotional responses. In *2024 11th International and the 17th National Conference on E-Learning and E-Teaching (ICeLeT)* (pp. 1–6). IEEE.

Darejeh, A., Mondal, M., Marcus, N., & Vassar, A. (2024b). Exploring the cognitive load effects of diverse virtual reality interaction methods in interactive educational platforms. *In 2024 IEEE International Symposium on Mixed and Augmented Reality Adjunct (ISMAR-Adjunct)* (pp. 435–438). IEEE.

Franco, A. A., Loup-Escande, E., Loiseaux, G., Chotard, J.-N., Zapata-Dominguez, D., Ciger, J., Leclere, A., Denisart, L., & Lelong, R. (2023). From battery manufacturing to smart grids: Towards a metaverse for the energy sciences. *Batteries & Supercaps*, *6*(1), e202200369.

Ghazali, A. K., Ab. Aziz, N. A., Ab. Aziz, K., & Tse Kian, N. (2024). The usage of virtual reality in engineering education. *Cogent Education, 11*(1), 2319441. https://doi.org/10.1080/2331186X.2024.2319441

Haryana, M. R. A., Warsono, S., Achjari, D., & Nahartyo, E. (2022). Virtual reality learning media with innovative learning materials to enhance individual learning outcomes based on cognitive load theory. *The International Journal of Management Education*, *20*(3), 100657.

Hasanah, U., Massango, H., Tsutaoka, T., & Shimizu, K. (2017). A teaching material for learning alternating current circuit using soundcard oscilloscope- experiments for the electrical resonance. *Unnes Science Education Journal*, *6*(2), 1596–1601.

Huggett, J. (2020). *Virtually real or really virtual: Towards a heritage metaverse. Studies in Digital Heritage*, 4(1), 1–15. Indiana University (IU). https://doi.org/10.14434/sdh.v4i1.26218

Jamil, S., Rahman, M., & Fawad. (2022). A comprehensive survey of digital twins and federated learning for industrial internet of things (iiot), internet of vehicles (iov) and internet of drones (iod). *Applied System Innovation*, *5*(3), 56.

Kirschner, P. A. (2002). *Cognitive load theory: Implications of cognitive load theory on the design of learning, 12*(1), 1–10. Elsevier. https://doi.org/10.1016/S0959-4752(01)00014-7

Liang, H., & Xiaoming, B. (2013). Application research of virtual reality technology in electronic technique teaching. *Intelligence Computation and Evolution- ary Computation: Results of 2012 International Conference of Intelligence Computation and Evolutionary Computation ICEC 2012 Held July 7, 2012 in Wuhan, China*, 153–159.

Liegmann, F., Schorge, L., Kelker, M., & Haubrock, J. (2022). Virtual reality grid control station for learning the operational management of real electrical grids on emergency situation. *2022 IEEE German Education Conference (GeCon)*, 1–6.

Lin, J., Li, Q., Wang, C., & Hu, Z. (2024). Product development and design frame- work based on interactive innovation in the metaverse perspective. *Applied System Innovation*, *7*(4), 58.

Long, J., Hartas, K., & An, A. (2005). An ac signal-generator and pc-oscilloscope test package for distance students in undergraduate electronics. *Proceedings of the 2005 ASEE/AaeE 4th Global Colloquium on Engineering Education*, 26–30.

Mazuryk, T., & Gervautz, M. (1995). Two-step prediction and image deflection for exact head tracking in virtual environments. *Computer Graphics Forum*, *14*(3), 29–41.

Orlosky, J., Itoh, Y., Ranchet, M., Kiyokawa, K., Morgan, J., & Devos, H. (2017). Emulation of physician tasks in eye-tracked virtual reality for remote di- agnosis of neurodegenerative disease. *IEEE Transactions on Visualization and Computer Graphics*, *23*(4), 1302–1311.

Paas, F., Renkl, A., & Sweller, J. (2003). Cognitive load theory and instructional design: Recent developments. *Educational Psychologist*, *38*(1), 1–4.

Salutina, T. Y., Platunina, G. P., & Frank, I. A. (2024, March). Development and application of interactive educational technologies in training students in electrical and electronic engineering specialties. In *2024 Systems of Signals Generating and Processing in the Field of on Board Communications* (pp. 1–5). IEEE.

Saparamadu, P. V. I. N., Sepasgozar, S., Guruge, R. N. D., Jayasena, H. S., Darejeh, A., Ebrahimzadeh, S. M., & Eranga, B. A. I. (2025). Optimising contract interpretations with large language models: A comparative evaluation of a vector database-powered Chatbot vs. ChatGPT. *Buildings*, *15*(7), 1144.

Sargazi Moghadam, T., Darejeh, A., Delaramifar, M., & Mashayekh, S. (2024). Toward an artificial intelligence-based decision framework for developing adaptive e-learning systems to impact learners' emotions. *Interactive Learning Environments*, *32*(7), 3665–3685.

Shi, Z., Zhao, S., & Zhu, S.-a. (2011). An internet-based electrical engineering virtual lab: Using modelica for unified modeling. *2011 IEEE 3rd International Conference on Communication Software and Networks*, 555–559.

Sidhu, M. S., Lee, C. K., Hassan, Z., Mousakhani, S., Sidhu, K. K., & Anuar, A. (2024a). Preliminary study on the perception of engineering students on metaverse. In *2024 IEEE 6th Symposium on Computers & Informatics (ISCI)* (pp. 12–17). IEEE.

Singh, G., Mantri, A., Sharma, O., & Kaur, R. (2021). Virtual reality learning environment for enhancing electronics engineering laboratory experience. *Computer Applications in Engineering Education*, *29*(1), 229–243.

Soliman, M., Pesyridis, A., Dalaymani-Zad, D., Gronfula, M., & Kourmpetis, M. (2021). The application of virtual reality in engineering education. *Applied Sciences*, *11*(6), 2879.

Valdez, M. T., Ferreira, C. M., Martins, M. J. M., & Barbosa, F. M. (2015). 3d virtual reality experiments to promote electrical engineering education. *2015 International Conference on Information Technology Based Higher Education and Training (ITHET)*, 1–4.

Van Merrienboer, J. J., & Ayres, P. (2005). Research on cognitive load theory and its design implications for e-learning. *Educational Technology Research and Development*, *53*(3), 5–13.

Wallace, S., Parsons, S., Westbury, A., White, K., White, K., & Bailey, A. (2010). Sense of presence and atypical social judgments in immersive virtual en- vironments: Responses of adolescents with autism spectrum disorders. *Autism*, *14*(3), 199–213.

Zhang, S., Guan, Z., Ci, Y., & Ding, G. (2024). Design and implementation of collaborative editing system based on physical simulation engine running state. *2024 3rd International Conference on Artificial Intelligence and Au- tonomous Robot Systems (AIARS)*, 864–869.

4 Metaverse and Virtual Reality Integration in Medical Science

Applications in Virtual Laboratories and Training

Lauren Huynh, Xiaohan Ji, Ali Darejeh, and Samad M. E. Sepasgozar

1 Introduction

Traditional medical education has long relied on theoretical instruction supported by hands-on training in clinical or laboratory settings. However, access to advanced medical laboratories and cadaver-based experiences is often limited due to financial constraints, logistical challenges, and ethical considerations. These barriers highlight the need for innovative, technology-enhanced solutions to enrich the learning process (Flynn & Frost, 2021).

The integration of virtual reality (VR) and metaverse technologies into medical science education presents a promising response to these limitations. By offering simulated, interactive, and fully immersive environments, these tools allow learners to repeatedly practice high-risk procedures in a safe setting—well before entering clinical practice. This approach not only mitigates access issues but also represents a transformative shift in how students engage with complex medical concepts (Singh, Vig, & Kaunert, 2024).

The metaverse extends the capabilities of VR by creating persistent, collaborative virtual spaces where students and instructors can interact regardless of geographical constraints (Kala, 2022; Zaidi et al., 2024). In medical education, both VR and metaverse-based technologies offer immersive learning experiences across various domains such as anatomy, surgical training, and clinical procedures. These environments allow students to explore human anatomy in three dimensions, interact with virtual patients, and practice high-risk surgical techniques without the limitations or risks associated with live-patient training (Ghaempanah et al., 2024; Kurul et al., 2020; Sandrone, 2022; Thomsen et al., 2016).

A significant advantage of these technologies is their ability to overcome barriers inherent in traditional laboratory-based instruction (Darejeh et al., 2024a). Physical training facilities often require in-person attendance, which can be a challenge for distance learners, international students, or those with limited access to medical resources. In contrast, metaverse-based virtual laboratories enable remote participation and offer a scalable solution that maintains high-quality educational experiences (Chen et al., 2020).

DOI: 10.1201/9781003557555-4

Moreover, several studies have highlighted the potential of VR and the metaverse to enhance both conceptual understanding and practical skills in medical education. By simulating real-world clinical environments, these technologies provide students with the opportunity to engage in interactive problem-solving and collaborative exercises. VR, in particular, introduces depth perception, spatial interaction, and real-time feedback—features that traditional learning tools often lack—thus fostering a more comprehensive understanding of medical principles (Pulijala et al., 2018).

One of the most significant advantages of VR and the metaverse in medical education is their ability to bridge the gap between theory and practice (Darejeh, 2023; Darejeh et al., 2024b). Many students struggle with abstract medical concepts when learning solely from textbooks and lectures. VR enables students to visualize and manipulate complex biological structures, leading to a deeper understanding of anatomical and physiological processes (Popov et al., 2024). For example, medical students can use VR to dissect virtual cadavers, observe cellular interactions at a microscopic level, and practice intricate surgical procedures in a risk-free environment. Studies have demonstrated that VR-based learning improves knowledge retention and engagement compared to traditional methods, making it a valuable tool in medical education (Aïm et al., 2016).

The incorporation of immersive technologies in medical education also enhances student motivation and engagement. Traditional instructional methods often involve passive learning, where students absorb information without direct interaction. VR transforms this experience into an active learning process, where students can manipulate virtual surgical instruments, simulate medical procedures, and diagnose virtual patients in real time. This hands-on approach fosters curiosity and encourages students to explore medical concepts in greater depth (Meyerbröker & Emmelkamp, 2010).

Moreover, VR has the potential to address cognitive load issues associated with learning complex medical subjects. By breaking down intricate anatomical and physiological concepts into interactive simulations, VR helps students process information more effectively. Immersive learning environments provide contextual cues that aid in knowledge retention, reducing the mental effort required to grasp difficult topics (Bueno-Vesga, Xu, & He, 2021). Furthermore, VR's multimodal approach—incorporating visual, auditory, and haptic feedback—caters to diverse learning styles, ensuring a more inclusive educational experience (Saposnik & Levin, 2011).

This chapter explores the rise of immersive technologies and their transformative impact on medical science education, with a particular focus on their application in clinical and medical science laboratory training. It begins with a comprehensive literature review on using VR and the metaverse in medical education, followed by a detailed presentation of our empirical study on simulating medical procedures in VR. By examining how these technologies enhance learning, this chapter aims to assess their potential for shaping the future of medical training and their suitability for integration into current curricula.

2 Literature Review

This section reviews existing literature on the application of VR and the metaverse across various domains of medical science, including laboratory techniques and experimental procedures, surgical training simulations, patient treatment, and VR-assisted tools.

2.1 VR in Laboratory Techniques and Experimental Procedures

The metaverse and VR have become increasingly valuable in medical laboratories and experimental training, revolutionizing how students and professionals develop proficiency in laboratory techniques and disease diagnosis. Traditional training methods often face limitations such as restricted access to specialized equipment, safety concerns, and logistical challenges. VR-based training addresses these issues by enabling repeated practice in a controlled, risk-free environment, thereby enhancing both diagnostic and experimental skills (Darejeh et al., 2024a).

These advantages are particularly evident in high-risk or resource-intensive areas of medical science. For example, in microbiology experiments, exposure to microorganisms presents inherent risks, and preparation can be time-consuming and costly. VR simulations mitigate these challenges by allowing students to conduct experiments in a virtual environment, reducing contamination risks and minimizing the use of consumables (Chitra et al., 2024). In their study, Chitra et al. (2024) developed a 3D VR laboratory featuring six microbiology experiments. Student feedback indicated that the VR environment was engaging and enjoyable and provided a safe space for iterative learning. The program's built-in error-prevention system ensured that students only progressed once they had accurately understood each step, thereby enhancing knowledge retention and procedural accuracy.

The benefits of VR extend beyond laboratory-based experiments to broader areas of medical training. For instance, Gunn et al. (2017) demonstrated the value of VR-based simulations in diagnostic radiology, where students interacted with virtual patients and radiation equipment to replicate real-world clinical scenarios. Their findings showed that students trained in VR achieved outcomes comparable to those trained through conventional methods, positioning VR as a highly effective supplementary tool.

Moreover, VR has proven useful in addressing training challenges within ethically sensitive areas of medical research, such as the handling of laboratory animals. Although proper training is critical for ensuring animal welfare and research validity, many institutions provide limited hands-on opportunities, relying instead on theoretical instruction. Tang et al. (2020) addressed this gap by designing a VR-based simulation for procedures like administering injections and positioning animals. Their study revealed that VR training not only improved the practical application of theoretical knowledge but also reinforced ethical practices. Additionally, the integration of real-time feedback and observational tools allowed instructors to more effectively assess and guide student performance.

Another critical application of VR in medical laboratories is its role in training healthcare professionals in proper medical waste management. Improper waste disposal poses significant environmental and public health risks, yet many healthcare workers receive inadequate training in this area. Wu et al. (2020) conducted a two-year pilot study in China to evaluate the impact of VR training on laboratory and clinical waste classification and disposal. Participants engaged in VR-based exercises where they identified and correctly disposed of different types of medical waste. The study utilized a pre- and post-training assessment model, demonstrating that VR significantly improved participants' understanding of waste classification and increased their confidence in applying correct disposal techniques in real-world settings. These findings underscore the potential of VR to enhance compliance with medical waste regulations while minimizing risks associated with improper disposal.

2.2 Surgery Training Simulation

VR has had a transformative impact on surgical training by simulating real-world surgical environments, allowing trainees to practice complex procedures in a controlled, risk-free setting. While traditional surgical education remains essential, it is often constrained by ethical concerns, limited patient availability, restricted access to cadaver-based practice, and rising costs (Kuhn et al., 2021; Ntakakis et al., 2023; O'Toole et al., 1999). VR addresses these limitations by offering repeatable, customizable, and cost-effective alternatives that reduce reliance on live patients and expensive physical resources.

Peters et al. (2023) and O'Toole et al. (1999) explored this potential by designing a virtual environment where trainees practiced suturing on a simulated large, flexible vessel. These systems provided real-time feedback on tool positioning and force application, enabling learners to refine their technique through iterative practice. Results demonstrated significant improvements among both surgeons and non-surgeons, reinforcing VR's broad applicability in enhancing procedural skills across varying levels of surgical expertise.

Beyond general surgical training, VR is gaining significant recognition in specialized surgical fields such as orthopedic, maxillofacial, and ophthalmologic surgery. These disciplines demand high levels of precision, dexterity, and decision-making—skills that VR platforms are well-equipped to support. For instance, Aïm et al. (2016) conducted a systematic review of nine studies on VR-based orthopedic surgery training and concluded that VR holds strong potential for enhancing procedural accuracy and clinical judgment in orthopedic education. Building on this, Cevallos et al. (2022) evaluated the efficacy of VR in training for pediatric orthopedic procedures, specifically the pinning of a slipped capital femoral epiphysis (SCFE). In their study, participants who received additional training through the Osso VR surgical trainer outperformed those using only traditional study guides. The VR-trained group completed procedures more quickly, achieved more accurate pin placement, and committed fewer technical errors—demonstrating VR's capacity to improve both efficiency and precision in surgical execution.

The benefits of VR extend similarly into maxillofacial surgery. Pulijala et al. (2018) developed a VR-based simulation for Le Fort I osteotomy, a complex procedure involving the repositioning of the upper jaw. Their findings showed that surgical residents who trained using the VR system reported higher levels of confidence and knowledge acquisition compared to those who underwent conventional training. Notably, first-year residents experienced the most substantial improvements, underscoring VR's particular value in early-stage training where structured and interactive environments support foundational skill development.

VR's impact is equally evident in ophthalmologic training, where microsurgical precision is critical. Thomsen et al. (2016) introduced a cataract surgery VR simulator to evaluate its effectiveness in improving surgical performance. Their results revealed minimal changes among experienced surgeons but showed significant skill development among novice trainees. These findings suggest that, across surgical specialties, VR is especially effective in supporting the learning curve of junior practitioners, providing a safe and repeatable environment for refining essential techniques.

2.3 VR in Patient Treatment

Beyond its applications in medical education and surgical training, VR has emerged as a powerful tool in patient treatment, offering innovative solutions for rehabilitation, psychological therapy, and doctor-patient interactions. From physical rehabilitation programs that enhance mobility and reduce pain to psychological interventions that help patients manage anxiety, phobias, and post-traumatic stress disorder (PTSD), VR is reshaping patient care. The sections below review recent studies and advancements in the use of VR for patient treatment, highlighting its clinical benefits and evolving role in therapeutic practices.

2.3.1 Physical Therapy

VR integration into physical therapy provides an innovative approach to exercise, pain management, and patient motivation. Traditional rehabilitation techniques often face challenges related to patient engagement and adherence, but VR has shown promise in addressing these limitations by creating interactive and immersive therapeutic experiences. A recent pragmatic randomized trial by Hemphill et al. (2021) compared VR therapy with standard physical therapy techniques across patients with varying rehabilitation needs, including injury recovery, postoperative rehabilitation, and chronic pain management. The study found that participants engaged in significantly more physical movement during VR-supplemented therapy sessions than standard physical therapy, regardless of the intervention order. Both therapists and patients reported positive feedback, highlighting VR's ability to enhance motivation and physical engagement, reinforcing its role as a valuable supplementary tool in rehabilitation settings.

Beyond movement facilitation, VR has also been explored as an adjunct for pain management in physical therapy (Goudman et al., 2022). Hoffman et al. (2000) conducted a controlled study evaluating VR's effectiveness in reducing pain during burn wound care. Participants underwent three minutes of traditional physical therapy followed by three minutes of VR-integrated therapy. The findings demonstrated a significant reduction in pain perception when VR was used, suggesting its efficacy as a distraction-based analgesic technique. Expanding on these results, Schmitt et al. (2011) investigated VR's role in pain management among hospitalized burn patients. Their study, involving 54 participants performing range-of-motion exercises, found that the use of VR significantly reduced pain on the first day of treatment. Additionally, patients expressed increased motivation, with many describing the VR experience as enjoyable. These findings underscore VR's potential as a non-pharmacological pain relief tool, particularly in populations undergoing painful rehabilitation therapies.

Recent research has further demonstrated VR's role in pain relief beyond traditional rehabilitation contexts. A randomized controlled trial conducted by Groninger et al. (2024) investigated the use of immersive VR for pain management in hospitalized cancer patients experiencing moderate to severe pain. The study compared a ten-minute VR distraction therapy session with a two-dimensional guided imagery (2DGI) intervention. While both interventions resulted in pain reduction, VR demonstrated significantly greater pain relief immediately post-intervention and sustained benefits 24 hours later. These findings suggest that VR can serve as an effective non-pharmacological analgesic intervention, even for patients with complex and severe pain conditions, such as those undergoing cancer treatment.

Furthermore, VR's effectiveness in stroke rehabilitation, particularly in motor function recovery, has been well-documented. A meta-analysis by Saposnik and Levin (2011) evaluated the impact of VR on arm motor recovery in stroke patients, concluding that VR, when integrated with conventional rehabilitation, resulted in greater functional improvements than traditional therapy alone. The immersive and interactive nature of VR enables stroke survivors to engage in task-specific, repetitive movements within a controlled environment—an approach that promotes neuroplasticity and facilitates motor learning.

Building on these benefits, researchers have also explored the application of VR in treating chronic musculoskeletal disorders, where persistent pain and mobility limitations are common. Kantha et al. (2023) conducted a comparative study on immersive VR's impact on pain reduction and functional disability in patients with chronic musculoskeletal conditions. Their findings indicated that VR therapy significantly reduced pain and psychological distress, outperforming traditional rehabilitation approaches in several key measures. These results suggest that VR-based rehabilitation improves physical function and enhances emotional well-being, highlighting its holistic benefits in patient recovery.

2.3.2 Psychological Therapy

VR in psychological therapy offers immersive and controlled environments that enable patients to confront and manage mental health challenges in a safe, structured manner. By replicating real-world experiences in a highly controlled setting, VR facilitates therapeutic interventions across a range of psychological conditions. A comprehensive review by Bell et al. (2024) underscores VR's expanding role in mental health treatment, particularly in anxiety, psychotic disorders, PTSD, eating disorders, depression, and stress management. Also, Liu et al. (2022) conducted a two-decade bibliometric analysis of VR-aided therapy, identifying four key research domains: PTSD, anxiety and fear-related disorders, diseases of the nervous system, and pain management. Their findings illustrate the evolution of VR research from early applications in PTSD treatment to broader psychological and neurological conditions, emphasizing the need for innovative and personalized therapeutic approaches.

Among these domains, PTSD therapy remains one of the most well-established and impactful applications of VR. Cuperus et al. (2017) found that VR-based PTSD therapy elicited emotional and psychological responses similar to those experienced in real-world trauma-related settings, leading to improved therapeutic outcomes. Similarly, Dibbets and Schulte-Ostermann (2015) demonstrated that VR exposure therapy enables therapists to identify trauma-related triggers and cognitive patterns better. These findings reinforce VR's potential as an effective tool for trauma therapy, particularly for individuals who struggle with conventional talk-based interventions.

Another well-established application of VR in psychological therapy is the treatment of phobias. VR-based exposure therapy enables individuals to gradually confront their fears under therapist-guided conditions, providing a structured and repeatable experience that is often more accessible and less distressing than real-world exposure therapy (Freeman et al., 2018). Unlike traditional exposure therapy, which may be challenging to control and replicate, VR allows for precise manipulation of stimuli, ensuring a safe yet immersive therapeutic process. A key advantage of VR exposure therapy is its ability to simulate complex environments that are difficult to recreate in clinical settings. Freeman et al. (2018) explored this capability in a study on VR-based interventions for acrophobia (fear of heights), where participants engaged in height-related virtual scenarios without physical risk. The results demonstrated a significant reduction in fear responses, highlighting VR's efficacy as an alternative to conventional phobia treatments. Similarly, Meyerbröker and Emmelkamp (2010) compared VR exposure therapy with cognitive-behavioral therapy (CBT) and found that VR provided comparable therapeutic outcomes, with the added benefit of increased patient engagement. By eliminating logistical barriers and offering highly personalized treatment options, VR broadens access to effective therapy for individuals struggling with various anxiety disorders.

Recent advancements have expanded VR's scope to include psychosis treatment. Freeman et al. (2023) conducted a clinical trial assessing the effectiveness of VR therapy for patients diagnosed with psychosis, focusing on satisfaction, side effects, and overall experience. The study revealed that most participants reported being either very satisfied or mostly satisfied with VR therapy, with only minor side effects such as mild concentration difficulties and temporary headaches. More importantly, VR therapy led to significant reductions in agoraphobic avoidance behaviors, particularly among patients with severe symptoms. These findings suggest that VR therapy can serve as a well-received and viable intervention for psychosis, especially for individuals experiencing high levels of social anxiety and avoidance.

In addition to its role in phobia and psychosis treatment, VR has demonstrated promising applications in addiction therapy. Controlled exposure to addiction-related triggers within a VR environment allows patients to develop and practice coping mechanisms in a safe and controlled setting. Segawa et al. (2020) conducted a systematic review on VR-based addiction therapy, concluding that VR simulations effectively replicate real-world cravings and high-risk environments. This controlled exposure enables individuals to refine their self-regulation strategies before encountering real-life triggers, potentially enhancing relapse prevention efforts and reinforcing long-term behavioral change.

Beyond clinical treatment, VR is increasingly being explored as a mental health intervention for broader populations, particularly the elderly. Kwan et al. (2023) examined the impact of VR therapy on elderly individuals in long-term care and found significant improvements in psychological well-being, including reductions in feelings of isolation, anxiety, and depression. The study highlights VR's potential not only as a clinical therapy tool but also as a holistic approach to enhancing quality of life in aging populations. By fostering cognitive engagement and social interaction, VR therapy offers an innovative way to support mental health and emotional resilience across diverse patient groups.

2.3.3 VR-Assisted Tools

Beyond direct treatment, another promising application of VR is its role as an assistive tool in medical settings, particularly in mitigating the trauma associated with aggressive treatments and surgeries. For young children, surgical procedures can be distressing, potentially leading to hyperalgesia, postoperative delirium, or long-term psychological disorders.

To address this challenge, Wu et al. (2022) developed a highly realistic virtual environment modelled after an operating room, enabling children to interact with their surroundings through an "adventure story" simulation of the surgical process. While this approach did not significantly reduce postoperative pain, results indicated that VR substantially alleviated preoperative anxiety, leading to greater satisfaction among parents.

In addition to reducing anxiety, VR technology can enhance communication between doctors and patients. Pandrangi et al. (2019) introduced a virtual image-viewing environment featuring a 3D model of an abdominal aortic aneurysm derived from CT scans. This interactive model enabled patients to understand their medical condition better. Feedback from patients suggested that they felt more informed and engaged in their care, reinforcing VR's effectiveness as a tool for patient education and improved doctor-patient interactions.

Furthermore, VR technology facilitates communication not only between doctors and patients but also among medical professionals. Yang et al. (2021) developed a mixed-reality environment featuring a 3D model of an acoustic neuroma. Surgeons wearing headsets could simultaneously view, enlarge, rotate, and annotate the model through an interactive system. This technology enabled surgeons to comprehensively assess the patient's condition and devise a more precise surgical plan. The interactive model clarified the condition, treatment options, and associated risks when shared with patients and their families. This approach not only streamlined communication but also reduced misunderstandings and disputes, ultimately fostering greater trust and satisfaction in the medical process.

While the literature highlights VR's broad applications in patient treatment, ranging from physical rehabilitation to psychological therapy and doctor-patient communication, its effectiveness in enhancing hands-on procedural learning remains an important area for further study. To further explore how VR can support skill acquisition in medical laboratory training—particularly in the use of high-tech equipment—the following section presents the development and experimental evaluation of our VR application, developed as part of the University of New South Wales (UNSW) metaverse initiative, as introduced in Chapter 1. This empirical study focuses on the impact of VR on learning complex laboratory procedures, with a specific emphasis on performing the polymerase chain reaction (PCR) using a PCR machine—a foundational technique in medical science education.

3 Empirical Study

In this study, two experiments were conducted to evaluate the effectiveness of VR in enhancing laboratory skills and conceptual understanding. In Experiment 1, participants were required to learn about and complete a PCR procedure in a VR laboratory. In Experiment 2, participants performed the same PCR procedure using both the VR and computer display versions of the simulation to enable a direct comparison between the two interfaces.

All participants in both experiments were recruited in accordance with a human research ethics protocol approved by the University of New South Wales and sanctioned by its institutional review board.

3.1 Experiment One: Evaluating VR for Teaching PCR Procedures

This experiment aimed to assess the effectiveness of VR-based training in teaching PCR procedures to learners of different expertise levels. The study aimed to determine whether VR enhances learners' confidence, conceptual understanding, and perceived realism. Additionally, it examined whether prior experience with PCR influenced learning outcomes in a VR environment.

3.1.1 Application Development

A VR-based medical science laboratory was developed using Unreal Engine 5 (UE5) and deployed on HTC VIVE Pro 2 headsets as part of UNSW's metaverse initiative. The system was designed to replicate UNSW real-life medical science laboratory environment (Figure 4.1), enabling participants to learn and perform PCR procedures interactively. Everything in the VR laboratory is functional, allowing users to interact with it and perform various tasks.

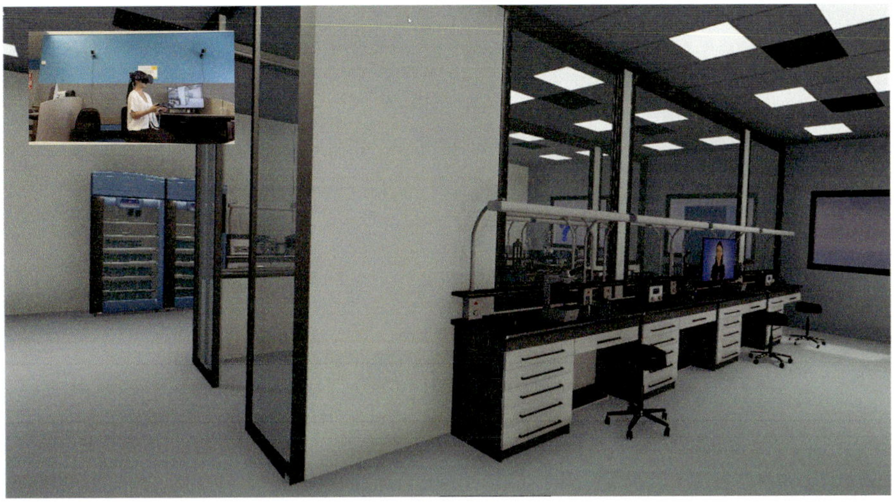

Figure 4.1 The simulated laboratory environment.

Upon initiating the application, users are immersed in the virtual laboratory environment, where they are first presented with safety instructions displayed on the wall. Within this space, users interact with personal protective equipment (PPE) and laboratory tools to become familiar with essential safety protocols and procedures (Figure 4.2).

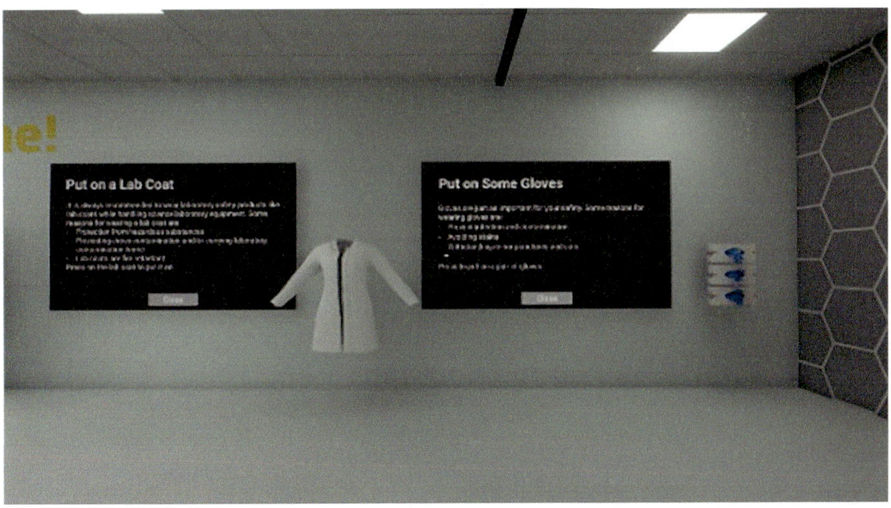

Figure 4.2 Laboratory safety instruction.

After completing the safety protocol section of the virtual laboratory, users encounter a poster detailing the PCR procedure, which outlines the fundamental principles and real-world applications of the process. These procedures are illustrated in Figure 4.3, which demonstrates how informational posters are integrated into the virtual environment. This approach enhances the richness of the virtual laboratory by integrating text or informational panels directly into the immersive space, supporting a more engaging learning experience.

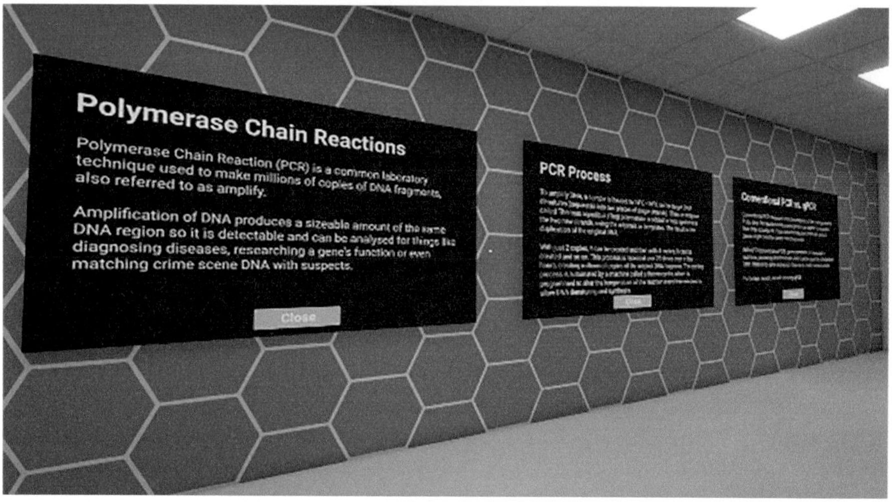

Figure 4.3 The introductions to the PCR procedure.

The final poster familiarized participants with the operation of PCR equipment by providing a step-by-step walkthrough of the essential procedures involved in performing a PCR reaction, as shown in Figure 4.4.

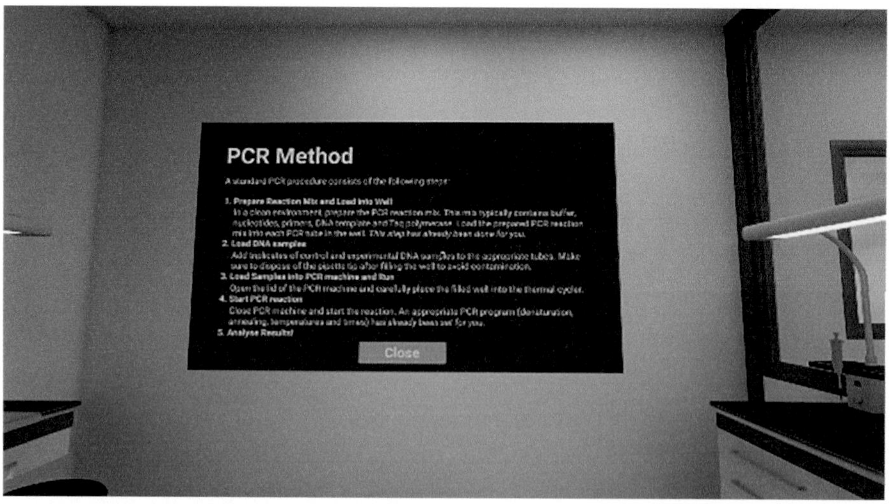

Figure 4.4 The guideline of the PCR procedure.

After reviewing the instructional content, users proceeded to conduct the PCR process by assembling the required samples and loading them into the PCR machine using the provided apparatus, as illustrated in Figure 4.5.

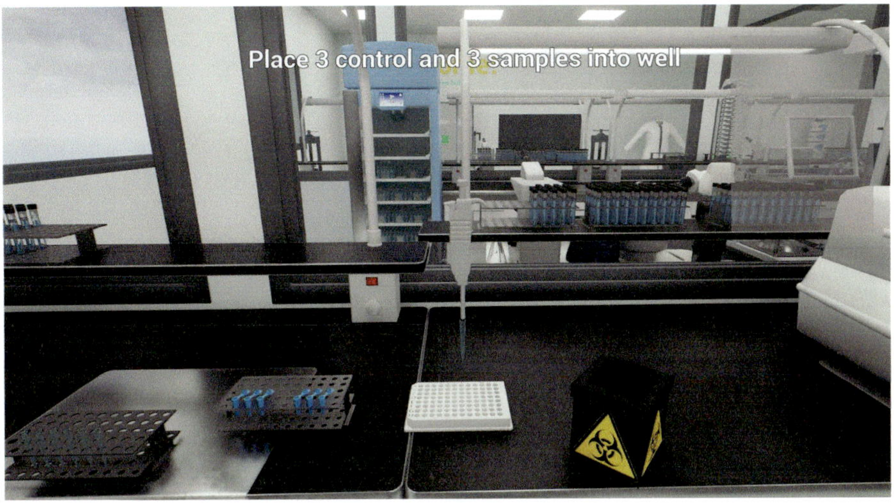

Figure 4.5 The process of preparing the sample.

Once the samples are prepared, users place them into the PCR machine, as illustrated in Figure 4.6, to observe its operation and become familiar with its functions within a simulated environment. This hands-on experience fosters confidence and understanding while reducing the risks commonly associated with first-time use of physical equipment.

Figure 4.6 The interactable PCR machine in the simulation.

Finally, the system presented an analysis of the results, enabling participants to interpret virtual data displayed on a simulated computer screen. The simulated output and accompanying instructional content are shown in Figure 4.7.

Figure 4.7 The analysis results of the sample.

3.1.2 Methodology for Conducting Experiment One

The study recruited six medical science students and professionals from UNSW. Participants were categorized into three expertise levels: novice, intermediate, and expert. Novices were defined as those who had completed between zero and three PCR procedures in real life, intermediates had completed between four and ten, while experts had conducted more than ten. Two participants were assigned to each expertise level. To ensure consistency, all participants were required to be at least 18 years old and have some prior experience with VR to minimize the impact of motion sickness.

The experimental procedure was divided into three phases. In the pre-experiment phase, participants provided informed consent after being briefed on the study's objectives and procedures and completed a demographic questionnaire assessing their prior experience with PCR and VR. A system demonstration was also conducted to familiarize them with the VR controls. The training phase required participants to wear a VR headset and engage independently with the instructional posters displayed on the walls of the virtual laboratory, allowing them to progress through the content at their own pace. They were encouraged to interact naturally with the environment.

Upon completion of the training phase, participants proceeded to the main phase of the experiment, which involved performing a PCR procedure within the virtual environment and interpreting the resulting data. In the post-experiment phase, participants completed a questionnaire evaluating their confidence, perceived realism, and usability of the VR system. Additionally, a short interview was conducted to gather qualitative feedback on their training experience.

3.1.3 Results from Experiment One

The quantitative analysis revealed significant differences in participant responses based on their level of expertise. A one-way ANOVA examined confidence levels in performing PCR procedures after VR training among novice, intermediate, and expert users, yielding a significant effect [$F(2,3) = 10.5$, $p < 0.05$]. Novices reported the highest confidence ($M = 4$, $SD = 0$), intermediates had moderate confidence ($M = 3.5$, $SD = 0.6$), while experts reported the lowest confidence ($M = 1.5$, $SD = 0.7$). Similarly, perceptions of the VR simulation's realism varied significantly [$F(2,3) = 10.5$, $p < 0.05$]. Novices rated the simulation as highly realistic ($M = 4$, $SD = 0$), intermediates provided slightly lower ratings ($M = 3.5$, $SD = 0.7$), and experts perceived it as considerably less realistic ($M = 1.5$, $SD = 0.5$).

However, no significant differences were found in the perceived difficulty of using VR technology [$F(2,3) = 0.6$, $p > 0.05$]. Intermediates found it somewhat difficult ($M = 3$, $SD = 0$), novices rated it as less challenging ($M = 2.5$, $SD = 0.7$), and experts perceived it as the least difficult ($M = 2$, $SD = 1.4$).

The qualitative analysis, conducted using NVivo software, revealed key insights. Novices and intermediates reported that VR was particularly useful in strengthening their understanding of PCR principles and procedures. Experts, however, commented on the lack of tactile fidelity in the simulation, suggesting that while VR was engaging, it did not fully replicate the intricacies of real-world PCR execution. Some participants suggested adding haptic feedback or additional tutorials to enhance their understanding of the process. Others noted that while VR provided an immersive experience, they felt that real-life PCR requires more fine motor skills, which are difficult to replicate in the simulation.

3.1.4 Insights from Experiment One

The outcomes of this experiment are consistent with findings from previous studies across various fields, indicating that VR is an effective tool for laboratory training, particularly for novice learners. The immersive environment reinforced fundamental PCR concepts and provided an engaging alternative to traditional learning methods. Interview responses indicated that novices appreciated the ability to practice in a risk-free setting, which helped build their confidence. However, expert learners believed that the additional instructions in the VR environment introduced unnecessary complexity.

The difference in effectiveness can be attributed to the role of prior knowledge and cognitive schemas. Novices, with limited prior experience in PCR procedures, benefited from VR's structured guidance, interactive feedback, and visual representations, which supported the construction of new schemas and enhanced knowledge retention. In contrast, experts, who already possessed well-developed mental models, found the additional interactivity unnecessary, leading to increased cognitive load and a perception of redundancy in the training. Additionally, experts may prefer traditional hands-on methods that align more closely with their existing procedural knowledge. These findings support the expertise reversal effect described in cognitive load theory, which suggests that instructional strategies effective for novice learners may be less beneficial—or even counterproductive—for experts due to differences in cognitive processing.

3.2 Experiment Two: Comparing VR and Traditional Computer-Based Learning for PCR Training

This experiment aimed to compare the effectiveness of VR-based training with traditional computer-based training in teaching PCR procedures to novice learners. The study aimed to determine whether VR provides superior learning outcomes, engagement, and cognitive load compared to a conventional 2D learning method using a computer display, keyboard, and mouse.

3.2.1 Methodology for Conducting Experiment Two

This study recruited 16 UNSW students with no prior experience in PCR and divided them into two groups of eight to compare VR-based and computer display-based PCR training. A crossover design was employed, in which each participant experienced both training methods. One group used the VR system first, followed by the computer display version, while the other group completed the tasks in the reverse order. The VR application used in this experiment was the same application developed for Experiment 1; however, a version for computer display was also created.

Similar to the first experiment, upon arrival, participants completed a consent form for the study, followed by a brief questionnaire assessing their prior experience with PCR. Based on their assigned group, participants then engaged with either the VR-based or computer display-based version of the simulation and completed a full PCR procedure using the assigned system.

To evaluate learning performance, participants completed a quiz on the PCR procedure immediately after completing the activity in their first assigned system (VR or computer display). This approach allowed for a between-subjects comparison of learning outcomes, ensuring that quiz scores reflected initial training effects without the influence of prior exposure. Following the quiz, participants also completed the NASA Task Load Index (NASA-TLX), a validated instrument for measuring cognitive load following training tasks (Park et al., 2024). The NASA-TLX was administered only once, after the first training session, to evaluate perceived cognitive demands associated with the initial training system. This ensured consistency in comparisons of both learning performance and cognitive load between groups—those who used the VR version first and those who began with the computer display.

After using the second system (either the VR or computer display version), participants took part in a semi-structured post-experiment interview. These interviews explored participants' perceptions of engagement, ease of navigation, and the effectiveness of each method in visualizing the PCR process.

3.2.2 Results from Experiment Two

The t-test results showed no statistically significant difference in learning performance between the two groups—those who used the VR version first and those who used the computer display version first—as indicated by quiz scores: VR group ($M = 4.8$, $SD = 0.5$) and display group ($M = 4.1$, $SD = 1.1$), $t(7) = 1.49$, $p > 0.05$. Similarly, cognitive load, measured using the NASA-TLX after the first training session, did not differ significantly between groups. Participants who used the VR version first reported slightly higher cognitive load ($M = 22.8$, $SD = 4.8$) compared to those who began with the computer display version ($M = 22.0$, $SD = 3.4$); however, this difference was not statistically significant, $t(7) = 0.76$, $p > 0.05$.

Although the quantitative data did not reveal a significant difference between participants who used the VR version first and those who started with the computer-based training, qualitative feedback from post-experiment interviews suggested that most participants found the VR version more engaging and immersive. Participants frequently used terms such as "interactive," "immersive," and "motivating" to describe the VR experience. Some noted that the VR training helped them better visualize the PCR process, whereas others found the computer-based simulation easier to navigate due to their familiarity with traditional interfaces.

3.2.3 Insights from Experiment Two

The results of this study indicate that both VR-based and computer-based training methods effectively teach PCR procedures to novice learners, with no clear advantage in knowledge retention. This suggests that the choice of medium may not significantly impact learning outcomes when the instructional content is consistent. However, VR training demonstrated notable benefits in enhancing user engagement and motivation. Participants generally found the immersive nature of VR to be more interactive and engaging, which may contribute to higher levels of interest and active participation during the training process.

Although VR training was perceived as more immersive, some participants preferred the computer-based version due to its familiarity and ease of navigation. This highlights the potential trade-off between engagement and usability, where highly immersive systems may introduce additional complexity for users accustomed to traditional interfaces. While VR introduced slightly higher cognitive demands, this did not appear to negatively impact the learning experience. These findings suggest that while VR enhances engagement, its usability should be further optimized to maximize accessibility and ease of use.

4 Discussion

The integration of the metaverse and VR into medical science education has demonstrated significant potential for enhancing conceptual understanding, student engagement, and hands-on learning. The study presented in this chapter reinforces these findings, aligning with existing literature that highlights how immersive technologies can effectively bridge the gap between theoretical knowledge and practical application (Birkheim, Calogiuri, & Martinsen, 2024; Koolivand et al., 2024; Yang et al., 2024). Specifically, the developed VR application was perceived to support students in visualizing and applying theoretical concepts related to the operation of a PCR machine and the analysis of its results. Building on this, similar simulations may enable students to engage with complex medical procedures and laboratory techniques in a safe, controlled environment—enhancing both confidence and skill acquisition. A particularly noteworthy advantage of VR in medical education is its capacity to promote deep engagement and immersion; participants in the study frequently described the learning experience as "engaging" and "immersive." These perceptions echo prior research demonstrating that immersive environments significantly boost learner motivation and active participation compared to traditional didactic methods (Freeman et al., 2018; Kala, 2022).

The study also indicated that novice learners are more likely to benefit from VR training than experts, as structured and interactive VR applications can support the development of foundational knowledge. This aligns with research by Flynn and Frost (2021) and Pulijala et al. (2018), which highlights the effectiveness of immersive learning environments in improving knowledge retention and comprehension of medical procedures, particularly for novice learners. However, expert learners exhibited signs of the expertise reversal effect, where VR training felt redundant due to their existing mental models. This suggests that while VR is highly effective for beginner and intermediate learners, expert users may require advanced simulations incorporating features like haptic feedback and high-fidelity modelling to replicate real-world complexities (Aïm et al., 2016). The adaptability of VR and metaverse-based training has been emphasized in the literature as a crucial factor in maintaining engagement and maximizing educational benefits across different levels of expertise (Ghaempanah et al., 2024).

Additionally, the cognitive load associated with VR-based learning varied among participants. While the immersive nature of VR reduced extraneous cognitive load by providing real-time feedback and interactive guidance, some users reported difficulties navigating the system, increasing their intrinsic cognitive load. These findings align with cognitive load theory (Paas et al., 2003), emphasizing the importance of designing virtual learning environments that optimize learning without overwhelming users. The challenge of balancing immersion with usability remains critical, particularly in ensuring that students from diverse backgrounds, including those from lower-middle-income countries, can effectively engage with VR-based education (Zaidi et al., 2024). Usability issues such as navigation difficulties and controller responsiveness were observed in the study. This is consistent with prior research suggesting that refining user interfaces and ensuring seamless integration with real-world medical tools can enhance the effectiveness of VR training environments and reduce users' cognitive load (Saposnik & Levin, 2011; Wei et al., 2025).

Although the application presented in the study is part of a broader metaverse platform, its collaborative features were not evaluated in this instance. However, it should be noted that these features offer significant potential for enhancing medical education through interactive, socially driven learning experiences. In the context of medical laboratory training—such as the PCR laboratory presented in the study—the metaverse facilitates social learning by enabling students to engage in shared virtual environments where they can collaboratively operate equipment and discuss procedures in real time. This digital interactivity fosters peer-to-peer instruction and reflective learning, particularly beneficial in resource-constrained settings where access to physical laboratories is limited (Zaidi et al., 2024).

In clinical training, metaverse platforms extend learning beyond physical constraints by allowing students to participate in team-based clinical simulations regardless of geographic location. These environments support structured role-play scenarios—such as multidisciplinary rounds or emergency responses—promoting the development of communication, coordination, and decision-making skills essential in real-world clinical practice (Popov et al., 2024). Additionally, instructors can join these sessions to provide live feedback, enhancing learner-instructor interaction and strengthening clinical reasoning (Lewis et al., 2024).

The impact on patient education is equally transformative. Through interactive metaverse spaces, patients can participate in group-based health education workshops, visualize complex medical conditions, and simulate treatment journeys in a safe and supportive setting. These social experiences are shown to reduce anxiety, increase health literacy, and encourage shared understanding, especially among individuals from underserved or remote communities (Ghaempanah et al., 2024; Kala, 2022).

Lastly, in the realm of rehabilitation therapy, metaverse environments support collaborative exercises, where patients can join virtual group therapy sessions. These sessions not only promote physical recovery but also reduce the sense of isolation often associated with long-term rehabilitation. Clinicians can supervise and adjust exercises in real time, while peer presence fosters mutual motivation and emotional support (Ahuja et al., 2023; Suh et al., 2023). The shared virtual settings thereby merge therapeutic effectiveness with human connection, making rehabilitation more holistic and patient-centered.

From an institutional perspective, integrating VR and the metaverse into medical curricula requires alignment with existing educational standards and clinical training protocols. Faculty training and technical support are essential to facilitate this transition. Future metaverse systems should incorporate AI-driven tutoring capabilities to provide personalized guidance and real-time feedback (Darejeh et al., 2024c; Darejeh, Marcus, & Sweller, 2022; Pulijala et al., 2018; Sargazi Moghadam et al., 2024), building on the success of large language models in domain-specific applications such as legal and engineering document interpretation (Saparamadu et al., 2025). By addressing these considerations, VR and the metaverse have the potential to transform medical education and healthcare training, ensuring students gain hands-on experience with cutting-edge medical technologies while fostering innovation in patient care (Ghaempanah et al., 2024; Kala, 2022).

Finally, it should be noted that the experiments presented in this chapter are intended to demonstrate an approach for developing effective VR applications within the context of medical education, rather than to produce definitive numerical outcomes. Due to the limitations related to sample size, the findings should be considered preliminary and not fully generalizable. Replication with larger and more diverse samples will be essential to ensure statistical validity and strengthen the evidence base. The primary aim of this study was to establish an observational research framework and to design a novel VR laboratory informed by cognitive load dimensions—thereby laying the foundation for future investigations.

Looking ahead, future studies should focus on increasing sample sizes and applying a broader range of analytical techniques to generate more robust and generalizable insights. Moreover, the metaverse component of this project warrants further evaluation to examine how collaborative participation among multiple students in a virtual medical science laboratory affects their learning performance and cognitive load. Such investigations will be critical in assessing the full potential of socially interactive, immersive environments in enhancing medical education.

5 Conclusion

The integration of the metaverse and VR in medical science education represents a paradigm shift in how students and professionals acquire and apply knowledge. Unlike traditional teaching methods, which are often constrained by physical resources and logistical barriers, immersive technologies offer a scalable, flexible, and interactive alternative that extends beyond the limits of conventional classrooms and laboratories. As institutions continue to explore digital transformations in education, the metaverse provides an opportunity to rethink medical training by fostering experiential learning, remote collaboration, and adaptive teaching strategies. This shift is not merely about supplementing existing educational models but redefining them to better align with the evolving demands of the medical field.

However, while the benefits of VR and the metaverse in medical education are evident, their successful adoption requires addressing challenges such as technological accessibility, curriculum integration, and long-term effectiveness. The financial and infrastructural demands of these systems raise questions about equity in education, particularly for institutions in resource-limited settings. Furthermore, as these technologies advance, ongoing research is necessary to assess their impact on skill retention, clinical competency, and patient outcomes. The future of medical education will likely be shaped by how well educators, researchers, and policymakers can balance innovation with practical implementation, ensuring that immersive learning environments are effective and inclusive. Ultimately, the metaverse holds the potential to revolutionize medical education, not just by introducing new tools but by fundamentally reshaping how knowledge is delivered, shared, and applied in the healthcare industry.

References

Ahuja, A. S., Polascik, B. W., Doddapaneni, D., Byrnes, E. S., & Sridhar, J. (2023). The digital metaverse: Applications in artificial intelligence, medical education, and integrative health. *Integrative Medicine Research*, *12*(1), 100917.

Aïm, F., Lonjon, G., Hannouche, D., & Nizard, R. (2016). Effectiveness of virtual reality training in orthopaedic surgery. *Arthroscopy: The Journal of Arthroscopic & Related Surgery, 32*(1), 224–232.

Bell, I. H., Pot-Kolder, R., Rizzo, A., Rus-Calafell, M., Cardi, V., Cella, M., … & Valmaggia, L. (2024). Advances in the use of virtual reality to treat mental health conditions. *Nature Reviews Psychology*, *3*(8), 552–567.

Birkheim, S. L., Calogiuri, G., & Martinsen, R. (2024). Advancing immersive virtual reality-based simulation practices: Developing an evidence-based and theory-driven pedagogical framework for VR-based simulations of non-technical skills among healthcare professionals. *Interactive Learning Environments*, *32*(7), 3579–3591.

Bueno-Vesga, J. A., Xu, X., & He, H. (2021, March). The effects of cognitive load on engagement in a virtual reality learning environment. In *2021 IEEE Virtual Reality and 3D User Interfaces (VR)* (pp. 645–652). IEEE.

Cevallos, N., Zukotynski, B., Greig, D., Silva, M., & Thompson, R. M. (2022). The utility of virtual reality in orthopedic surgical training. *Journal of Surgical Education*, *79*(6), 1516–1525.

Chen, S., Zhu, J., Cheng, C., Pan, Z., Liu, L., Du, J., …& Pan, H. (2020). Can virtual reality improve traditional anatomy education programmes? A mixed-methods study on the use of a 3D skull model. *BMC Medical Education*, *20*, 1–10. https://doi.org/10.1186/s12909-020-02255-6

Chitra, E., Mubin, S. A., Nadarajah, V. D., Se, W. P., Sow, C. F., Er, H. M., …& Davamani, F. (2024). A 3-D interactive microbiology laboratory via virtual reality for enhancing practical skills. *Scientific Reports*, *14*(1), 12809.

Cuperus, A. A., Klaassen, F., Hagenaars, M. A., & Engelhard, I. M. (2017). A virtual reality paradigm as an analogue to real-life trauma: Its effectiveness compared with the trauma film paradigm. *European Journal of Psychotraumatology*, *8*(sup1), 1338106. https://doi.org/10.1080/20008198.2017.1338106

Darejeh, A. (2023). Empowering education through eerp: A customizable educational vr escape room platform. In *2023 IEEE International Symposium on Mixed and Augmented Reality Adjunct (ISMAR-Adjunct)* (pp. 764–766). IEEE.

Darejeh, A., Huynh, L., Ji, X., Marcus, N., & Polly, P. (2024a). Assessing the efficacy of virtual reality-based laboratories versus real-life settings for PCR procedures, within medical science laboratories. In *2024 IEEE International Symposium on Mixed and Augmented Reality Adjunct (ISMAR-Adjunct)* (pp. 489–492). IEEE.

Darejeh, A., Marcus, N., & Sweller, J. (2022). Increasing learner interactions with E-learning systems can either decrease or increase cognitive load depending on the nature of the interaction. *L'Année psychologique*, *122*(3), 405–437.

Darejeh, A., Moghadam, T. S., Delaramifar, M., & Mashayekh, S. (2024c). A framework for AI-powered decision making in developing adaptive e-learning systems to impact learners' emotional responses. In *2024 11th International and the 17th National Conference on E-Learning and E-Teaching (ICeLeT)* (pp. 1–6). IEEE.

Darejeh, A., Mondal, M., Marcus, N., & Vassar, A. (2024b). Exploring the cognitive load effects of diverse virtual reality interaction methods in interactive educational platforms. In *2024 IEEE International Symposium on Mixed and Augmented Reality Adjunct (ISMAR-Adjunct)* (pp. 435–438). IEEE.

Dibbets, P., & Schulte-Ostermann, M. A. (2015). Virtual reality, real emotions: A novel analogue for the assessment of risk factors of post-traumatic stress disorder. *Frontiers in Psychology*, *6*, 681. https://doi.org/10.3389/fpsyg.2015.00681.

Flynn, C., & Frost, P. (2021). Making VR a Reality in the Classroom. [online] Available at: https://er.educause.edu/articles/2021/4/making-vr-a-reality-in-the-classroom.

Freeman, D., Haselton, P., Freeman, J., Spanlang, B., Kishore, S., Albery, E., Denne, M., Brown, P., Slater, M., & Nickless, A. (2018). Automated psychological therapy using immersive virtual reality for treatment of fear of heights: A single-blind, parallel-group, randomised controlled trial. *The Lancet Psychiatry*, *5*(8), 625–632. https://doi.org/10.1016/s2215-0366(18)30226-8.

Freeman, D., Rosebrock, L., Waite, F., Loe, B. S., Kabir, T., Petit, A., …& Lambe, S. (2023). Virtual reality (VR) therapy for patients with psychosis: Satisfaction and side effects. *Psychological Medicine*, *53*(10), 4373–4384.

Ghaempanah, F., Moasses Ghafari, B., Hesami, D., Hossein Zadeh, R., Noroozpoor, R., Moodi Ghalibaf, A., & Hasanabadi, P. (2024). Metaverse and its impact on medical education and health care system: A narrative review. *Health Science Reports*, *7*(9), e70100.

Goudman, L., Jansen, J., Billot, M., Vets, N., De Smedt, A., Roulaud, M., …& Moens, M. (2022). Virtual reality applications in chronic pain management: Systematic review and meta-analysis. *JMIR Serious Games*, *10*(2), e34402.

Groninger, H., Violanti, D., & Mete, M. (2024). Virtual reality for pain management in hospitalized patients with cancer: A randomized controlled trial. *Cancer*, *130*(14), 2552–2560.

Gunn, T., Jones, L., Bridge, P., Rowntree, P., & Nissen, L. (2017). The use of virtual reality simulation to improve technical skill in the undergraduate medical imaging student. *Interactive Learning Environments*, *26*(5), 613–620. https://doi.org/10.1080/10494820.2017.1374981

Hemphill, S., Rodriguez, S., Wang, E., Koeppen, K., Aitken-Young, B., Jackson, C., Simons, L., & Caruso, T. J. (2021). Virtual reality augments movement during physical therapy. *American Journal of Physical Medicine & Rehabilitation*, *101*(3), 229–236. https://doi.org/10.1097/phm.0000000000001779.

Hoffman, H. G., Patterson, D. R., & Carrougher, G. J. (2000). Use of virtual reality for adjunctive treatment of adult burn pain during physical therapy: A controlled study. *The Clinical Journal of Pain*, *16*(3), 244–250. https://doi.org/10.1097/00002508-200009000-00010

Kala, N. (2022). Revolutionizing medical education with metaverse. *International Journal of Scientific Research in Computer Science, Engineering and Information Technology*, *8*(4), 26–32.

Kantha, P., Lin, J. J., & Hsu, W. L. (2023). The effects of interactive virtual reality in patients with chronic musculoskeletal disorders: A systematic review and meta-analysis. *Games for Health Journal*, *12*(1), 1–12.

Koolivand, H., Shooreshi, M. M., Safari-Faramani, R., Borji, M., Mansoory, M. S., Moradpoor, H., …& Azizi, S. M. (2024). Comparison of the effectiveness of virtual reality-based education and conventional teaching methods in dental education: A systematic review. *BMC Medical Education*, *24*(1), 8.

Kuhn, S., Huettl, F., Deutsch, K., Kirchgaessner, E., Huber, T., & Kneist, W. (2021). Surgical education in the digital age-virtual reality, augmented reality and robotics in the medical school. *Zentralblatt fur Chirurgie*, *146*(1), 37–43.

Kurul, R., Ögün, M. N., Neriman Narin, A., Avci, Ş., & Yazgan, B. (2020). An alternative method for anatomy training: Immersive virtual reality. *Anatomical Sciences Education*, *13*, 648–656. https://doi.org/10.1002/ase.1959

Kwan, R. Y. C., Ng, F., Lam, L. C. W., Yung, R. C., Sin, O. S. K., & Chan, S. (2023). The effects of therapeutic virtual reality experience to promote mental well-being in older people living with physical disabilities in long-term care facilities. *Trials, 24*(1), 558. https://doi.org/10.1186/s13063-023-07592-7

Lewis, K. O., Popov, V., & Fatima, S. S. (2024). From static web to metaverse: Reinventing medical education in the post-pandemic era. *Annals of Medicine*, *56*(1), 2305694.

Liu, Z., Ren, L., Xiao, C., Zhang, K., & Demian, P. (2022). Virtual reality aided therapy towards health 4.0: A two-decade bibliometric analysis. *International Journal of Environmental Research and Public Health*, *19*(3), 1525.

Meyerbröker, K., & Emmelkamp, P. (2010). Virtual reality exposure therapy in the treatment of anxiety disorders. *International Journal of Anxiety Disorders*, *29*(4), 345–356. https://doi.org/10.1002/da.20734

Ntakakis, G., Plomariti, C., Frantzidis, C., Antoniou, P. E., Bamidis, P. D., & Tsoulfas, G. (2023). Exploring the use of virtual reality in surgical education. *World Journal of Transplantation*, *13*(2), 36.

O'Toole, R. V., Playter, R. R., Krummel, T. M., Blank, W. C., Cornelius, N. H., Roberts, W. R., Bell, W. J., & Raibert, M. (1999). Measuring and developing suturing technique with a virtual reality surgical simulator. *Journal of the American College of Surgeons, 189*(1), 114–128. https://doi.org/10.1016/S1072-7515(99)00076-9

Paas, F., Tuovinen, J. E., Tabbers, H., & Van Gerven, P. W. (2003). Cognitive load measurement as a means to advance cognitive load theory. In F. Paas, A. Renkl, & J. Sweller (Eds.), *Cognitive Load Theory* (pp. 63–71). Routledge.

Pandrangi, V. C., Gaston, B., Appelbaum, N. P., Albuquerque Jr, F. C., Levy, M. M., & Larson, R. A. (2019). The application of virtual reality in patient education. *Annals of Vascular Surgery*, *59*, 184–189. https://doi.org/10.1016/j.avsg.2019.01.015

Park, S., Yoo, J., Lee, Y., DeGuzman, P. B., Kang, M. J., Dykes, P. C., …& Cha, W. C. (2024). Quantifying emergency department nursing workload at the task level using NASA-TLX: An exploratory descriptive study. *International Emergency Nursing*, *74*, 101424.

Peters, P., Lemos, M., Bönsch, A., Ooms, M., Ulbrich, M., Rashad, A., …& Puladi, B. (2023). Effect of head-mounted displays on students' acquisition of surgical suturing techniques compared to an e-learning and tutor-led course: A randomized controlled trial. *International Journal of Surgery*, *109*(8), 2228–2240.

Popov, V., Mateju, N., Jeske, C., & Lewis, K. O. (2024). Metaverse-based simulation: A scoping review of charting medical education over the last two decades in the lens of the 'marvelous medical education machine'. *Annals of Medicine*, *56*(1), 2424450.

Pulijala, Y., Ma, M., Pears, M., Peebles, D., & Ayoub, A. (2018). Effectiveness of immersive virtual reality in surgical training: A randomized control trial. *Journal of Oral and Maxillofacial Surgery*, *76*(5), 1065–1072. https://doi.org/10.1016/j.joms.2017.10.002

Sandrone, S. (2022). Medical education in the metaverse. *Nature Medicine*, *28*(12), 2456–2457.

Saparamadu, P. V. I. N., Sepasgozar, S., Guruge, R. N. D., Jayasena, H. S., Darejeh, A., Ebrahimzadeh, S. M., & Eranga, B. A. I. (2025). Optimising contract interpretations with large language models: A comparative evaluation of a vector database-powered Chatbot vs. ChatGPT. *Buildings*, *15*(7), 1144.

Saposnik, G., & Levin, M. (2011). Virtual reality in stroke rehabilitation. *Stroke*, *42*(5), 1380–1386. https://doi.org/10.1161/strokeaha.110.605451.

Sargazi Moghadam, T., Darejeh, A., Delaramifar, M., & Mashayekh, S. (2024). Toward an artificial intelligence-based decision framework for developing adaptive e-learning systems to impact learners' emotions. *Interactive Learning Environments*, *32*(7), 3665–3685.

Schmitt, Y. S., Hoffman, H. G., Blough, D. K., Patterson, D. R., Jensen, M. P., Soltani, M., Carrougher, G. J., Nakamura, D., & Sharar, S. R. (2011). A randomized, controlled trial of immersive virtual reality analgesia, during physical therapy for pediatric burns. *Burns, [online]*, *37*(1), 61–68. https://doi.org/10.1016/j.burns.2010.07.007.

Segawa, T., Baudry, T., Bourla, A., Blanc, J. V., Peretti, C. S., Mouchabac, S., & Ferreri, F. (2020). Virtual Reality (VR) in Assessment and Treatment of Addictive Disorders: A Systematic Review. Frontiers in Neuroscience, [online] 13. https://doi.org/10.3389/fnins.2019.01409

Singh, B., Vig, K., & Kaunert, C. (2024). Modernizing healthcare: Application of augmented reality and virtual reality in clinical practice and medical education. In K. K. Hiran, R. Doshi, & M. Patel (Eds.), *Modern Technology in Healthcare and Medical Education: Blockchain, IoT, AR, and VR* (pp. 1–21). IGI Global Scientific Publishing.

Suh, I., McKinney, T., & Siu, K. C. (2023). Current perspective of metaverse application in medical education, research and patient care. In *Virtual Worlds*, *2*(2), 115–128. MDPI.

Tang, F. M. K., Lee, R. M. F., Szeto, R. H. L., Cheung, J. C. T., & Ngan, O. M. Y. (2020). Experiential learning with virtual reality: Animal handling training. *Innovation and Education*, *2*(1), 1–9. https://doi.org/10.1186/s42862-020-00007-3

Thomsen, A. S. S., Bach-Holm, D., Kjærbo, H., Højgaard-Olsen, K., Subhi, Y., Saleh, G. M., Park, Y. S., la Cour, M., & Konge, L. (2016). Operating room performance improves after proficiency-based virtual reality cataract surgery training. *Ophthalmology, 124*(4), 524–531. https://doi.org/10.1016/j.ophtha.2016.11.015

Wei, J., Siegel, E., Sundaramoorthy, P., Gomes, A., Zhang, S., Vankipuram, M., …& Bailenson, J. (2025, March). Cognitive load inference using physiological markers in virtual reality. *In 2025 IEEE Conference Virtual Reality and 3D User Interfaces (VR)* (pp. 759–769). IEEE.

Wu, S. H., Yang, Y. Y., Huang, C. C., Liu, C. W., Yang, L. Y., Chen, C. H., Kao, S. Y., & Lee, F. Y. (2020). Virtual reality simulation increases Chinese physicians' and lab technicians' familiarity and confidence regarding proper clinical wastes segregation/disposal: A 2-year pilot study. *BMJ Simulation and Technology Enhanced Learning*, *7*(1), 38–40. https://doi.org/10.1136/bmjstel-2019-000554

Wu, Y., Chen, J., Ma, W., Guo, L., & Feng, H. (2022). Virtual reality in preoperative preparation of children undergoing general anesthesia: A randomized controlled study. *Die Anaesthesiologie*, *71*(Suppl 2), 204–211. https://doi.org/10.1007/s00101-022-01177-w

Yang, C., Zhang, J., Hu, Y., Yang, X., Chen, M., Shan, M., & Li, L. (2024). The impact of virtual reality on practical skills for students in science and engineering education: A meta-analysis. *International Journal of STEM Education*, *11*(1), 28.

Yang, J., Li, X., Wang, T., Li, C., Li, Z., Hu, H., & Sun, X. (2021). Application of mixed reality technology in surgical resection of vestibular schwannoma and doctor-patient communication. *Cancer Research on Prevention and Treatment*, *48*(8), 788–793. https://doi.org/10.3971/j.issn.1000-8578.2021.21.0407

Zaidi, S. S. B., Adnan, U., Lewis, K. O., & Fatima, S. S. (2024). Metaverse-powered basic sciences medical education: Bridging the gaps for lower middle-income countries. *Annals of medicine*, *56*(1), 2356637.

5 Metaverse and Virtual Reality Integration in Computer Science

Applications in Programming Education

Ali Darejeh, Kanit Srihakorth, Sara Mashayekh, and Samad M. E. Sepasgozar

1 Introduction

The field of programming has experienced remarkable growth over the past decade, with computer science driving innovations across numerous sectors, from artificial intelligence (AI) to data science and software engineering (Segura et al., 2020). Despite this rapid advancement and its transformative impact, computer science education continues to face persistent challenges, particularly in attracting and retaining students in programming-related disciplines. Many students encounter significant obstacles when learning programming due to its abstract nature, particularly in areas like object-oriented programming (OOP), which serves as a cornerstone for modern software development (Jusas et al., 2022). These difficulties often stem from the complexity of concepts, the requirement for logical thinking, and the inability to relate abstract ideas to tangible outcomes. For novice learners, these challenges can lead to frustration and disengagement, especially when timely support is unavailable. As a result, a lack of motivation frequently becomes a barrier to sustained learning and mastery of advanced topics (Zhao et al., 2018).

To mitigate these issues, researchers and educators have explored innovative teaching methodologies to make programming more engaging, accessible, and intuitive for students. One such approach is the use of serious games, which have gained popularity for their ability to transform complex educational content into interactive and enjoyable experiences (Kumar & Sharma, 2018). Serious games are explicitly designed with educational objectives in mind, creating environments where learners can solve challenges, explore scenarios, and acquire both theoretical knowledge and practical skills. These games are not merely entertaining; they have been shown to positively influence learners' moods and increase their intrinsic motivation, which in turn enhances their academic performance (Zhonggen, 2019). For instance, platforms like ProLounge provide a dynamic learning environment that leverages problem-solving tasks, quizzes, and matching exercises to help students grasp complex programming topics more effectively. By integrating gameplay mechanics with educational content, such platforms create a structured yet enjoyable framework that fosters deeper engagement and learning retention (Kumar & Sharma, 2018).

DOI: 10.1201/9781003557555-5

Beyond serious games, the concept of educational escape rooms has emerged as a highly engaging and immersive teaching tool that promotes collaborative, interactive learning while fostering critical thinking, problem-solving, creativity, and practical skills across diverse educational disciplines (Rushdan et al., 2025; Yang, Chang, & Jen, 2023). These interactive environments, which challenge students to solve puzzles and complete tasks to escape within a set time frame, have been widely adopted in fields such as medicine and engineering for their ability to create immersive, hands-on learning experiences (Ho, 2018). In programming education, escape rooms have proven particularly effective in teaching abstract concepts through practical application. For example, escape room activities might involve debugging code, solving algorithmic puzzles, or identifying logical errors in a program, all requiring students to apply their knowledge in real time (López-Pernas et al., 2019a).

Another transformative tool in modern education is virtual reality (VR), which has rapidly expanded its applications across disciplines due to its unique ability to create immersive and interactive environments. These immersive experiences enhance learners' understanding of complex topics and provide opportunities to experiment and learn from mistakes without real-world consequences (Darejeh et al., 2024a). VR's potential in programming education lies in its capacity to visualize abstract concepts and create 3D environments where students can interact with code, objects, and algorithms in intuitive ways (Di Natale et al., 2020; Huang et al., 2023).

Within programming education, VR has been explored as a solution to the challenges of teaching both foundational and advanced programming concepts. Virtual environments have leveraged storytelling, gamification, and 3D visualizations to help learners connect abstract ideas—such as OOP—to tangible, relatable representations (Bouali et al., 2019; Díaz et al., 2008). Systems like VEnvI (Parmar et al., 2017) and Imikode (Bouali et al., 2019) demonstrate the potential of immersive environments to engage students and make programming more approachable. While current VR platforms have laid a strong foundation, many remain limited in scope—often focusing narrowly on introductory algorithms or specific tasks—and lack the robustness required to support comprehensive programming curricula. Moreover, they frequently do not incorporate advanced collaborative features aligned with Metaverse principles, which emphasize persistent, shared virtual spaces that support real-time interaction and teamwork (Jun, 2020; Nevelsteen, 2018).

To address existing gaps in programming education, this chapter explores the transformative potential of VR, particularly through the lens of gamified escape room experiences. It reviews the current state of VR in programming education and presents a novel VR-based escape room platform designed to enhance teaching and learning. This next-generation platform integrates gamification, storytelling, and collaborative elements within a fully immersive virtual environment (Darejeh, 2023). Its design emphasizes challenging puzzles that map directly to key programming concepts, encouraging critical thinking, problem-solving, and the practical application of knowledge. Furthermore, the platform aligns with Metaverse principles, enabling real-time collaboration in a shared virtual space.

By examining existing research, identifying critical gaps, and presenting a detailed empirical study, this chapter offers insights into how immersive technologies can help overcome persistent challenges and pave the way for more effective, engaging, and inclusive programming education.

2 Literature Review

In this section, related studies on the application of VR in learning programming and the application of the escape room concept in teaching computer science will be discussed. At the end, a conclusion will be drawn, and the literature gap will be indicated.

2.1 Applications of VR in Learning Programming

Over the years, various VR platforms have been developed to introduce fundamental programming concepts, including OOP, and these platforms demonstrate diverse levels of interactivity, gamification, and educational effectiveness (Agbo et al., 2021). One notable example is the system developed by Stigall and Sharma (2017), which employed a vehicle-themed approach to teach OOP. In this platform, students could build and visualize vehicles by selecting predefined classes. For instance, they could choose vehicle types such as "fly," "drive," or "land," and pair them with usage categories like "commercial," "military," or "service." While constructing these vehicles, the system displayed Unified Modeling Language (UML) diagrams alongside 3D models, enabling students to visualize the relationships between objects and their components. To further engage learners, the system incorporated quizzes where students assigned access modifiers (e.g., public, private, or protected) to variables like "speed" and "color," earning points for correct answers. Despite the platform's innovative approach, its educational impact was not comprehensively assessed, leaving questions about its effectiveness in improving learning outcomes.

Similarly, Bouali et al. (2019) introduced Imikode, a VR-based learning environment designed to teach OOP concepts. In an empty farm environment, users interacted with elements to construct objects such as houses, trees, and animals using OOP keywords. While the platform encouraged hands-on learning, it also lacked an evaluation of its impact on student performance—highlighting a recurring issue in the field.

Other VR systems have emphasized gamification and immersive learning experiences. Segura et al. (2020) developed VR-OCKS, a maze-based VR game where students solved puzzles using action blocks, such as loops and conditional statements, to guide characters through mazes. This approach proved effective in boosting students' motivation and learning performance. Similarly, Horst et al. (2019) created FunPlogs, another maze-based game focused on algorithm simplification. Students used advanced action blocks like "if" and "while" to streamline their algorithms and solve puzzles more efficiently. Both VR-OCKS and FunPlogs demonstrated the potential of gamified learning to influence students' attitudes toward programming positively. Building on this trend, Grivokostopoulou et al. (2017) designed a maze game to teach search algorithms. By following a numbered tree view of algorithm elements, students navigated through mazes, which significantly improved their learning effectiveness and motivation. In a similar vein, Darejeh et al. (2024b) developed a VR application for teaching programming that presented a maze and a robot positioned at its starting point. Users were required to assemble code blocks—either through drag-and-drop or keyboard input—to guide the robot to the end of the maze. The results indicated a positive student attitude toward the system and an increased understanding of algorithmic concepts in programming. These studies collectively highlight the role of gamification in enhancing the appeal and effectiveness of VR platforms for programming education.

Several VR platforms targeted broader programming concepts through gamification and interactive activities. For example, Sharma and Ossuetta (2017) developed a mini-game system to teach decision-making, loops, and arrays. One game required students to shoot ducks of specific colors, illustrating decision-making processes, while another game focused on understanding arrays by having students identify specific elements within a matrix. Berns et al. (2019) introduced MYR, a browser-based platform that allowed students to write code on a computer screen to generate 3D animated characters, viewable through VR headsets. Nguyen et al. (2020) expanded on this idea with VRASP, a platform that integrates coding and output visualization entirely within a VR environment. While demonstrating potential for enhancing motivation, these systems also revealed significant usability issues and highlighted the need for design improvements.

Beyond gamification, several studies have extended VR applications to incorporate embodied activities and advanced 3D visualizations. Parmar et al. (2017) introduced VEnvI, a VR platform where students created dance routines using programming constructs such as variables, loops, and conditional statements. By linking visual programming with pseudocode, VEnvI helped bridge the gap between abstract concepts and traditional programming languages, making the learning process more intuitive. Similarly, Chandramouli et al. (2014) developed a VR system that visualized OOP concepts—such as classes, methods, inheritance, and encapsulation—using 3D text and diagrams. However, this system lacked interactivity and served only as a visual representation rather than an interactive learning environment. Building on this, Ortega et al. (2017) developed 3D-VPL, a VR environment that enabled users to engage with block-based programming (BBP). In this system, visual elements like cubes and rectangles were used to represent programming constructs such as objects, functions, and properties. While these systems demonstrated the potential of visualization in programming education, they also revealed notable usability challenges, particularly for novice learners.

Collectively, these studies underscore the transformative potential of VR-based platforms in programming education. By leveraging gamification, interactivity, and immersive experiences, VR systems can enhance student engagement, motivation, and learning performance. However, the lack of consistent evaluations and the limited scope of many systems suggest further research.

2.2 Applications of Escape Rooms in Teaching Computer Science

Although educational escape rooms are a relatively recent topic in academia, numerous studies have demonstrated their effectiveness across various disciplines. In medical education, escape rooms have been shown to facilitate learning in diverse areas, including infection prevention (Adams et al., 2018), cardiopulmonary resuscitation, blood sample collection (Gómez-Urquiza et al., 2019), and laboratory result evaluation (Brown, Darby, & Coronel, 2019). These escape rooms immerse learners in realistic settings, such as surgery rooms (Kinio et al., 2019), dermatology clinics (Guckian, Sridhar, & Meggitt, 2020), radiology labs (Jambhekar, Pahls, & Deloney, 2020), physiotherapy rooms (Jiménez Sánchez et al., 2017), and pathophysiology labs (Monaghan & Nicholson, 2017), where they must complete task sequences based on scenarios to resolve challenges and escape the room. Similarly, positive effects of educational escape rooms have been reported in pharmacy education, covering topics such as medication error identification (Clauson et al., 2019), dose calculation (Hermanns et al., 2017), diabetes knowledge (Eukel, Frenzel, & Cernusca, 2017), and infectious diseases (Cotner et al., 2018). These escape rooms combine puzzle mechanics, such as unlocking objects and locating hidden elements, with domain-specific tasks, including performing calculations, consulting references, interpreting results, and operating laboratory equipment.

In computer science education, escape rooms are less frequently studied but have been applied to teach Information and Communication Technology (ICT) competence (Musil et al., 2019), programming (López-Pernas et al., 2019a, 2019b), computer security (Beguin et al., 2019), computer networks (Borrego et al., 2017), and cryptography (Ho, 2018). Musil et al. (2019) designed a physical escape room to teach ICT concepts, including hardware, software, and communication principles. Tasks involved connecting cables, configuring files, and transferring data, with pre-test and post-test surveys indicating increased ICT knowledge among participants. Similarly, López-Pernas et al. (2019a) developed an educational escape room for web programming, where students stopped a simulated bomb explosion by completing a website's HTML, CSS, and JavaScript code. Survey results highlighted increased motivation and learning performance. In a related study, López-Pernas et al. (2019b) introduced a more advanced escape room for web development, featuring HTML, CSS, JavaScript, Node.js, and SQL. This escape room tasked students with generating an antivirus code to cure a virus-infected scientist, demonstrating higher learning performance and motivation than traditional methods.

In the domain of computer security, Beguin et al. (2019) developed two escape rooms where students took on defender or hacker roles. Defenders secured computers by addressing weak passwords and sensitive information, while hackers bypassed security measures to retrieve critical files. Positive survey results underscored participants' engagement and learning. Borrego et al. (2017) designed an escape room for computer networks and information security, incorporating concepts such as TCP conversations and RSA schemes. Despite its innovative scenarios, no evaluation of its learning effectiveness was conducted. Similarly, Ho (2018) created an escape room to teach cryptography, where students solved riddles using

Sage Math to unlock boxes, with interviews reflecting a positive attitude toward motivation and learning.

Escape rooms have also been explored in VR environments for programming education, albeit to a limited extent. Bolivar et al. (2019) developed a 3D VR escape room game where users indirectly engaged with programming concepts. For example, loops were taught by repeating actions, and conditional statements were explored by weighing apples to find the heaviest. However, the system did not explicitly teach programming, and no formal evaluation of its effectiveness was conducted.

Overall, the literature highlights the efficacy of escape rooms as an engaging teaching strategy in computer science. These environments increase students' engagement with content, foster creativity, and enhance problem-solving and decision-making skills—essential pillars of programming education (McFadden & Porter, 2018; Pan, Lo, & Neustaedter, 2017). Despite these promising findings, few studies have examined the technical feasibility and design considerations necessary to explore the potential of VR escape rooms for enhancing programming education, highlighting a clear need for further research.

In summary, the existing literature on the use of VR and escape rooms for teaching programming reveals that none of the current systems offer comprehensive coverage of all essential elements of fundamental programming. Most existing platforms focus on isolated components, such as variables or object models in OOP, rather than delivering a complete instructional experience. To address this gap, the empirical study presented in the following section introduces our platform—a holistic system designed to encompass all critical aspects of both fundamental programming and OOP. This approach aims to provide a more comprehensive, engaging, and effective learning experience for students.

3 Empirical Study: VR Escape Room for Learning the Fundamentals of Programming

This study evaluates the impact of a custom-developed escape room on students' motivation and learning of fundamental programming and OOP concepts.

3.1 Application Development

The gamified environment developed for this research project was built using Unity version 2020.3.25f1, with C# as the primary programming language, and was deployed on Meta Quest 2, Meta Quest 3, and HTC VIVE Pro headsets. Designed as an escape room game, the environment comprises two modules: the first introduces users to puzzles related to OOP, while the second focuses on more advanced programming concepts, including conditional statements and loops. Both modules incorporate narrative-style gameplay, which has been shown to be an effective instructional method for reducing cognitive load and enhancing learner engagement (Darejeh, Marcus, & Sweller, 2021).

As detailed in Chapter 1 of this book, this application was created as part of a broader metaverse platform initiative at the University of New South Wales

(UNSW). The application enables real-time peer collaboration and multiplayer puzzle-solving, enriching the learning experience and simulating real-world programming teamwork.

3.1.1 Module One

Module one consists of ten levels, and players must successfully solve the puzzle within each level to unlock the door to the next room and progress in the game. The puzzles are based on the story of creatures traveling from another planet in their spaceship to Earth and getting trapped in their spaceship. Users should solve programming-related puzzles at each level to open doors and progress to the next level. When users open the application, they receive instructions on interacting with the system. Figure 5.1 displays the landing screen of the first module of the application as it appears on the VR headsets.

In the initial stages of the game, we introduce users to fundamental programming concepts, such as variables, variable types, functions, and classes. These early rooms serve as tutorials and playgrounds, allowing users to familiarize themselves with the basics of programming. To represent each programming concept, we have employed cartoon characters and integrated them into our distinctive theme, replacing traditional blocks commonly used in programming environments.

For example, Figures 5.2A and 5.2B depict an instance of teaching how to modify object properties in OOP. Following the instructions located at the bottom of the screen (see Figures 5.2A and 5.2B), users are instructed to modify certain properties of the pot object (see Figure 5.2A). They should change the pot color to green and increase the pot's size to 2, allowing them to execute the 'Grow' method (see Figure 5.2B).

Figure 5.1 The introductory screen of the application on the VR headset.

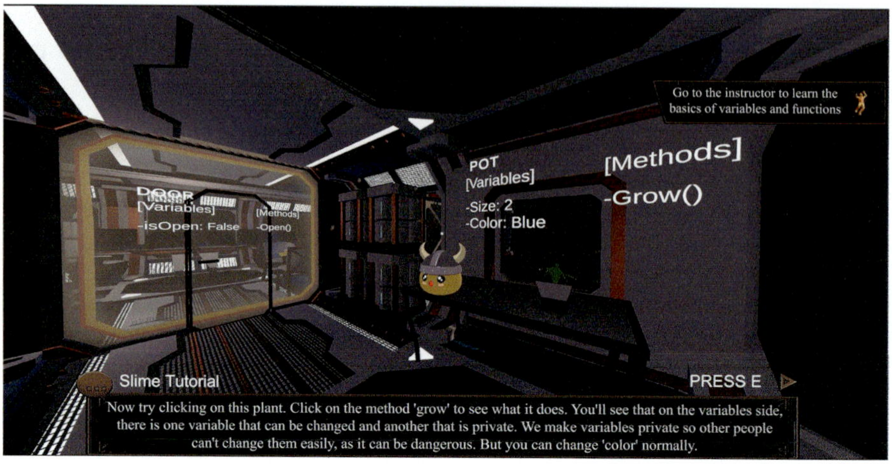

Figure 5.2A The instruction says the pot size should be two, and the color should be green.

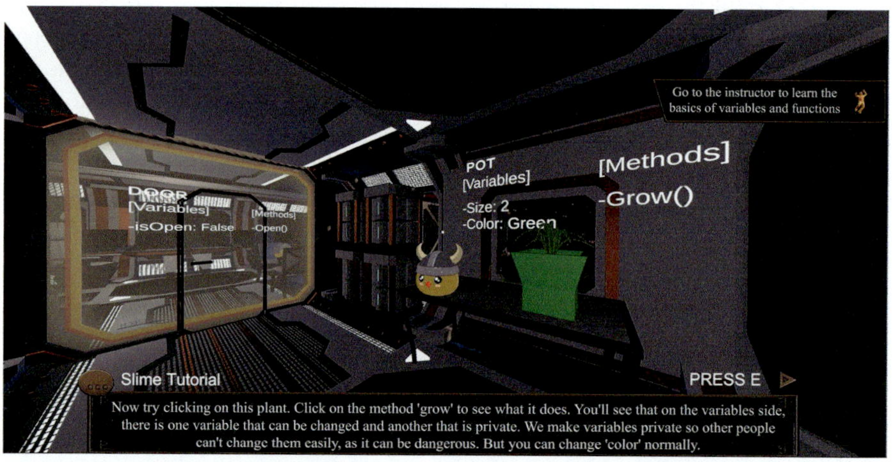

Figure 5.2B The pot color is green, and the size equals two.

This is a requirement for opening the door. As can be observed in Figures 5.3A and 5.3B, once users have made the necessary modifications to the pot object, they can change the value of the 'Is Open' variable to true, thus enabling them to open the door.

As users advance through the game, they will encounter increasingly challenging puzzles. The progressive nature of the game ensures that as users advance, they are presented with more complex and intricate puzzles. This deliberate design choice aims to strike a balance between keeping the game engaging and fostering a sense of approachability for newcomers to programming. By gradually introducing new challenges, users can comfortably build upon their existing knowledge while expanding their understanding of programming concepts.

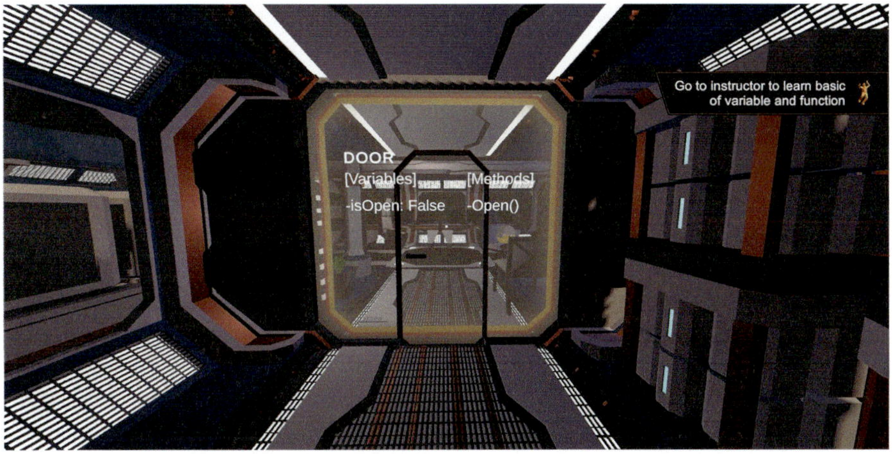

Figure 5.3A The door is closed.

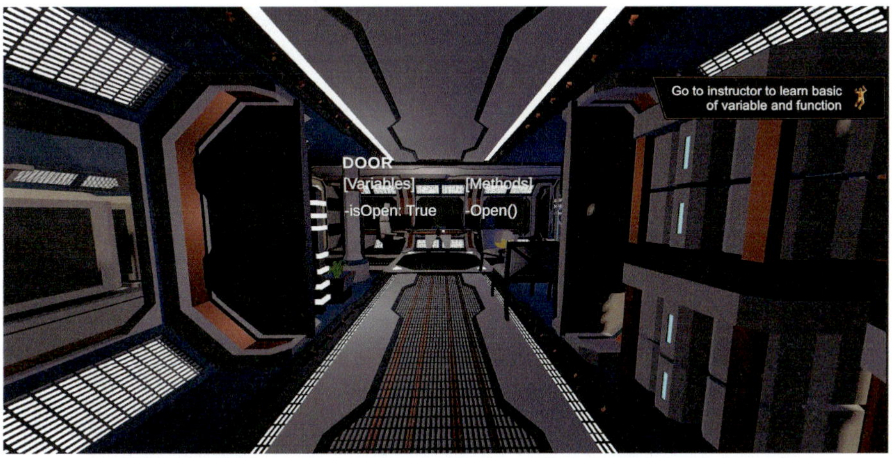

Figure 5.3B The door is open.

In the subsequent levels, users are introduced to the fundamental concepts of methods, functions, and classes, which serve as building blocks for structuring code in OOP. In this stage, users assemble variables and methods to construct classes, demonstrating the practical application of key OOP concepts.

3.1.2 Module Two

Module two consists of five levels designed to teach algorithms and programming concepts such as conditional statements and loops. This module is implemented as an escape room set in a virtual house with multiple rooms. Users must solve a series of puzzles to progress through each room, ultimately leading them to escape the house upon completing all the levels. Each puzzle is carefully crafted to reinforce specific programming concepts, ensuring learners engage with the material practically and interactively.

When users begin this module, they find themselves inside the living room of a virtual building (Figure 5.4), designed to resemble a warm and immersive home environment. This space acts as the central hub of the experience. Using VR controllers, users can interact with objects in the environment, initiating the first stage of the escape room challenge. To proceed to the second floor, learners must solve a series of programming-based puzzles, starting with an algorithm pathfinding challenge (Figure 5.5). In this task, users help a green circle navigate a maze by assembling directional commands using basic control structures such as loops. This reinforces sequencing and logic building, core to algorithmic thinking.

The next task is a matching puzzle focused on programming keywords, where users must correctly associate control flow elements—such as loops (for, while) and conditional statements (if, switch)—with their visual representations and expected behavior (Figure 5.6). The drag-and-drop interface enhances interactivity, encouraging active recall of syntax and functionality. As users progress, they encounter a challenge centered on understanding the switch-case control structure (Figure 5.7). A series of case conditions are displayed, and users must analyze the input-output behavior of the function. The goal is to enter the correct output for the switch-case statement using the password board mounted on the wall.

Figure 5.4 The landing room of the escape room.

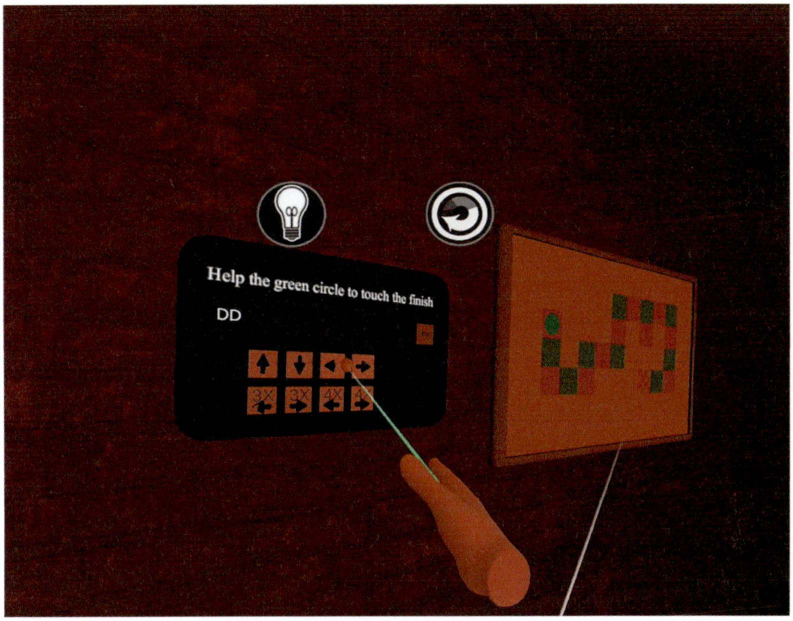

Figure 5.5 Learn the fundamentals of algorithms.

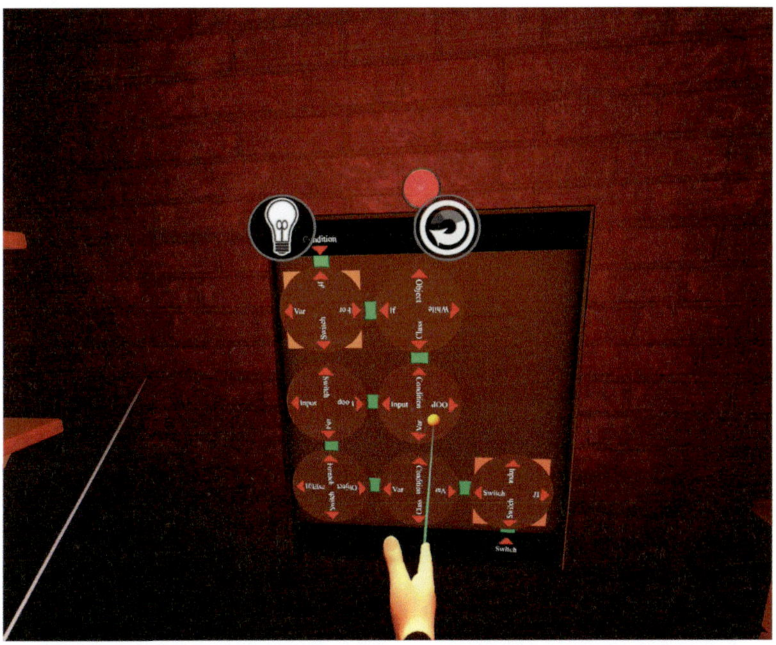

Figure 5.6 Learning keywords related to conditional statements and loops.

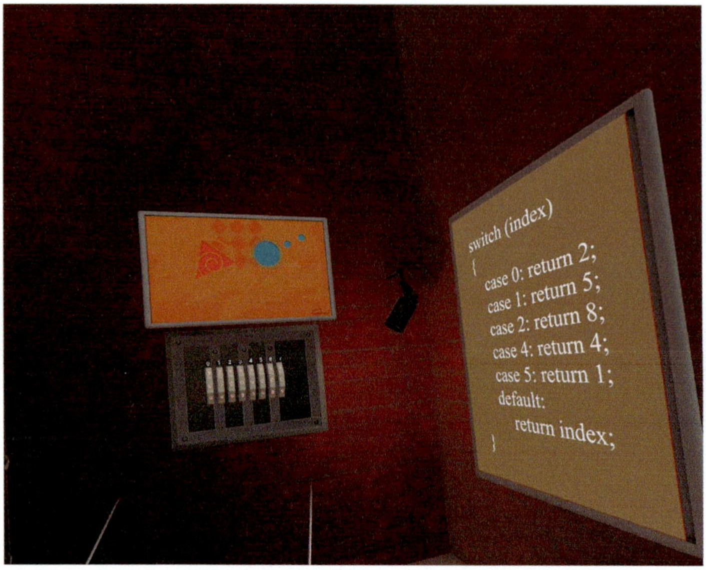

Figure 5.7 Learning a conditional statement.

Following this, users face a code-matching puzzle that presents various code snippets, each representing a distinct function (Figures 5.8a and 5.8b). The task requires users to match these snippets based on their output behavior, using logical reasoning and their understanding of conditionals and loops. This reinforces both functional decomposition and code comprehension.

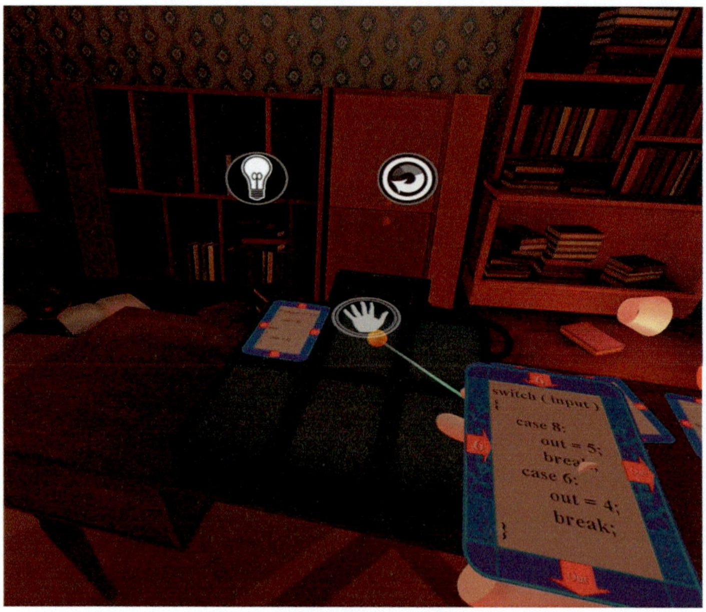

Figure 5.8A Practicing a combination of conditional statements and loops.

Figure 5.8B Practicing a combination of conditional statements and loops.

3.2 Methodology of Conducting the Study

The study recruited 40 students from UNSW to evaluate whether a gamified learning platform could enhance students' perceptions of their learning performance and motivation in OOP. Participants were selected based on specific criteria: they were over 18 years of age, had a satisfactory level of English proficiency, and had no or minimal prior experience with programming. This ensured a homogeneous group of novice users, aligning with the study's objective of assessing the platform's usability and educational value among beginners.

The experiment was conducted in accordance with the university-approved human research ethics protocol, as authorized by the institutional ethics review board. All participants provided informed consent, voluntarily agreeing to take part in the study and permitting the use of their data for research purposes within the project. To protect confidentiality, participant identities were anonymized, and no personally identifiable information was disclosed in the reporting of results or discussion.

The experiment was conducted in two main phases. In the first phase, upon arrival, participants were briefed about the study's purpose and procedure and asked to sign an informed consent form. They then completed a pre-questionnaire using a 0–9 Likert scale to assess their background knowledge in fundamental and OOP, motivation to learn these subjects, and confidence in the potential effectiveness of the gamified platform. Following this, participants received a demonstration on how to use the VR headset and platform. They were then allowed to interact with the gamified system at their own pace, solving puzzles embedded within the game to learn programming concepts.

After gameplay, participants completed a post-questionnaire containing the same questions as the pre-test, designed to measure knowledge, motivation, and confidence changes. Furthermore, a semi-structured interview was conducted to gather deeper insights. Participants were asked about their prior experiences with traditional programming learning methods, preferences between those methods and the gamified platform, elements they found most engaging, and how the platform affected their motivation to learn fundamental and OOP.

In the second phase of the experiment, participants interacted with a computer display version of the escape room used in the VR system. This version was navigated using a mouse and keyboard and mirrored the VR experience. Participants were given 15 minutes to explore this version, after which they answered four Likert-scale questions comparing the VR and non-VR versions in terms of engagement, motivation, and personal preference.

3.3 Results from the Study

3.3.1 First Phase of the Experiment

The findings from the first phase of the study were obtained through two Likert scale questionnaires (a pre-test and a post-test) and an interview conducted at the conclusion of this phase. The first two questions of the Likert scale questionnaires asked participants about their knowledge level of fundamental programming and OOP. Participants reported a significant improvement in their knowledge level, transitioning from low familiarity in the pre-test Question one: $M = 2.00$, $SD = 0.72$; Question two: $M = 1.70$, $SD = 0.76$ to medium-level familiarity in the post-test Question one: $M = 4.7$, $SD = 0.88$; Question two: $M = 5.17$, $SD = 0.79$. The paired samples t-test showed a significant difference between the pre-test and post-test questionnaire for these two questions: Question one $t(39) = -14.83$, $p < 0.001$ and Question two $t(39) = -17.94$, $p < 0.001$.

The third question asked participants about their motivation to learn fundamental and OOP. Participants reported an average motivation level of $M = 3.72$, $SD = 0.74$ in the pre-test and $M = 5.34$, $SD = 0.94$ in the post-test, indicating increased motivation after using the VR system. The paired samples t-test showed a significant difference for Question three $t(39) = -6.43$, $p < 0.001$. The fourth question asked about their confidence in the platform's ability to assist them in self-learning fundamental and OOP. Participants reported an average confidence level of $M = 4.40$, $SD = 0.75$ in the pre-test and $M = 5.93$, $SD = 0.85$ in the post-test, indicating increased perceived usefulness of the gamified platform. A significant difference was found for Question four $t(39) = -9.45$, $p < 0.001$.

As discussed in the methodology section at the end of phase one, the participants of this study were interviewed to better understand their preferred method of learning programming, the aspects of the gamified learning platform they liked, and whether the platform was successful in increasing their motivation to learn fundamental programming and OOP.

The interview conducted with the participants revealed their unanimous preference for a gamified learning platform over traditional methods of learning programming, such as reading books or watching videos. Several common themes emerged from their comments, highlighting the reasons behind their preference. Firstly, they found the visual elements and interactive nature of the learning method to be enjoyable and fun. This interactive approach engaged the participants actively and enhanced their self-study experience, making it more effective.

The ability to instantiate objects infinitely allowed them to experiment and explore coding ideas, providing a practical understanding of programming and OOP. The participants also expressed that the presented method was more engaging and satisfying, and kept them interested throughout the learning process. It was described as 'lively', 'interesting', 'dynamic', 'less intimidating', and an 'enjoyable way to acquire knowledge'. While nine participants acknowledged the value of traditional methods, such as reading, they still emphasized the overall superiority of the gamified learning platform in terms of engagement, satisfaction, and knowledge acquisition. The participants appreciated the practicality and comprehensiveness of the learning method, particularly its visual elements, interactivity, and the ability to experiment with coding ideas. This approach provided them with a practical understanding of programming concepts, allowing them to see how coding ideas functioned in practice and reinforcing their grasp of key principles.

The findings also indicate that the gamified learning platform had a positive impact on their motivation to learn fundamental programming and OOP. Participants reported that the platform was 'quite stimulating' and 'provided a sense of motivation', particularly for someone who had 'never learned programming before'. They expressed that programming seemed easier to grasp and understand through the interactive and engaging nature of the platform.

3.3.2 Second Phase of the Experiment

In evaluating the engagement level and participant preferences of the escape room VR version compared to its computer display counterpart to determine whether the sense of immersion provided by VR enhances user engagement and motivation, we analyzed the Likert-scale results of the second phase of the experiment. The findings are as follows:

The first two questions on the Likert scale assessed participants' perceptions of engagement and motivation when using the escape room's computer display version versus its VR counterpart. Participants reported an average score of $M = 4.86$, $SD = 0.84$ for the computer display version and $M = 6.19$, $SD = 0.59$ for the VR version. A paired samples t-test revealed a statistically significant difference between the two conditions, $t(39) = -8.91$, $p < 0.001$, indicating that the VR version was perceived as significantly more engaging and motivating than the computer display version.

Similarly, the third and fourth questions gauged participants' preferences for using the escape room in either format. Participants gave an average score of $M = 3.24$, $SD = 0.90$ for the computer display version and $M = 6.23$, $SD = 0.73$ for the VR version. The paired samples t-test again demonstrated a significant difference, $t(39) = -16.21$, $p < 0.001$, reflecting a strong preference for the VR version in terms of engagement and overall appeal.

3.4 Insights from Both Experiments

The results of this study affirm the effectiveness of VR-based escape rooms in enhancing novice learners' motivation and perceived competence in programming. Participants showed statistically significant improvements in their knowledge of both fundamental and OOP concepts, along with increased motivation and confidence in self-learning. These findings are consistent with existing research suggesting that immersive and gamified environments foster deeper engagement and promote positive learning outcomes. Participants not only found the VR platform more enjoyable and less intimidating than traditional methods but also appreciated its interactive, exploratory design, which allowed them to visualize and apply programming concepts in context.

Furthermore, the comparison between the VR and computer display versions revealed a clear preference for the immersive VR experience, highlighting the added value of spatial presence and interaction in maintaining attention and enhancing motivation. However, as with other emerging educational technologies, some usability challenges were noted, particularly for users unfamiliar with VR equipment. These insights underscore the importance of iterative, user-centered design when developing VR learning tools.

4 Discussion

This chapter explored the intersection of VR, gamification, and programming education by combining a comprehensive review of the literature with an empirical study on a VR-based escape room platform. The integration of these components provided a more holistic view of how immersive technologies are reshaping the teaching and learning of programming, particularly for novice learners.

The existing literature highlighted a growing trend toward using VR to support programming education, demonstrating its ability to visualize abstract concepts, support interactive learning, and increase student engagement. Studies such as those by Bouali et al. (2019), Parmar et al. (2017), and Segura et al. (2020) emphasized the effectiveness of VR environments in simplifying complex programming topics through visualization, embodied learning, and gamified tasks. However, these platforms often focused narrowly on specific programming elements—such as loops, conditional statements, or object models—without offering a comprehensive instructional experience or systematic evaluation of learning outcomes. Our study addresses these limitations by presenting a structured, multi-module VR environment that incorporates both fundamental programming and OOP concepts through scaffolded, narrative-driven gameplay.

The study's empirical findings confirmed many of the theoretical benefits reported in prior research. Participants exhibited significant improvements in perceived knowledge, motivation, and confidence when learning with the VR-based escape room. These outcomes reinforce existing evidence that gamified and immersive platforms can enhance cognitive engagement, emotional connection, and motivation in educational contexts (Darejeh et al., 2024a; Kumar & Sharma, 2018; Zhonggen, 2019). Moreover, participants' strong preference for the VR version over the computer display version aligns with broader studies emphasizing the value of spatial presence, embodiment, and interactive feedback in immersive learning environments (Di Natale et al., 2020; Huang et al., 2023).

Importantly, this chapter also contributes to a deeper understanding of educational escape rooms in computer science. While prior studies have shown their value in disciplines such as medicine and pharmacy (Adams et al., 2018; Clauson et al., 2019), their application in programming education remains underexplored. Our findings echo the promising outcomes reported in earlier escape room studies (e.g., Borrego et al., 2017; López-Pernas et al., 2019a, 2019b), extending them by demonstrating how VR escape rooms can integrate problem-solving, code comprehension, and algorithmic thinking into a cohesive and motivating experience. Participants not only appreciated the game-like format but also benefited from tasks that required them to construct and manipulate programming structures such as variables, classes, loops, and switch-case statements—bridging theoretical knowledge with hands-on practice.

These motivational gains are especially important in the context of programming education, where learners often encounter barriers due to abstract syntax, debugging challenges, and a lack of immediate feedback (Jusas et al., 2022; Zhao et al., 2018). Interview data from the study further reinforced this finding—participants found the gamified VR experience "less intimidating," "enjoyable," and "effective," reflecting the broader pedagogical impact of emotional engagement in immersive learning. This echoes the findings of Servotte et al. (2020) and Daassi and Debbabi (2021), who emphasized the role of sensory immersion, emotional presence, and realism in enhancing learning outcomes in VR contexts.

Importantly, our study also demonstrated a significant preference for the VR version over the computer display version of the escape room. Participants reported higher engagement, motivation, and satisfaction with the VR interface, attributing their preference to the sense of presence, interactivity, and realism. These findings align with prior literature emphasizing VR's unique capacity to promote spatial awareness and embodied cognition (Di Natale et al., 2020; Huang et al., 2023). While platforms like Bolivar et al. (2019) and Ortega et al. (2017) have explored immersive environments for programming, they often lack the systematic evaluation and narrative depth provided by our platform.

Although the platform developed for this project goes beyond introductory content by covering conditional statements and loops, it still lacks comprehensive modules on advanced topics such as data structures, algorithms, concurrent and parallel programming, and functional programming. This limitation reflects a broader trend seen in earlier systems, which often address isolated topics without offering a cohesive or scalable curriculum (Berns et al., 2019; Bouali et al., 2019). Future development should aim to integrate these complex domains, including compiler construction, distributed computing, and cybersecurity, to support learners at both novice and intermediate levels.

Another key area for enhancement is the inclusion of adaptive learning mechanisms. Currently, the system is primarily system-paced and does not dynamically respond to user behavior. Recent research suggests that integrating AI-driven instructional agents could significantly improve learning outcomes by offering real-time feedback, identifying misconceptions, adjusting task difficulty, and even monitoring emotional states to personalize the learning experience (Darejeh, Marcus, & Sweller, 2022; Darejeh et al., 2024c; Saparamadu et al., 2025; Sargazi Moghadam et al., 2024). Such enhancements would help transition the platform into a more intelligent, self-directed learning tool, particularly beneficial for remote or underserved learners.

The platform's collaborative capabilities also remain underexplored. While our evaluated application was part of a metaverse platform and included multiplayer features, the study focused on individual interaction. Future experiments could explore the application's metaverse capabilities more fully, particularly its support for real-time peer collaboration in puzzle-solving tasks (Jun, 2020; Nevelsteen, 2018). This would enable programming learners to collaborate within shared virtual spaces across global institutions, enhancing both social interaction and learning outcomes.

In addition to these conceptual extensions, practical refinements such as improved user interfaces, onboarding processes, and haptic feedback should be considered to enhance accessibility and usability for diverse learners. Moreover, while the current study demonstrates a use case centered on OOP, the limited participant pool restricts the generalizability of findings. Broader empirical validation may be warranted in future work, though the primary aim of this chapter is to present a focused use case rather than provide exhaustive behavioral analysis.

5 Conclusion

The integration of immersive technologies in programming education is not merely a pedagogical enhancement—it reflects a broader redefinition of how future learners engage with abstract thinking, logical reasoning, and collaborative problem-solving. This chapter's exploration of a VR-based escape room platform reveals that such environments are uniquely suited to fostering these skills in ways that traditional instruction cannot. As educational systems strive to accommodate increasingly diverse learners, immersive platforms stand out for their potential to deliver equitable, flexible, and context-rich learning experiences, regardless of learners' geographical or socioeconomic background.

Looking ahead, the convergence of VR, AI, and metaverse technologies presents a powerful opportunity to redefine the boundaries of computer science education. When thoughtfully implemented, the Metaverse can serve as more than just a delivery medium—it can evolve into an ecosystem for continuous learning, experimentation, and innovation. By embedding these technologies into educational practice, institutions can move beyond traditional knowledge transmission to create learning environments that are immersive, adaptive, and truly transformative. As programming becomes a foundational skill across disciplines, there is a growing need for platforms that not only teach code but also foster creativity, collaboration, and computational thinking.

References

Adams, V., Burger, S., Crawford, K., & Setter, R. (2018). Can you escape? Creating an escape room to facilitate active learning. *Journal for Nurses in Professional Development*, *34*(2), E1–E5.

Agbo, F. J., Sanusi, I. T., Oyelere, S. S., & Suhonen, J. (2021). Application of virtual reality in computer science education: A systemic review based on bibliometric and content analysis methods. *Education Sciences, 11*(3), 142. https://doi.org/10.3390/educsci11030142

Beguin, E., Besnard, S., Cros, A., Joannes, B., Leclerc-Istria, O., Noel, A., et al. (2019). Computer-security-oriented escape room. *IEEE Security & Privacy, 17*(4), 78–83.

Berns, C., Chin, G., Savitz, J., Kiesling, J., & Martin, F. (2019). MYr: A web-based platform for teaching coding using VR. In *Proceedings of the 50th ACM Technical Symposium on Computer Science Education* (pp. 77–83). ACM. https://doi.org/10.1145/3287324.3287482

Bolivar, S., Perez, D., Carrasquillo, A., Williams, A. S., Rishe, N. D., & Ortega, F. R. (2019). 3D interaction for computer science educational VR game. In *International Conference on Human-Computer Interaction* (pp. 408–419). Springer. https://doi.org/10.1007/978-3-030-23560-4_30

Borrego, C., Fernández, C., Blanes, I., & Robles, S. (2017). Room escape at class: Escape games activities to facilitate the motivation and learning in computer science. *Journal of Technology and Science Education, 7*(2), 162–171.

Bouali, N., Nygren, E., Oyelere, S. S., Suhonen, J., & Cavalli-Sforza, V. (2019). Imikode: A VR game to introduce OOP concepts. In *Proceedings of the 19th Koli Calling International Conference on Computing Education Research* (pp. 1–2). ACM Press.

Brown, N., Darby, W., & Coronel, H. (2019). An escape room as a simulation teaching strategy. *Clinical Simulation in Nursing*, *30*, 1–6.

Chandramouli, M., Zahraee, M., & Winer, C. (2014). A fun-learning approach to programming: An adaptive virtual reality (VR) platform to teach programming to engineering students. In *IEEE International Conference on Electro Information Technology* (pp. 581–586). https://doi.org/10.1109/EIT.2014.6871829

Clauson, A., Hahn, L., Frame, T., Hagan, A., Bynum, L. A., Thompson, M. E., & Kiningham, K. (2019). An innovative escape room activity to assess student readiness for advanced pharmacy practice experiences (APPEs). *Currents in Pharmacy Teaching and Learning, 11*(7), 723–728. IEEE. https://doi.org/10.1109/EIT.2014.6871829

Cotner, S., Smith, K. M., Simpson, L., Burgess, D. S., & Cain, J. (2018, November). 1311. incorporating an 'escape room'game design in infectious diseases instruction. *Open Forum Infectious Diseases*, *5*(1), S401. Oxford University Press.

Daassi, M., & Debbabi, S. (2021). Intention to reuse AR-based apps: The combined role of the sense of immersion, product presence and perceived realism. *Information & Management*, *58*(4), 103453.

Darejeh, A. (2023). Empowering education through eerp: A customizable educational vr escape room platform. In *2023 IEEE International Symposium on Mixed and Augmented Reality Adjunct (ISMAR-Adjunct)* (pp. 764–766). IEEE.

Darejeh, A., Huynh, L., Ji, X., Marcus, N., & Polly, P. (2024a). Assessing the efficacy of virtual reality-based laboratories versus real-life settings for PCR procedures, within medical science laboratories. In *2024 IEEE International Symposium on Mixed and Augmented Reality Adjunct (ISMAR-Adjunct)* (pp. 489–492). IEEE.

Darejeh, A., Marcus, N., & Sweller, J. (2021). The effect of narrative-based E-learning systems on novice users' cognitive load while learning software applications. *Educational Technology Research and Development*, *69*(5), 2451–2473.

Darejeh, A., Marcus, N., & Sweller, J. (2022). Increasing learner interactions with E-learning systems can either decrease or increase cognitive load depending on the nature of the interaction. *L'Année psychologique*, *122*(3), 405–437.

Darejeh, A., Moghadam, T. S., Delaramifar, M., & Mashayekh, S. (2024c). A framework for AI-powered decision making in developing adaptive e-learning systems to impact learners' emotional responses. In *2024 11th International and the 17th National Conference on E-Learning and E-Teaching (ICeLeT)* (pp. 1–6). IEEE.

Darejeh, A., Mondal, M., Marcus, N., & Vassar, A. (2024b). Exploring the cognitive load effects of diverse virtual reality interaction methods in interactive educational platforms. In *2024 IEEE International Symposium on Mixed and Augmented Reality Adjunct (ISMAR-Adjunct)* (pp. 435–438). IEEE.

Díaz, G. J., Albarrán, M. G., Gómez-Martín, M. A., & González-Calero, P. A. (2008). Visualization and role-play to teach object-oriented programming. In A. J. Mendes, I. Pereira, & R. Costa (Eds.), *Computers and Education* (pp. 167–177). Springer. https://doi.org/10.1007/978-1-84628-929-3_17

Di Natale, A. F., Repetto, C., Riva, G., & Villani, D. (2020). Immersive virtual reality in K-12 and higher education: A 10-year systematic review of empirical research. *British Journal of Educational Technology*, *51*(6), 2006–2033.

Eukel, H. N., Frenzel, J. E., & Cernusca, D. (2017). Educational gaming for pharmacy students–design and evaluation of a diabetes-themed escape room. *American Journal of Pharmaceutical Education*, *81*(7), 6265. https://doi.org/10.5688/ajpe8176265.

Gómez-Urquiza, J. L., Gómez-Salgado, J., Albendín-García, L., Correa-Rodríguez, M., González-Jiménez, E., & Cañadas-De la Fuente, G. A. (2019). The impact on nursing students' opinions and motivation of using a "Nursing Escape Room" as a teaching game: A descriptive study. *Nurse Education Today*, *72*, 73–76.

Grivokostopoulou, F., Perikos, I., & Hatzilygeroudis, I. (2017). An innovative educational environment based on virtual reality and gamification for learning search algorithms. In *Proceedings of the IEEE 8th International Conference on Technology for Education* (pp. 110–115). IEEE. https://doi.org/10.1109/T4E.2016.029

Guckian, J., Sridhar, A., & Meggitt, S. J. (2020). Exploring the perspectives of dermatology undergraduates with an escape room game. *Clinical and Experimental Dermatology*, *45*(2), 153–158.

Hermanns, M., Deal, B., Hillhouse, S., Opella, J. B., Faigle, C., & Campbell IV, R. H. (2017). *Using an "Escape Room" toolbox approach to enhance pharmacology education. Journal of Nursing Education and Practice*, *8*(4), 89–95. Sciedu Press. https://doi.org/10.5430/jnep.v8n4p89

Ho, A. M. (2018). Unlocking ideas: Using escape room puzzles in a cryptography classroom. *Primus, 28*(9), 835–847.

Horst, R., Naraghi-Taghi-Off, R., Diez, S., Uhmann, T., Müller, A., & Dörner, R. (2019). A serious puzzle mini-game for learning FunPlogs fundamental programming principles using visual scripting. *Lecture Notes in Computer Science, 11844*, 494–504. https://doi.org/10.1007/978-3-030-33720-9_38

Huang, X., Huss, J., North, L., Williams, K., & Boyd-Devine, A. (2023). Cognitive and motivational benefits of a theory-based immersive virtual reality design in science learning. *Computers and Education Open, 4*(6), 1–14. 100124. https://doi.org/10.1016/j.caeo.2023.100124

Jambhekar, K., Pahls, R. P., & Deloney, L. A. (2020). Benefits of an escape room as a novel educational activity for radiology residents. *Academic Radiology, 27*(2), 276–283.

Jiménez Sánchez, C., Lafuente Ureta, R., Ortiz Lucas, M., Bruton, L., & Millán Luna, V. (2017). Room Escape: A Gamification Proposal for the Physiotherapy Degree. In En In-Red 2017. *III National Congress on Educational Innovation and Online Teaching* (pp. 537–551). Editorial Universitat Politècnica de València. https://doi.org/10.4995/INRED2017.2017.6855

Jun, G. (2020). Virtual reality church as a new mission frontier in the metaverse: Exploring theological controversies and missional potential of virtual reality church. *Transformation, 37*(4), 297–305.

Jusas, V., Barisas, D., & Jančiukas, M. (2022). Game elements towards more sustainable learning in object-oriented programming course. *Sustainability, 14*(4), 2325.

Kinio, A. E., Dufresne, L., Brandys, T., & Jetty, P. (2019). Break out of the classroom: The use of escape rooms as an alternative teaching strategy in surgical education. *Journal of Surgical Education, 76*(1), 134–139.

Kumar, B., & Sharma, K. (2018). A gamified approach to achieve excellence in programming. In *Proceedings of the IEEE* (pp. 107–114). IEEE. https://doi.org/10.1109/ICCS.2018.00026

López-Pernas, S., Gordillo, A., Barra, E., & Quemada, J. (2019a). Examining the use of an educational escape room for teaching programming in a higher education setting. *IEEE Access, 7*, 31723–31737.

López-Pernas, S., Gordillo, A., Barra, E., & Quemada, J. (2019b). Analyzing learning effectiveness and students' perceptions of an educational escape room in a programming course in higher education. *IEEE Access, 7*, 184221–184234.

McFadden, C., & Porter, S. (2018). Augmented reality escape rooms as high-engagement educational resources. In *Proceedings of the ICERI* (pp. 4361–4365). IATED.

Monaghan, S. R., & Nicholson, S. (2017). Bringing escape room concepts to pathophysiology case studies. *HAPS Educator, 21*(2), 49–65.

Musil, B., Gartner, S., Pesek, I., & Krašna, M. (2019, May). ICT competences assessment through ICT escape room. In *2019 42nd International Convention on Information and Communication Technology, Electronics and Microelectronics (MIPRO)* (pp. 622–626). IEEE.

Nevelsteen, K. J. (2018). Virtual world, defined from a technological perspective and applied to video games, mixed reality, and the Metaverse. *Computer Animation and Virtual Worlds, 29*(1), e1752.

Nguyen, V. T., Zhang, Y., Jung, K., Xing, W., & Dang, T. (2020). Vrasp: A virtual reality environment for learning answer set programming. In *International Symposium on Practical Aspects of Declarative Languages* (pp. 82–91). Springer. DOI:10.1007/978-3-030-39197-3_6

Ortega, F. R., Bolivar, S., Bernal, J., Galvan, A., Tarre, K., Rishe, N., & Barreto, A. (2017). Towards a 3D virtual programming language to increase the number of women in computer science education. In *2017 IEEE Virtual Reality Workshop on K-12 Embodied Learning through Virtual and Augmented Reality* (pp. 1–6). IEEE.

Pan, R., Lo, H., & Neustaedter, C. (2017). Collaboration, awareness, and communication in real-life escape rooms. In *Proceedings of the 2017 Conference on Designing Interactive Systems* (pp. 1353–1364). ACM. https://doi.org/10.1145/3064663.3064767

Parmar, D., Babu, S. V., Lin, L., Jörg, S., D'Souza, N., Leonard, A. E., & Daily, S. B. (2017). Can embodied interaction and virtual peer customization in a virtual programming environment enhance computational thinking? In *2016 IEEE Research on Equity and Sustained Participation in Engineering, Computing, and Technology* (pp. 1–8). IEEE. DOI:10.1109/RESPECT.2016.7836179

Rushdan, E. E., Mohamed, M. A. E. S., Abdelhalim, G. E., El-Ashry, A. M., & Ali, H. F. M. (2025). Effect of an escape room as a gamification evaluation tool on clinical reasoning and teamwork skills among nursing students: A quasi-experimental study. *Nurse Education in Practice*, *82*, 104188.

Saparamadu, P. V. I. N., Sepasgozar, S., Guruge, R. N. D., Jayasena, H. S., Darejeh, A., Ebrahimzadeh, S. M., & Eranga, B. A. I. (2025). Optimising contract interpretations with large language models: A comparative evaluation of a vector database-powered Chatbot vs. ChatGPT. *Buildings*, *15*(7), 1144.

Sargazi Moghadam, T., Darejeh, A., Delaramifar, M., & Mashayekh, S. (2024). Toward an artificial intelligence-based decision framework for developing adaptive e-learning systems to impact learners' emotions. *Interactive Learning Environments*, *32*(7), 3665–3685.

Segura, R. J., del Pino, F. J., Ogáyar, C. J., & Rueda, A. J. (2020). VR-OCKS: A virtual reality game for learning the basic concepts of programming. *Computer Applications in Engineering Education, 28*(1), 31–41.

Servotte, J. C., Goosse, M., Campbell, S. H., Dardenne, N., Pilote, B., Simoneau, I. L., … & Ghuysen, A. (2020). Virtual reality experience: Immersion, sense of presence, and cybersickness. *Clinical Simulation in Nursing*, *38*, 35–43.

Sharma, S., & Ossuetta, E. (2017). Virtual reality instructional modules in education based on gaming metaphor. *IS and T International Symposium on Electronic Imaging Science and Technology*, 11–18. https://doi.org/10.2352/ISSN.2470-1173.2017.3.ERVR-090

Stigall, J., & Sharma, S. (2017). Virtual reality instructional modules for introductory programming courses. *ISEC 2017- Proceedings of the 7th IEEE Integrated STEM Education Conference,* 00(c), 34–42. https://doi.org/10.1109/ISECon.2017.7910245

Yang, C. L., Chang, C. Y., & Jen, H. J. (2023). Facilitating undergraduate students' problem-solving and critical thinking competence via online escape room learning. *Nurse Education in Practice*, *73*, 103828.

Zhao, D., Chis, A. E., Muntean, G. M., & Muntean, C. H. (2018). A large-scale pilot study on game-based learning and blended learning methodologies in undergraduate programming courses. In *Proceedings of EDULEARN* (pp. 3716–3724). IATED.

Zhonggen, Y. (2019). A meta-analysis of the use of serious games in education over a decade. *International Journal of Computer Games Technology*, 2019, 797032. https://doi.org/10.1155/2019/4797032

6 Metaverse and Virtual Reality Integration in Filmmaking Education

Applications in Cinematography and Artificial Intelligence Feedback

David Liusun, Ali Darejeh, and Samad M. E. Sepasgozar

1 Introduction

The field of filmmaking education is undergoing a paradigm shift, driven by the metaverse and incorporating technologies such as virtual reality (VR), generative artificial intelligence (GenAI), and game engines. Traditional film education—which depends on dedicated studio facilities, costly equipment, and hands-on training—faces significant challenges related to accessibility, financial constraints, and logistical complexity (Liu et al., 2023; Song et al., 2023; Xu et al., 2025).

The metaverse and VR offer a transformative alternative to traditional filmmaking education by providing an immersive, interactive, collaborative, and scalable platform where students can experiment with lighting, framing, camera angles, and mise-en-scène in virtual spaces free from physical constraints (Xu et al., 2025). This shift not only reduces financial barriers but also enhances creative freedom, enabling students to engage with a variety of film techniques without the limitations of physical production environments. By utilizing digital replicas of professional film sets, students can practice industry-standard techniques in a flexible and cost-effective manner. Moreover, VR enables institutions to broaden their curricula by incorporating specialized courses in virtual production, aligning with the film industry's increasing adoption of digital workflows. These developments are consistent with studies highlighting VR's ability to replicate real-world learning environments, including laboratories and cinematography settings, thereby allowing students to engage in iterative, risk-free practice (Darejeh et al., 2024a, 2024b; Darejeh, 2023; Wei et al., 2023; Xu et al., 2025).

Additionally, the integration of artificial intelligence (AI)-driven tools such as ChatGPT-4o can significantly enhance filmmaking education by providing real-time feedback. A combination of computer vision and machine learning models evaluates cinematographic elements such as framing, lighting balance, tonal consistency, composition, narrative coherence, and lighting precision, offering objective, data-driven feedback that bridges creative intuition and technical execution (Awidi, 2024; El-Keilany et al., 2022; Hassoulas et al., 2023; Saparamadu et al., 2025; Zhang et al., 2023).

DOI: 10.1201/9781003557555-6

The integration of VR and AI addresses key challenges in cinematography education, which has traditionally required extensive hands-on practice (Alton, 2013; Keating, 2009). By analyzing compositions based on established film principles, AI-powered systems can identify strengths and areas for improvement, assisting educators in providing tailored guidance without the constraints of time-intensive manual review. Furthermore, AI's capacity to process large datasets enables comparative analysis against industry benchmarks, fostering a deeper understanding of professional filmmaking techniques. Together, these technologies make filmmaking education more dynamic, responsive, and personalized, supporting the development of a critical eye for film aesthetics (Awidi, 2024; Brako & Mensah, 2023; Hassoulas et al., 2023; Xu et al., 2025; Zhang et al., 2023).

This chapter provides a comprehensive literature review on the metaverse and VR applications in film production education. We also present an empirical study detailing our innovative application for cinematic training. This study explores how VR and AI-driven platforms have been developed to address specific challenges in film education, such as accessibility, real-time feedback, and cost reduction. Finally, we discuss broader implications, challenges, and future directions for these technologies in filmmaking education, highlighting how continued advancements can shape the future of cinematic training by making it more immersive, accessible, and technologically integrated.

2 Literature Review

This literature review examines the historical foundations of cinematographic education, the emerging role of metaverse technologies and VR, the benefits and limitations of real-time rendering engines, and the impact of AI-driven feedback systems in filmmaking training. Furthermore, it explores the ethical considerations associated with adopting these technologies, ensuring a balanced perspective on their potential to reshape filmmaking education.

2.1 Traditional Cinematographic Education

Traditional cinematographic education has long relied on hands-on instruction in physical environments, such as soundstages. Early film schools focused heavily on practical camera work, requiring students to master manual exposure settings, film stock selection, and traditional editing techniques. However, historical studies have explored how shifts in cinematographic technology have influenced film education, particularly the transition from celluloid to digital production methodologies (Babbar, 2024). With the advent of digital cinematography, many technical aspects have become automated, prompting a new approach to teaching that emphasizes creative composition and digital post-production. This evolution has sparked debates about whether traditional cinematographic skills remain essential or whether education should shift toward digital tools and techniques that align with contemporary industry practices (Keating, 2009).

While pedagogical approaches have evolved, traditional cinematographic education continues to face substantial challenges, including high costs, limited accessibility, and logistical constraints (Liu et al., 2023; Song et al., 2023). Many institutions struggle to equip students with the necessary resources to master cinematography lighting techniques, and opportunities for hands-on engagement remain limited, particularly in time-constrained educational settings. As class sizes grow, instructors often find it difficult to provide individualized guidance, creating gaps in practical training. Furthermore, financial limitations and geographical barriers restrict many students' access to high-quality film equipment, exacerbating disparities in learning opportunities.

In addition to these challenges, traditional film education often lags in integrating emerging trends in digital cinematography, such as computer-generated imagery (CGI), real-time rendering, and virtual production. These technologies rapidly transform the industry, yet many film schools lack the infrastructure and expertise to teach them effectively. Consequently, students may graduate with limited exposure to the advanced tools and workflows currently shaping professional film production, leaving them at a disadvantage in an increasingly technology-driven industry (Rodrigues & Loureiro, 2024).

2.2 *The Role of Metaverse and VR in Film Education*

Immersive learning, facilitated by the metaverse and VR, allows students to develop practical filmmaking skills within a controlled virtual environment. Unlike traditional classroom-based training, VR enables hands-on engagement with cinematic concepts in an immersive, interactive, and flexible setting (Liu et al., 2023). Using metaverse-based environments, students can collaborate in settings that closely replicate real-world film production, where different crew members work together behind the scenes. Within these virtual spaces, students can repeatedly rehearse scenes, test camera locations and angles, and experiment with lighting setups—eliminating the need for physical set construction and associated costs (Harkema & Rosendaal, 2020).

The following sections explore the potential of VR as an alternative to conventional filmmaking education and examine the development of emerging systems.

2.2.1 *Immersive Learning and Creativity in Cinematography*

The ability of VR to emulate real-world environments makes it an effective tool for teaching complex and abstract subjects such as cinematography, lighting, and framing. Lighting and framing are fundamental components of visual storytelling, shaping not only the aesthetic quality of a scene but also its psychological and emotional impact on the audience (Huttunen, 2022). Researchers have long recognized the psychological effects of lighting and framing in film, emphasizing their role in enhancing audience engagement and emotional resonance (Xiang et al., 2023). Additionally, research on digital game worlds reinforces this perspective, demonstrating that different lighting conditions affect both players' affective states and in-game performance, underscoring the importance of lighting as a key element in immersive virtual experiences (Knez & Niedenthal, 2008).

Metaverse and VR not only support the teaching of lighting and framing but also play a vital role in fostering creativity in filmmaking education. Thornhill-Miller and Dupont (2016) identify five key mechanisms through which VR enhances creativity, making it an effective pedagogical approach: altering self-perception through avatars, optimizing interactions and collaboration, controlling environmental conditions, facilitating creative problem-solving, and integrating cognitive enhancement technologies. These mechanisms allow students to engage in innovative learning experiences beyond traditional filmmaking methods.

By providing interactive and immersive learning environments, VR empowers students to experiment with virtual film sets, pre-visualization techniques, and character design, simultaneously developing their creative and technical skills (Liu et al., 2023). Additionally, VR within the metaverse enhances collaboration by enabling real-time, multi-user interaction in a shared virtual space. This functionality allows students to work together on scene composition, lighting adjustments, and camera placements, mirroring real-world production workflows (Xu et al., 2025; Ziemsen & Fenuta, 2023).

2.2.2 Bridging the Gap between Theory and Practice

Traditional film production training methods—such as storyboards, camera plan views, and miniature models—often fail to fully capture real-world cinematographic scenarios. These techniques can be overly abstract or lack the level of control necessary for industry-standard training (Liu et al., 2023). VR's unique ability to integrate theoretical learning and practical training within a single virtual environment addresses these limitations, eliminating the traditional divide between classroom instruction and hands-on practice.

One of the most significant advantages of VR-based film education is its cost-effectiveness (Liu et al., 2023; Wei et al., 2023; Xu et al., 2025). By reducing reliance on expensive cameras, lighting setups, and production equipment, VR makes filmmaking education more financially viable. Although VR technology requires an initial investment, the savings in physical film production gear often offset its costs. Additionally, VR hardware is versatile, serving multiple educational purposes compared to the single-use nature of traditional film equipment.

Customization is another key advantage of VR-based learning environments. Virtual simulations can be tailored to include various lighting setups, camera rigs, and digital actors, creating an adaptable training experience. This flexibility allows students to explore film genres and experiment with innovative storytelling techniques. Research by Mateer (2017) suggests that VR-based training methods provide a sense of presence and realism that closely mimics traditional filmmaking, aligning with industry trends that increasingly rely on virtual production workflows and digital pre-visualization.

2.2.3 *Enhancing Repetitive Practice and Safety*

A fundamental benefit of VR-based training is facilitating repetitive practice without time constraints. Traditional film production often operates under tight schedules and limited equipment access, restricting students' opportunities to observe, practice, and refine their skills. In contrast, VR enables students to revisit scenarios, adjust compositions, and continuously improve their craft without logistical limitations (Harkema & Rosendaal, 2020).

Additionally, VR minimizes safety concerns associated with traditional film sets, where heavy equipment, electrical hazards, and complex rigging setups pose potential risks. In a virtual environment, learners can experiment with cinematographic techniques in a secure and controlled setting, reducing the likelihood of injury or damage to costly equipment (Grassini & Laumann, 2020).

2.2.4 *Existing VR Applications Developed for Film Education*

Several initiatives have developed VR applications for filmmaking training, aiming to explore how immersive environments can enhance the teaching of technical, creative, and collaborative skills essential to filmmaking education. Among these initiatives are Meta Shadow (Wei et al., 2023) and Art Mirror (Xu et al., 2025), both built on the Unity game engine and focused specifically on cinematography lighting education. Meta Shadow provides a virtual studio equipped with a range of lighting fixtures—such as Tungsten, LED, and Daylight—allowing students to manipulate parameters like intensity and color in real time. Similarly, Art Mirror offers interactive lighting equipment manipulation and visual cues for lighting direction within a virtual soundstage environment.

Beyond lighting-focused applications, other VR initiatives have expanded into broader aspects of filmmaking education. For example, Cine DESK (Hendry et al., 2023) is a virtual production tool developed using the Unreal Engine that supports pre-visualization, scene blocking, camera staging, and editing practice. It features a touchscreen interface that allows for rapid scene prototyping, making it particularly useful for early-stage production planning. In a different approach, Cine Voyage (Kal & Samur, 2024) adopts a game-based framework, emphasizing foundational cinematic literacy, film grammar, and storyboard creation. Through narrative-driven levels, students engage with historical filmmakers like Méliès and Kubrick, deepening their understanding of cinematic history. Additionally, Liu et al.'s (2023) collaborative VR platform focuses on developing filmmaking skills through multi-role training and real-time feedback. This research uses a three-minute animated film segment as a learning context to simulate real-world production dynamics, offering students experiential learning opportunities that mirror professional workflows.

The strengths of these VR applications are evident in their ability to enhance learning engagement and facilitate rapid prototyping. Studies on Meta Shadow and Art Mirror demonstrate that VR-based instruction significantly improves user engagement, presence, and creative flexibility compared to traditional lecture-based approaches (Wei et al., 2023; Xu et al., 2025). The Cine DESK offers the advantage of enabling projects to be completed in days rather than weeks, making it highly effective for visual storytelling training (Hendry et al., 2023). Cine Voyage fosters student motivation and emotional engagement by allowing learners to interact virtually with historical cinematic pioneers (Kal & Samur, 2024). Furthermore, Liu et al.'s (2023) collaborative VR system supports teamwork and communication skills, which are essential in professional filmmaking environments.

However, despite these strengths, VR in film education presents several limitations. Technological barriers, such as the high-end VR hardware requirements of Art Mirror and Cine Voyage, restrict accessibility and scalability. Additionally, VR struggles to replicate the tactile sensations of handling physical lighting equipment, limiting its effectiveness for certain technical training aspects (Hendry et al., 2023). Health concerns, such as nausea and visual fatigue, also impact usability, with platforms like Cine Voyage capping play sessions at 30 minutes to mitigate these issues (Kal & Samur, 2024).

2.3 Benefits of Real-Time Rendering Engines for Film Production and Education

In addition to VR technology, the emergence of real-time rendering technologies has significantly transformed the landscape of virtual film production and filmmaking education. Game engines such as Unreal Engine and Unity play a pivotal role in creating dynamic, interactive virtual environments, allowing filmmakers to refine cinematographic elements, such as set design and lighting, at various stages of production. Compared to traditional film production frameworks, these tools offer a more flexible workflow, enabling real-time modifications and iterative creative decisions. Willment, Swords, and Thomas (2023) conducted an extensive study analyzing over 600 secondary sources and industry interviews, concluding that virtual production enhances efficiency by streamlining previsualization, motion tracking, and digital asset integration.

When applied to education, this technology allows students to visualize virtual mise-en-scène components in real-time, providing an immersive and practical learning experience. Additionally, virtual production facilitates camera movement experimentation, enabling students to practice techniques such as dolly shots, handheld simulations, and aerial perspectives without requiring heavy physical camera rigs. This ability to experiment in a controlled, digital environment bridges the gap between theoretical knowledge and hands-on practice (Dooley & Emery, 2023).

The continual advancement of game engine technology has improved the realism of virtual scenes and expanded opportunities for students to explore practical cinematography within a digital space. A particularly significant development is the integration of metahumans and AI-driven avatars, which allow students to practice scene composition, character blocking, and performance direction without needing a physical cast. These tools provide an invaluable training ground for developing essential filmmaking skills while reducing the logistical constraints of live-action production (An, 2022).

Real-time photorealistic lighting simulation further enhances virtual production by minimizing reliance on expensive lighting setups and time-consuming configurations. With realistic dynamic lighting simulations, students can experiment with brightness, shadows, and color tones without needing physical light sources. A breakthrough in this field came with NVIDIA's 2018 introduction of hardware-supported ray tracing via the DirectX Raytracing (DXR) API, which expanded real-time rendering capabilities previously limited to offline film production (Skorobogatova, 2022). This advancement has made high-fidelity lighting and reflection effects more accessible in virtual environments, significantly improving the visual quality of real-time cinematic training.

One of the most notable innovations in real-time rendering is the Unreal Engine 5 Lumen system, which combines software and hardware ray tracing to handle dynamic scenes with realistic global illumination, reflections, and ambient occlusion. However, while Lumen offers significant advancements, it is not without limitations. Software-based ray tracing struggles with complex geometries, while hardware-accelerated modes demand high computational power, limiting accessibility for lower-end systems (Skorobogatova, 2022). Additionally, the automation of lighting parameters, while reducing the workload for artists and students, can sometimes restrict creative control. As Skorobogatova (2022) notes, due to the limited range of available options, users may struggle to achieve precise control when configuring lighting scenarios essential for game production. These constraints highlight the ongoing challenge of balancing automation with creative flexibility in virtual production education.

Beyond the technological limitations, infrastructure challenges also hinder the effective integration of real-time rendering technologies into educational settings. Pires et al. (2022) highlight that many educational institutions lack the technical infrastructure and faculty expertise necessary to fully adopt these emerging tools. Nevertheless, despite these challenges, the continual evolution of real-time rendering and virtual production technologies presents promising new opportunities for advancing both game development and filmmaking education.

2.4 Computer Vision AI-Driven Feedback in Filmmaking Education

AI has the potential to revolutionize filmmaking education by automating the evaluation of aesthetic cinematographic elements and providing students with real-time, data-driven feedback to enhance their performance. These AI-driven systems enable dynamic, personalized challenges tailored to students' skill levels, testing their understanding of key cinematic principles such as camera movements, lighting techniques, and character blocking. By evaluating visual elements like the rule of thirds, shot composition, and scene coherence, AI enhances students' ability to refine visual storytelling while ensuring adherence to professional cinematography standards (Brako & Mensah, 2023).

Advanced GenAI tools such as ChatGPT-4 and computer vision-based analytics have demonstrated significant potential in cinematographic analysis, particularly in their ability to assess multiple visual elements simultaneously. These technologies can analyze frame composition, lighting quality, and color balance, comparing student work to industry-standard datasets to identify issues and offer corrective suggestions (Xu et al., 2025; Zhang et al., 2023). ChatGPT-4, in particular, provides detailed explanations of visual content, making it a valuable tool for evaluating fundamental aspects such as shot composition, lighting, and blocking in filmmaking education (Brako & Mensah, 2023; Zhang et al., 2023). However, despite its strengths in analyzing general cinematographic structure, GPT-4 struggles with fine-grained technical details and tends to overestimate visual quality compared to human experts (Zhang et al., 2023). This limitation highlights the need for a hybrid approach, where AI offers foundational feedback while human mentors address more advanced technical nuances. When appropriately prompted, GPT-4's reference-free evaluation capabilities prove useful for assessing basic compositional elements and narrative coherence. Nonetheless, its role remains supplementary rather than primary, requiring well-defined guidelines to optimize its effectiveness (Awidi, 2024; Brako & Mensah, 2023; Hassoulas et al., 2023; Zhang et al., 2023).

The methodological rigor of large-scale computational film analysis further underscores the potential and challenges of AI-driven assessment. A study by El-Keilany et al. (2022) utilized computer vision to decode visual patterns by tracking color palettes, brightness gradients, and emotional tone across narrative arcs. By quantifying stylistic shifts—such as progressive darkening correlating with increasing plot seriousness—their research demonstrates how algorithmic analysis can reveal the intricate relationship between visual language and storytelling, a core principle of cinematography training. More than just a technical tool, their methodology reframes cinematography as a structured visual syntax, where lighting gradients and chromatic rhythms function as grammatical constructs within a film's language. This shift from qualitative critique to data-driven hermeneutics aligns with AI's role in democratizing cinematography education, making abstract concepts such as visual tension and tonal coherence more tangible for learners.

Despite its potential, standard computer vision models—trained largely on real-world imagery—struggle to analyze stylized or fantastical cinematography. Ethereal lighting and exaggerated color grading often challenge these models, highlighting the need for domain-adapted AI trained on artistic, rather than documentary, visual styles (El-Keilany et al., 2022). This issue extends to pedagogical contexts: while emotion recognition algorithms perform well on overt facial expressions, they often falter when analyzing obscured or stylized visages (Darejeh et al., 2024c; Sargazi Moghadam et al., 2024). Similarly, GPT-4 demonstrates limitations in evaluating nuanced technical clarity, reflecting broader difficulties AI systems face in interpreting creative intent (Awidi, 2024; El-Keilany et al., 2022; Hassoulas et al., 2023; Zhang et al., 2023). Although color analysis could, in theory, support assessments of students' narrative use of hue and saturation, current models still require specialized training to accurately interpret artistic decisions.

For educators, this duality presents both a challenge and an opportunity. AI can democratize access to foundational critique by evaluating compositional balance, tonal consistency, and visual coherence, while human mentors focus on advanced technical refinement and interpretive guidance. Future AI assessment systems may bridge this gap by integrating domain-adapted models that combine multiple analytical approaches. Drawing from El-Keilany et al.'s (2022) multi-method framework—incorporating brightness metrics, emotion mapping, and color progression—AI-driven tools could evolve into holistic assessment platforms that respect both technical precision and artistic vision. By refining these technologies, educators can leverage AI not only as a tool for technical evaluation but also as a means to enhance students' creative decision-making in cinematography education.

2.5 Ethical Considerations of Metaverse Technologies in Filmmaking Education

The integration of metaverse technologies, VR, and AI into filmmaking education offers transformative potential but also raises important ethical considerations that must be addressed within academic curricula. These considerations center around equity of access, the preservation of creative diversity, and the responsible use of AI-generated content.

A primary ethical challenge lies in balancing the democratization of access with persistent systemic inequities. While virtual production tools and immersive platforms reduce geographical and financial barriers, they do not inherently eliminate disparities in training opportunities, mentorship, and industry recognition. Students from underrepresented or economically disadvantaged backgrounds may still lack the necessary resources or networks to fully benefit from metaverse-based educational tools (Fair, 2023). Addressing this issue requires the design of equity-driven curricula that ensure all students—regardless of background—develop competencies in both analog and digital production techniques. Initiatives such as Ulster University's virtual production studio exemplify this approach by combining practical XR training with critical discussions on workforce inclusion and strategies for bridging digital divides (An, 2022).

In parallel, the growing reliance on prefabricated digital assets and AI-generated recommendations poses a significant risk to creative diversity. Standardized virtual sets, AI-based cinematography suggestions, and procedurally generated scenes can unintentionally prioritize algorithmic conformity over artistic innovation (An, 2022; Yamazaki, 2024). To counter this trend, filmmaking education must emphasize the role of human intuition and creativity, positioning AI tools as supports rather than substitutes for original storytelling. Assignments that encourage students to critically engage with, subvert, or creatively modify AI-generated outputs can reinforce the irreplaceable value of human artistic judgment within virtual production environments.

Finally, as regulatory frameworks such as the EU's AI Act and emerging U.S. deepfake legislation continue to evolve (Yamazaki, 2024), filmmaking education must proactively prepare students to navigate the complex ethical and legal landscapes surrounding the use of metaverse and AI technologies.

In the next section, we report on the development and outcomes of our project, which integrates VR and GenAI to enhance filmmaking education. Developed as part of the University of New South Wales (UNSW) Metaverse initiative introduced in Chapter 1, this project was designed to provide students with immersive, hands-on training while leveraging AI-driven feedback to refine their technical skills in cinematography.

3 Empirical Study: GenAI-Driven VR-Based Film Production Training System

This study explores the development and evaluation of our VR-based film production training system designed to enhance cinematographic education. By integrating VR and GenAI, the system provides students with an immersive, hands-on learning environment, enabling them to practice lighting setup, camera placement, and composition analysis in a virtual setting. The purpose of this experiment is to evaluate student learning performance while practicing through the immersive environment, as well as the effectiveness of the feedback system in enhancing their understanding and helping them identify and correct mistakes when setting up a scene.

3.1 Application Development for the Study

The VR training system was developed using Unreal Engine 5 and deployed on HTC VIVE Pro 2 headsets, supported by high-performance computing systems. It replicates professional production studios and outdoor filming locations, offering a realistic, risk-free, and cost-effective alternative to physical film sets. The virtual environment is designed to address traditional limitations in filmmaking education by simulating professional production spaces where students can experiment with lighting setups, camera angles, and character blocking without logistical constraints.

The environment includes both indoor and outdoor scenarios for cinematographic practice. The lighting system replicates industry-standard equipment, such as softboxes and point lights, with full manipulation capabilities, including adjustments to position, height, and angle. This supports the broader goal of providing hands-on technical training while mitigating the safety risks associated with traditional lighting setups. Two movable MetaHuman mannequins are included for character blocking and composition practice, alongside a fully adjustable virtual camera system that allows modifications to height, rotation, and focal length. These features directly address pedagogical challenges identified in contemporary film education research. Figure 6.1 illustrates an outdoor setting with natural sunlight; on the left is a movable MetaHuman mannequin, while on the right is a set of industry-standard equipment replicating real-world counterparts.

Figure 6.1 An outdoor setting with natural sunlight.

Users can interact with scene objects through a menu interface that dynamically displays the settings of the nearest equipment, allowing for precise adjustments of object properties. Figure 6.2 shows the menu that allows interaction with equipment settings.

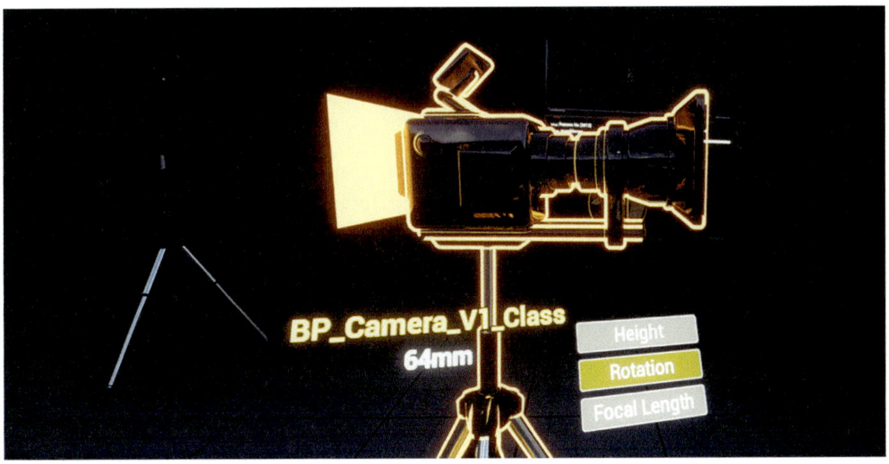

Figure 6.2 The menu that allows interaction with equipment settings.

The system also features an intelligent assessment engine powered by OpenAI's APIs, utilizing a combination of GPT-4o and GPT-4o Mini models for vision-based assessment, enabling immediate feedback loops. First, it generates specific cinematographic requirements based on established film techniques and genres.

Second, it develops comprehensive assessment criteria for fundamental cinematographic elements—such as lighting, framing, and composition—that align with industry standards and theoretical aesthetic principles. Third, after users apply the necessary settings based on the scenario, it provides detailed feedback and a numerical evaluation using computer vision analysis of user-created setups, offering immediate and constructive assessment of student work. Figure 6.3 illustrates how ChatGPT generates requirements, feedback, and progress tracking.

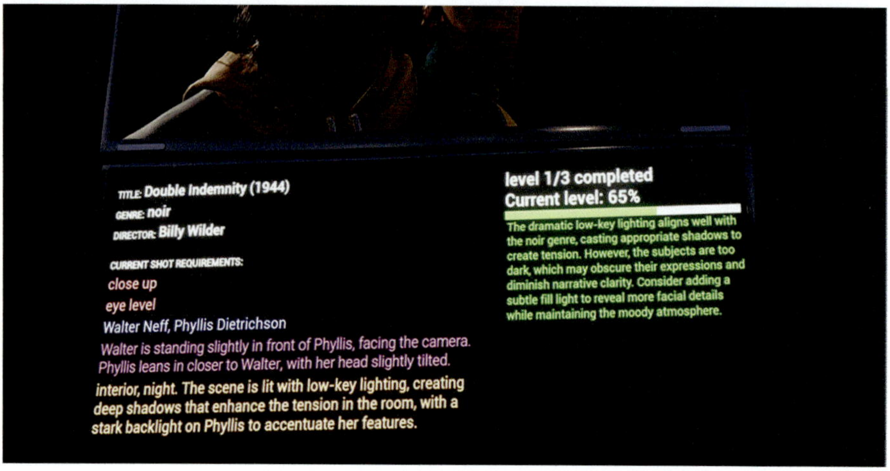

Figure 6.3 ChatGPT generates requirements, feedback, and progress tracking.

Here are key features of the VR training system:

- Customizable Lighting Setups: Users can manipulate light intensity, direction, and color to achieve different cinematic effects, simulating industry-standard lighting equipment.
- Interactive Virtual Cameras: Adjustable parameters such as height, rotation, and focal length enable students to explore various framing techniques.
- Realistic Scene Composition: Metahuman-based character models provide lifelike subject placement for shot planning.
- AI-Driven Cinematography Feedback: AI-driven assessment tools to provide real-time feedback on cinematographic techniques, analyzing lighting balance, camera angles, shot composition, and thematic coherence. The system offers qualitative assessments through a structured rubric and generates a score highlighting strengths and areas for improvement while providing alternative scene suggestions to refine the filmmaker's creative vision. Additionally, it evaluates the emotional impact of lighting and framing choices, helping students develop a deeper understanding of cinematographic storytelling. Figure 6.4 shows the overall application interface including the AI feedback.

Figure 6.4 The overall application interface, including the AI feedback.

3.2 Application Functionality and Workflow

The learning experience begins with the educator/administrator setting requirement variables in the system, whereupon ChatGPT generates contextually appropriate cinematography challenges. These challenges encompass comprehensive shot requirements, including composition parameters, camera positioning, character blocking, and lighting specifications, all derived from established cinematic conventions. Imagine a student wants to learn how to create a horror movie scene set in a dark environment. The instructor first configures the system settings and defines different requirements that will be assigned to students as tasks, as shown in Figure 6.5.

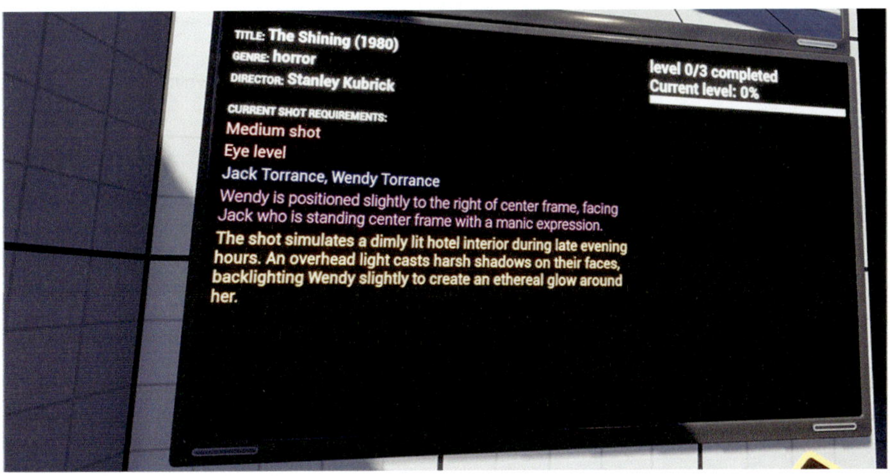

Figure 6.5 Configurations that are entered by the instructor.

Based on the task defined by the instructor, students must adjust various film production elements, including camera positioning, focal length (Figure 6.6), stage lights (Figure 6.7), and the manipulation of metahuman character models (Figure 6.8).

Figure 6.6 Camera controlling within the scene.

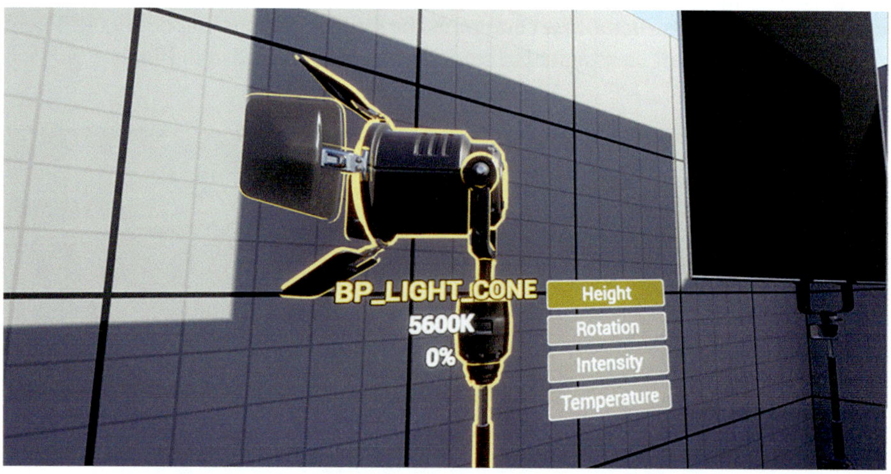

Figure 6.7 Lighting settings within the scene.

Figure 6.8 User-controlled human mannequin.

After making adjustments, students can view a live preview of their changes through the camera lens (Figure 6.9).

Figure 6.9 Live preview of changes through the camera lens.

Finally, the system generates a score and written feedback based on the accuracy of the student's adjustments (see Figures 6.10a and 6.10b). The assessment mechanism is powered by a multi-stage, AI-driven feedback loop designed to evaluate both technical proficiency and narrative coherence. After students complete their scene setup—adjusting camera angles, lighting configurations, and character placement—the system captures the final frame and analyzes it using computer vision models trained on cinematic principles.

The feedback engine extracts key visual features from the user-created frame, including spatial composition, lighting distribution, symmetry, subject positioning, and depth of field. These extracted features are then compared against a dynamic rubric generated from scene-specific requirements, which include genre conventions (e.g., horror, drama), lighting mood (e.g., high contrast, soft ambient), and framing rules (e.g., rule of thirds, leading lines).

The system outputs two types of feedback:

- A quantitative score (out of 100), which reflects how closely the student's configuration matches the predefined aesthetic and technical parameters.
- A qualitative critique that provides a narrative explanation of the student's performance, highlighting strengths and offering suggestions for improvement. This critique focuses on areas such as compositional balance, visual coherence, and the emotional impact of lighting and framing choices.

If a student's score exceeds 80, the system triggers a narrative contextualization module that unlocks additional backstory elements relevant to the scene. This serves both as a reward mechanism and as a scaffold for the next cinematographic challenge, reinforcing skill progression. This tiered structure ensures that students master fundamental cinematography techniques before advancing to more complex storytelling tasks.

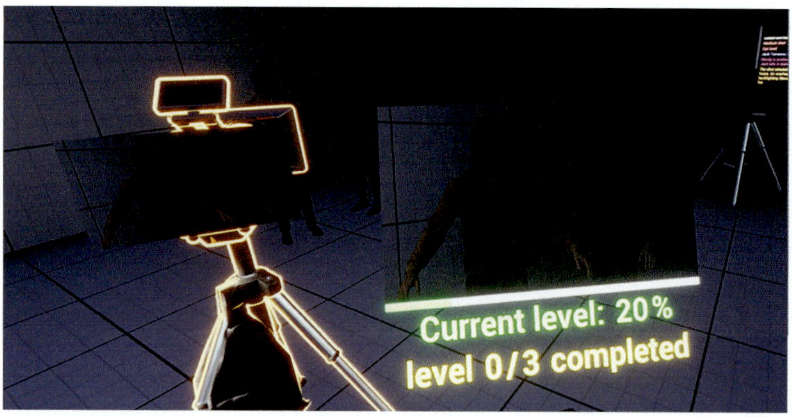

Figure 6.10a Scoring system based on adjustments.

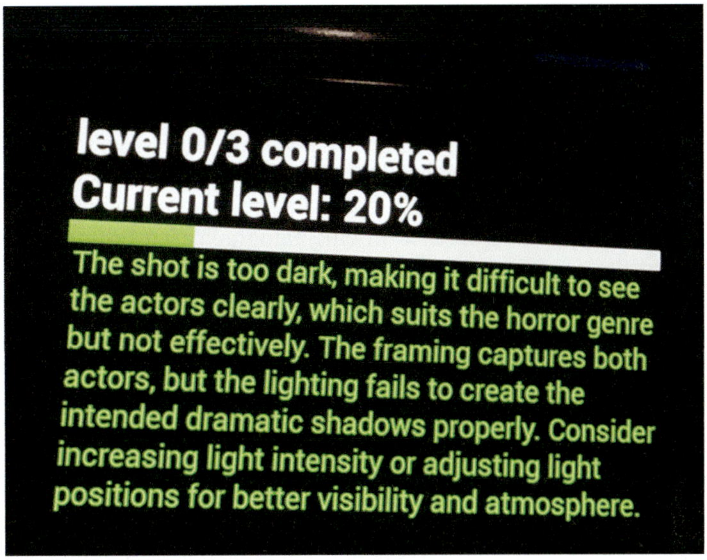

Figure 6.10b System feedback based on adjustments.

As the feedback loop is iterative, when students revise their scene based on the provided feedback, the system re-evaluates the setup and updates the performance report accordingly (see Figure 6.11). This encourages reflective practice

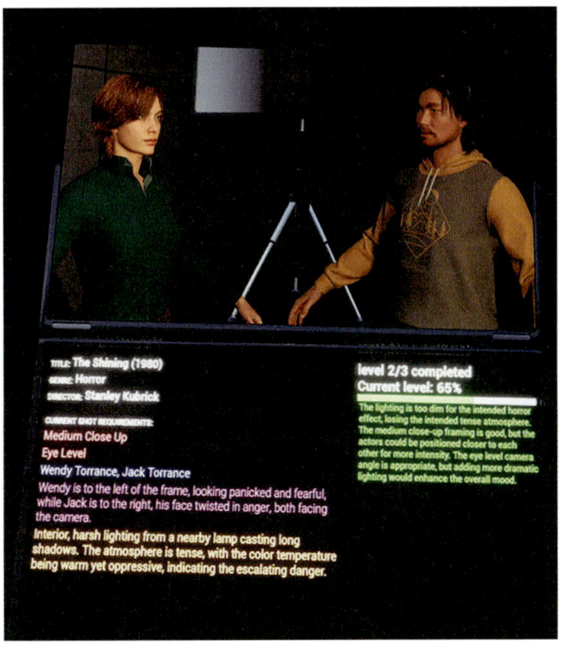

Figure 6.11 The updated score and feedback based on students' performance.

and iterative refinement, closely mirroring real-world filmmaking processes where multiple revisions lead to polished outputs.

3.3 Methodology for Conducting the Study

The study involved 10 participants, all of whom were UNSW students, to evaluate the effectiveness of the developed VR-based film production training platform. All participants had no prior experience in cinematography, ensuring that the evaluation reflected the perspectives of novices rather than experienced practitioners. Participants were selected based on a university-approved human research ethics protocol, as granted by the institutional review board. Eligible participants were volunteers aged 18 years or older. All participants provided informed consent, voluntarily agreeing to participate in the study and permitting the use of their anonymized data for research purposes. To keep confidentiality, the identities and names of participants were anonymized throughout the chapter, with no personally identifiable information disclosed in the reporting of results or discussions.

The evaluation process was divided into three phases: pre-assessment, interaction with the system, and post-assessment. Upon arrival, participants completed a pre-test using a Likert-scale questionnaire (ranging from 0 = no knowledge to 9 = professional knowledge) designed to assess their understanding of key cinematographic concepts, such as camera positioning, lighting techniques, composition, and focal length. This questionnaire focused on foundational knowledge in key areas such as lighting techniques, framing and composition, camera angles, and narrative visual coherence.

Following the pre-test, participants interacted with the VR system through a guided session introducing them to the platform's functionalities, including lighting setup, camera positioning, shot composition, and the AI-driven assessment engine. Throughout the session, participants received system-generated feedback, made adjustments based on the feedback, and were encouraged to iteratively refine their cinematic scenes.

After completing the hands-on VR activities, participants completed the same Likert-scale questionnaire as a post-test to evaluate changes in their cinematography knowledge and confidence. The pre- and post-test results were then statistically compared to determine whether using the VR platform significantly enhanced users' understanding of filmmaking techniques.

Additionally, participants engaged in a semi-structured interview immediately after the training session, which focused on several key dimensions. These included the usability and interface design of the system to evaluate its intuitiveness and ease of navigation; perceived learning effectiveness to assess how the platform contributed to participants' understanding of cinematographic techniques; AI feedback and assessment to explore users' views on the usefulness, accuracy, and clarity of the AI-generated feedback; and immersion and realism to evaluate how closely the virtual environment replicated real-world film production scenarios. The qualitative data collected through these interviews were analyzed using thematic analysis to identify common themes, user perceptions, challenges, and potential areas for system improvement.

3.4 Results from the Study

A paired-samples t-test was conducted to compare participants' self-reported cinematography knowledge before and after using the system, based on their pre-test and post-test questionnaire responses. The pre-test score was ($M = 2.3$, $SD = 1.25$), while the post-test score increased to ($M = 4.4$, $SD = 1.10$). The t-test revealed a statistically significant improvement: $t(9) = 7.22$, $p < 0.0001$, indicating a substantial learning gain.

The interview results showed that participants generally had a positive experience with the VR-based cinematography training system, describing it as realistic and engaging. They noted that the immersive virtual environment allowed them to practice cinematographic techniques in an intuitive and accessible way, increasing their confidence; however, some expressed uncertainty about whether they could effectively transfer these techniques to real-world settings involving actual cameras and lighting systems.

Several usability challenges were also identified, particularly regarding the complexity of the user interface. Participants suggested that a more intuitive control system and the inclusion of step-by-step tutorials would significantly enhance the learning experience. Also, the lack of haptic feedback was seen as a major limitation, with recommendations for future versions to incorporate tactile interfaces to better replicate the physical aspects of film production.

The interviews also revealed mixed perceptions of the AI-driven feedback system. While many participants found the AI-generated assessments useful for tracking their progress and refining technical skills, some expressed concerns about the accuracy of the scoring mechanism, particularly when creative or unconventional cinematographic choices were made.

3.5 Insights from the Study

The findings from this study demonstrate that VR-based cinematography training can significantly enhance students' understanding of fundamental filmmaking concepts, providing an engaging and risk-free environment for technical skill development. The notable improvement in participants' self-reported cinematography knowledge highlights the platform's potential as an effective educational tool. Although their understanding increased from a score of two to four out of nine on the self-reported learning scale, this result is promising considering the limited exposure time—less than 30 minutes of system use. Participants particularly valued the freedom to experiment with lighting, framing, and camera positioning without the constraints associated with traditional film production environments.

However, the results also suggest that VR alone may not fully replace real-world experience, as several participants expressed uncertainty about transferring virtual skills to practical filmmaking settings. Usability challenges, especially related to the complexity of the interface, were frequently reported, indicating a need for more intuitive controls and guided onboarding features. Feedback on the AI-driven assessment system was mixed; while participants found it helpful for tracking progress, concerns about score transparency and the interpretation of creative work highlighted the need for a hybrid evaluation approach combining AI and instructor feedback. The lack of haptic feedback also emerged as a key limitation, reducing the authenticity of the training experience. Overall, the system shows strong promise but would benefit from targeted refinements to maximize its educational impact.

4 Discussion

The adoption of metaverse and VR technologies in filmmaking education has led to significant advancements in accessibility, interactivity, and experiential learning. VR provides students with immersive environments that allow them to engage with filmmaking concepts in a way that traditional methods cannot replicate (Liu et al., 2023; Xu et al., 2025). VR fosters an iterative learning process that enhances cinematographic understanding by enabling hands-on experimentation with lighting, camera angles, and mise-en-scène (Wei et al., 2023). The ability to manipulate digital film sets in real-time democratizes access to high-quality production tools, particularly for students in institutions with limited resources (Song et al., 2023).

Moreover, AI-driven feedback systems offer real-time evaluation of cinematographic elements, bridging the gap between creative intuition and technical execution (El-Keilany et al., 2022; Zhang et al., 2023). AI-powered tools like ChatGPT-4o provide instant feedback on composition, lighting balance, and narrative coherence, allowing students to refine their work efficiently (Xu et al., 2025). These technologies also offer adaptive learning opportunities, allowing students to engage with filmmaking concepts at their own pace and skill level (Alton, 2013).

Our empirical study results further validate these benefits, demonstrating that students using VR-based cinematography tools appreciated the risk-free training environment, allowing them to experiment with camera movements, lighting setups, and scene composition without fear of damaging expensive equipment. Participants particularly valued the AI-driven feedback system, which helped them refine their cinematographic techniques. The structured critiques and real-time visual feedback provided clearer insights into the impact of lighting choices on mood and storytelling.

As observed from the interview results, while immersive learning environments offer an alternative to costly, resource-intensive film production training, their effectiveness ultimately depends on their ability to translate into real-world filmmaking scenarios. Although students can experiment with camera angles and lighting setups in VR, they may encounter difficulties replicating these skills in practical film settings. This suggests that, although VR effectively facilitates conceptual learning, it does not fully bridge the gap between theoretical understanding and hands-on cinematographic execution (Rodrigues & Loureiro, 2024).

A key factor contributing to this gap is the absence of physical interaction with equipment. While the system accurately simulated lighting and camera setups visually, the inability to physically manipulate lighting rigs and camera controls diminished the authenticity of the experience. Participants emphasized that the lack of tactile interaction made it difficult to develop the muscle memory essential for mastering real-world cinematographic techniques (Xu et al., 2025).

Building on these insights, participant feedback indicated that increasing the likelihood of successfully transferring VR-acquired skills to practical settings requires designing a more intuitive interface that closely mirrors real-world object manipulation. Moreover, integrating haptic feedback—another major concern raised—could significantly enhance the realism and effectiveness of the learning experience.

In addition to the challenges related to physical interaction, participants also raised concerns about the system's AI-driven feedback mechanisms. While these tools promise to enhance cinematography assessment by providing objective, data-driven evaluations of composition, lighting, and shot coherence, questions remain regarding their reliability and interpretability. The study revealed inconsistencies in AI-generated assessments, particularly in evaluating subjective elements such as mood, thematic coherence, and artistic intent, highlighting the need for a more nuanced and transparent feedback approach (Hashemi, Darejeh, & Cruz, 2024; Zhang et al., 2023).

One of the major criticisms of AI in film education is its reliance on pre-existing datasets, which may reinforce conventional cinematic norms while limiting the exploration of experimental or unconventional filmmaking techniques (Brako & Mensah, 2023). This raises concerns about whether AI fosters genuine creativity or merely standardizes film composition based on historical trends. Additionally, AI's opaque scoring mechanisms make it difficult for students to understand or contest their evaluations, potentially leading to over-reliance on automated feedback rather than critical engagement with their work (Awidi, 2024; Hassoulas et al., 2023).

Another challenge is AI's inherent bias, which reflects the data it has been trained on. For example, if AI models prioritize Hollywood-style cinematography, alternative storytelling techniques from diverse cultural perspectives may be undervalued or misinterpreted. This limitation suggests the need for a hybrid assessment approach where AI is used to complement, rather than replace, human feedback in cinematography education (Brako & Mensah, 2023; Zhang et al., 2023).

5 Conclusion

This chapter explored the integration of metaverse and VR technologies in filmmaking education, highlighting their potential to enhance accessibility, interactivity, and experiential learning. The study demonstrated that VR-based environments can significantly improve students' understanding of cinematographic concepts, providing a safe space for experimentation with lighting, framing, and camera movements. Participants responded positively to the immersive training platform and AI-driven feedback system, reporting increased confidence in their technical skills. However, the results also revealed important challenges, including the complexity of the user interface, the lack of haptic feedback, and concerns regarding the transparency and reliability of AI-generated assessments.

A major limitation of this study was the small sample size, with only 10 participants involved, all of whom were novice filmmaking students. While the findings provide valuable initial insights, the limited number of participants restricts the generalizability of the results. Future studies should involve a larger and more diverse cohort, including students with varying levels of filmmaking experience, to better understand the broader applicability of VR-based training tools. Additionally, longer-term studies are needed to assess whether skills acquired in virtual environments effectively transfer to real-world filmmaking practice.

Future work should focus on refining the system's user interface to reduce cognitive load, integrating haptic technologies to enhance realism, improving AI feedback mechanisms to better accommodate creative variability, and adding and testing the effects of avatar-based feedback compared to text-based feedback to determine which approach yields better learning performance (Darejeh, Marcus, & Sweller, 2022). Additionally, developing hybrid assessment models that combine AI-driven analysis with instructor oversight could provide more nuanced and reliable evaluations. Another promising direction is to evaluate the collaborative features of the application, enabling groups of students to assume different behind-the-scenes roles and collaboratively design, stage, and refine cinematic missions. This would more closely mirror real-world production workflows and further enhance the system's educational value. Overall, while the metaverse offers transformative possibilities for cinematography education, ongoing system refinement and larger-scale evaluations are essential to fully realize its potential in preparing students for the demands of the professional film industry.

References

Alton, J. (2013). *Painting with light*. University of California Press.

An, D. (2022). Technology-driven virtual production: The advantages and new applications of game engines in the film industry. *FAMECOS Magazine*, *29*(1), e43370. https://doi.org/10.15448/1980-3729.2022.1.43370

Awidi, I. T. (2024). Comparing expert tutor evaluation of reflective essays with marking by generative artificial intelligence (AI) tool. *Computers and Education: Artificial Intelligence*, *6*, 100226.

Babbar, I. (2024). Evolution of cinema. *International Journal for Multidisciplinary Research (IJFMR)*, *6*(2),825–835. https://doi.org/10.36948/ijfmr.2024.v06i02.17578

Brako, D. K., & Mensah, A. K. (2023). Robots over humans? The place of artificial intelligence in the pedagogy of art direction in film education. *Journal of Emerging Technologies*, *3*(2), 51–59.

Darejeh, A. (2023). Empowering education through eerp: A customizable educational vr escape room platform. In *2023 IEEE International Symposium on Mixed and Augmented Reality Adjunct (ISMAR-Adjunct)* (pp. 764–766). IEEE.

Darejeh, A., Huynh, L., Ji, X., Marcus, N., & Polly, P. (2024a). Assessing the efficacy of virtual reality-based laboratories versus real-life settings for PCR procedures, within medical science laboratories. In *2024 IEEE International Symposium on Mixed and Augmented Reality Adjunct (ISMAR-Adjunct)* (pp. 489–492). IEEE.

Darejeh, A., Marcus, N., & Sweller, J. (2022). Increasing learner interactions with E-learning systems can either decrease or increase cognitive load depending on the nature of the interaction. *L'Année psychologique*, *122*(3), 405–437.

Darejeh, A., Moghadam, T. S., Delaramifar, M., & Mashayekh, S. (2024c). A framework for AI-powered decision making in developing adaptive e-learning systems to impact learners' emotional responses. In *2024 11th International and the 17th National Conference on E-Learning and E-Teaching (ICeLeT)* (pp. 1–6). IEEE.

Darejeh, A., Mondal, M., Marcus, N., & Vassar, A. (2024b). Exploring the cognitive load effects of diverse virtual reality interaction methods in interactive educational platforms. In *2024 IEEE International Symposium on Mixed and Augmented Reality Adjunct (ISMAR-Adjunct)* (pp. 435–438). IEEE.

Dooley, K., & Emery, S. (2023). Creating screen stories with game engines: Challenges and opportunities for students and researchers working collaboratively across disciplines. *Media Practice and Education*, *24*(1), 21–34.

El-Keilany, A., Schmidt, T., & Wolff, C. (2022). Distant viewing of the *Harry Potter* movies via computer vision. *Digital Humanities in the Nordic and Baltic Countries, 4*(1), 33–49. https://doi.org/10.5617/dhnbpub.11273

Fair, J. (2023). Virtual production and the potential impact on regional filmmaking: Where do we go from here? *DBS Business Review*, *5*, Article 89. https://doi.org/10.22375/dbr.v5i.89

Grassini, S., & Laumann, K. (2020, November). Evaluating the use of virtual reality in work safety: A literature review. In *Proceedings of the 30th European Safety and Reliability Conference and the 15th Probabilistic Safety Assessment and Management Conference* (Vol. 1, pp. 779–786). Springer.

Harkema, G. J., & Rosendaal, A. (2020). From cinematograph to 3D model: How can virtual reality support film education hands-on?. *Early Popular Visual Culture*, *18*(1), 70–81.

Hashemi, M., Darejeh, A., & Cruz, F. (2024, October). A User-Centric Exploration of Axiomatic Explainable AI in Participatory Budgeting. In *Companion of the 2024 on ACM International Joint Conference on Pervasive and Ubiquitous Computing* (pp. 126–130). Association for Computing Machinery (ACM), New York, NY, USA.

Hassoulas, A., Powell, N., Roberts, L., Umla-Runge, K., Gray, L., & Coffey, M. (2023). Investigating marker accuracy in differentiating between university scripts written by students and those produced using ChatGPT. *Journal of Applied Learning & Teaching*, *6*(2), 71–77. https://doi.org/10.37074/jalt.2023.6.2.13

Hendry, M., Kottmann, N., Huber, V., Schaerer, M., & Iseli, C. (2023). Simulate this! Virtual production and cinematic education with cineDESK. In *Proceedings of Avanca Cinema 2023* (pp. 552–558). Avanca.

Huttunen, S. (2022). Ecological approach to cinematographic lighting of the human face – a pilot study. *Baltic Screen Media Review*, *10*(2), 274–291. https://doi.org/10.2478/bsmr-2022-0020

Kal, O., & Samur, Y. (2024). Educational virtual reality game design for film and animation. In N. Lee (Ed.), *Encyclopedia of Computer Graphics and Games*. (pp. 621–636). Springer. https://doi.org/10.1007/978-3-031-23161-2_168

Keating, P. (2009). *Hollywood lighting from the silent era to film noir*. Columbia University Press.

Knez, I., & Niedenthal, S. (2008). Lighting in digital game worlds: Effects on affect and play performance. *Cyberpsychology & Behavior, 11*(2), 129–137. https://doi.org/10.1089/cpb.2007.0006

Liu, Z., Jin, Y., Ma, M., & Li, J. (2023). A comparison of immersive and non-immersive VR for the education of filmmaking. *International Journal of Human–Computer Interaction*, *39*(12), 2478–2491. https://doi.org/10.1080/10447318.2022.2078462

Mateer, J. (2017). Directing for cinematic virtual reality: How the traditional film director's craft applies to immersive environments and notions of presence. *Journal of Media Practice, 18*(1), 14–25. https://doi.org/10.1080/14682753.2017.1305838

Pires, F., Silva, R., & Raposo, R. (2022). A survey on virtual production and the future of compositing technologies. *Avanca|Cinema, 13*, 692–699. https://doi.org/10.37390/avancacinema.2022.a447

Rodrigues, M. B., & Loureiro, S. M. C. (2024). X-reality film innovation circle framework and future directions. *Creativity and Innovation Management, 33*(3), 338–368. https://doi.org/10.1111/caim.12592

Saparamadu, P. V. I. N., Sepasgozar, S., Guruge, R. N. D., Jayasena, H. S., Darejeh, A., Ebrahimzadeh, S. M., & Eranga, B. A. I. (2025). Optimising contract interpretations with large language models: A comparative evaluation of a vector database-powered Chatbot vs. ChatGPT. *Buildings, 15*(7), 1144.

Sargazi Moghadam, T., Darejeh, A., Delaramifar, M., & Mashayekh, S. (2024). Toward an artificial intelligence-based decision framework for developing adaptive e-learning systems to impact learners' emotions. *Interactive Learning Environments, 32*(7), 3665–3685.

Skorobogatova, A. (2022). *Real-time global illumination in Unreal Engine 5* [Master's thesis, Masaryk University, Faculty of Informatics].

Song, L., Pan, X., & Fan, P. (2023). The practice of virtual reality technology in the teaching of film and television art. *Journal of Education and Educational Research, 5*(3), 220–226. https://doi.org/10.54097/jeer.v5i3.13724

Thornhill-Miller, B., & Dupont, J. (2016). Virtual reality and the enhancement of creativity and innovation: Under recognized potential among converging technologies? *Journal of Cognitive Education and Psychology, 15*(1), 102–121. https://doi.org/10.1891/1945-8959.15.1.102

Wei, Z., Xu, X., Lee, L.-H., Tong, W., Qu, H., & Hui, P. (2023). Feeling present! From physical to virtual cinematography lighting education with Metashadow. *In Proceedings of the 31st ACM International Conference on Multimedia* (pp. 1127–1136). ACM. https://doi.org/10.1145/3581783.3612580

Willment, N., Swords, J., & Thomas, B. (2023). *The evolution of virtual production: Issues and opportunities.* University of York. Retrieved from https://xrstories.co.uk/wp-content/uploads/2023/11/The-Evolution-of-Virtual-Production_final.pdf

Xiang, C., Yan, Y., Hu, T., Lv, Y., & Zeng, Y. (2023). A review of the effect of the light environment of the VDT workspace on the "learning to learn" effect of video game training. *Frontiers in Neuroscience, 17*. https://doi.org/10.3389/fnins.2023.1093602

Xu, X., Tong, W., Wei, Z., Xia, M., Lee, L. H., & Qu, H. (2025). Transforming cinematography lighting education in the metaverse. *Visual Informatics, 9*(1), 1–17. https://doi.org/10.1016/j.visinf.2024.11.003

Yamazaki, A. (2024). Digital replicas and democracy: Issues raised by the Hollywood actors' strike. *Humanities and Social Sciences Communications, 11*, 1661. https://doi.org/10.1057/s41599-024-04204-w

Zhang, X., Lu, Y., Wang, W., Yan, A., Yan, J., Qin, L., Wang, H., Yan, X., Wang, W. Y., & Petzold, L. R. (2023). *GPT-4V(ision) as a generalist evaluator for vision-language tasks.* arXiv preprint. https://arxiv.org/abs/2311.01361

Ziemsen, E., & Fenuta, E. (2023). Film production and architecture education in the metaverse. In Benítez Rojas, R., V. (Ed.), *The Future of Digital Communication* (pp. 131–143). CRC Press.

7 Metaverse-Enabled Virtual Museums

A Systematic Review and Empirical Study on Immersive Cultural Heritage Applications

Tianxing Xu, Ali Darejeh, and Samad M. E. Sepasgozar

1 Introduction

Virtual museums have emerged as an innovative approach to cultural heritage preservation and dissemination, offering visitors immersive and engaging experiences beyond traditional museum settings. These digital spaces provide expanded exhibit areas, simulations of inaccessible artifacts and environments, and safe exploration of hazardous locations. Moreover, virtual museums enhance accessibility by enabling remote engagement, allowing a broader audience to experience cultural heritage without the constraints of physical travel (Lepouras et al., 2004).

The rapid advancement of metaverse technologies—a concept that enables collaborative workspaces in 3D digital environments (Mystakidis, 2022) and integrates virtual humans into immersive settings (Spyrou, Hurst, & Krampe, 2025)—has accelerated the development of virtual museums. Metaverse and mixed reality (MR) technologies, including virtual reality (VR) and augmented reality (AR), are increasingly recognized as critical enablers for creating immersive experiences that showcase the diverse cultural, historical, and geographic features of heritage sites (Anwar et al., 2025).

By leveraging these technologies, virtual museums offer dynamic and interactive environments, allowing users to engage with cultural artifacts in ways that traditional museums cannot facilitate. This shift gained particular significance during the COVID-19 pandemic, profoundly affecting worldwide museum operations (Berros, 2025; Burke, Jørgensen, & Jørgensen, 2020). The necessity for social distancing and travel restrictions led to an increased reliance on virtual platforms, sparking a surge in research and innovation surrounding virtual museums as an alternative to physical visits (Lu et al., 2022).

Efforts to integrate virtual museums into the metaverse date back to the 1990s. Tsichritzis and Gibbs (1991) proposed an early prototype based on a client-server model allowing multiple users to interact with digital artifacts through descriptions and multimedia content. However, technological limitations at the time prevented the full realization of these ideas. Early interaction tools such as the EyePhone VR headset and the DataGlove (Blanchard et al., 1990) were also envisioned to enhance user engagement, but their capabilities were limited.

DOI: 10.1201/9781003557555-7

By the 2000s, the integration of virtual museums within metaverse platforms gained momentum, particularly within Second Life, a popular virtual world launched in 2003 (Varvello et al., 2011). Hazan and Hamaagal (2008) examined virtual museums in Second Life, highlighting key examples such as the Second House of Sweden, which showcased Swedish cultural heritage, and the Second Louvre Museum, renowned for its detailed 3D modeling. Similarly, the Old Masters Gallery in Dresden created an online representation of its 750 masterpieces to attract virtual visitors to its physical museum.

More recently, Steam, a leading digital content distribution platform developed by Valve, has become a prominent space for hosting virtual museum applications (Werning, 2019). Since its launch in 2003, Steam has become a major hub for game developers and interactive applications, including virtual museum experiences. The availability of virtual museums on Steam provides an opportunity to analyze their technological features, user engagement strategies, and overall impact on digital cultural heritage preservation.

Building on this opportunity, this chapter conducts two systematic reviews to explore the technologies, user experiences (UXs), and design strategies that define contemporary virtual museum spaces. The chapter is structured as follows: Section 2 presents a systematic review of existing literature on virtual museums, examining their technological foundations, benefits, and best practices for design and curation. Section 3 provides a second systematic review focused specifically on virtual museum applications available on Steam, analyzing their distribution by release year, employed technologies, and key features. This section also includes hands-on evaluations of selected applications, identifying limitations and areas for improvement. Section 4 presents an empirical study on virtual museums developed as part of the University of New South Wales (UNSW) Metaverse platform. Finally, Section 5 concludes the chapter by summarizing key insights for the design and development of virtual museums, highlighting emerging trends, outlining future research directions, and discussing broader implications for the field.

2 Literature Review

This literature review section presents two systematic reviews aimed at exploring the features and shortcomings of virtual museums: one based on existing research studies (Section 2.1) and the other based on available applications on Steam (Section 2.2).

2.1 Exploring Virtual Museums Through a Review of Existing Research

As the first step in conducting the systematic review, we established three research questions to guide our comprehensive analysis of the existing literature on virtual museums.

- RQ1: What technologies and tools are commonly employed in developing virtual museums?

- RQ2: What are the potential benefits of virtual museums?
- RQ3: What techniques should be used to design an efficient virtual museum?

Question 1 examines the technological aspects of virtual museums, focusing on the software frameworks and tools used in their creation. Questions 2 and 3 are closely related to human–computer interaction (HCI). Specifically, Question 2 investigates the benefits of virtual museums in different areas. In contrast, Question 3 examines the techniques that can be incorporated into the design and curation of efficient virtual museums, providing insights for their future development.

2.1.1 Inclusion Criteria for Studies

To address these questions, we established the following criteria for selecting relevant literature:

1 Studies should focus on virtual museums, which can be:

  a Either 3D reconstructions of physical museums or entirely imaginary environments where cultural artifacts are displayed (Styliani et al., 2009).
  b Designed to enhance visitors' learning experiences through interactive exhibits or immersive storytelling.
  c A software framework or solution for building virtual museums.

2 The virtual museums examined in the study must integrate at least one metaverse feature, including but not limited to:

  a Immersive experiences using technologies such as VR, AR, and MR (Darejeh et al., 2024a, 2024b; Mozumder et al., 2022).
  b A three-dimensional online environment where users, represented by avatars, interact in virtual spaces independent of the real world (Ritterbusch & Teichmann, 2023).
  c Features that enhance virtual interactions and simulate the experience of being physically together (Van der Land et al., 2011).

3 Publications should fall into the following categories, preferably with empirical studies or project examples:

  a Technical or Application-Oriented Studies: These publications describe the technical stacks, system design, and implementation processes in detail, showcasing technical innovations and applications.
  b HCI Studies: These publications focus on UX, interactive quality, and the system's impact on users, aiming to improve usability and user satisfaction.

4 Only the latest publication will be included if multiple similar studies exist by the same or closely related authors.
5 Publications must be written in English.

2.1.2 Search Method

An extensive search was conducted on Google Scholar using keywords such as "virtual museum," "metaverse," "VR," and "simulated." An example search query used was "Virtual Museum AND (VR OR Metaverse)." The search results were filtered based on the relevance of the titles and abstracts to the research questions. The search was conducted in January 2025 and was limited to publications from 2015 to 2025 to ensure the inclusion of recent studies.

During the search process, publications that did not align with the topic were excluded based on their titles and portions of the full text in the search results. For publications initially deemed relevant, the Zotero application was used to download the full texts and record their metadata. At this stage, 47 publications were downloaded.

Subsequently, these publications were further examined in Zotero, leading to additional exclusions: 6 publications were excluded based on their relevance to the topic, 1 publication was excluded based on its abstract, 12 publications were excluded after reviewing the full text, and 2 identical or older studies by the same or similar authors were excluded. After applying these exclusion criteria, 26 publications were selected. The exclusion and selection percentages are detailed in Figure 7.1.

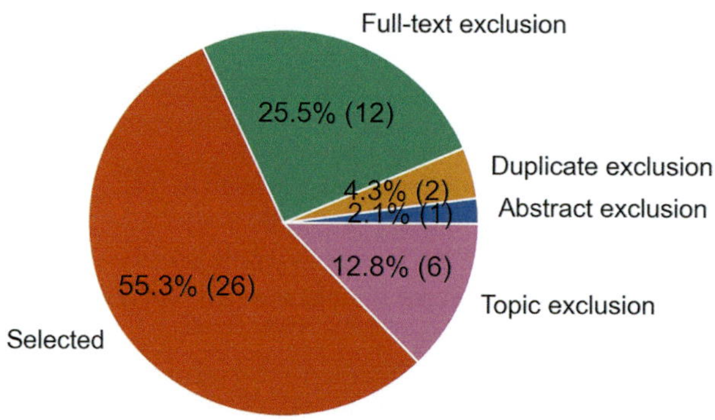

Figure 7.1 Percentages of excluded and selected publications.

2.1.3 Results and Findings

2.1.3.1 RESULTS OF THE SEARCH

Figure 7.2 illustrates the distribution of the selected publications by year. Most of these publications were published between 2020 and 2024, reflecting the growing interest in virtual museums and metaverse technologies in recent years, driven by advancements in VR, AR, and other immersive technologies.

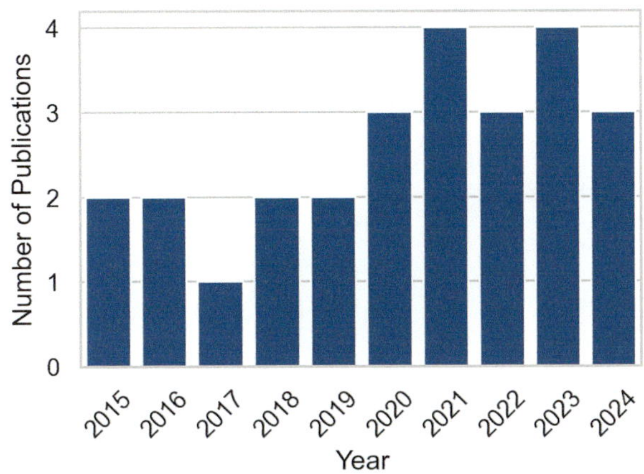

Figure 7.2 Distribution of selected publications by publication year.

Figure 7.3 illustrates the distribution of publication types derived from the "Entry Type" field of the BibTeX citation file exported from Zotero. The largest category, comprising 57.7% (15 instances), consists of journal articles. The second-largest category, accounting for 38.5% (10 instances), represents conference proceedings. The sole instance classified as "misc" is a preprint from arXiv.

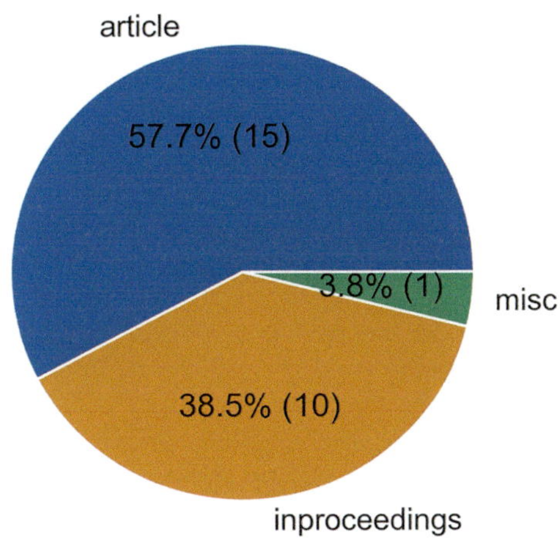

Figure 7.3 Distribution of selected publications by type.

Figure 7.4 illustrates the distribution of the selected publications by category. The "Technical or Application" and "HCI" categories were equally represented among the selected publications, each with 9 instances. Additionally, 8 publications covered both categories.

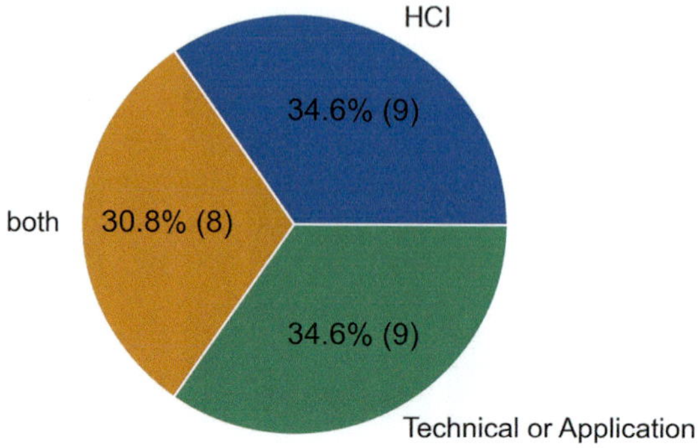

Figure 7.4 Distribution of selected publications by category.

Given that this chapter focuses on virtual museums—an area potentially influenced by the cultural constraints of each country—we aimed to illustrate the international distribution of authorship in the selected papers. To achieve this, we defined the number of authors in a paper as N and assigned a contribution of 1/N to the country of each author's affiliated institution. Figure 7.5 presents the distribution of author contributions by country. China, Italy, and Greece show the highest levels of contribution, followed by Japan, Korea, the United Kingdom, and the United States.

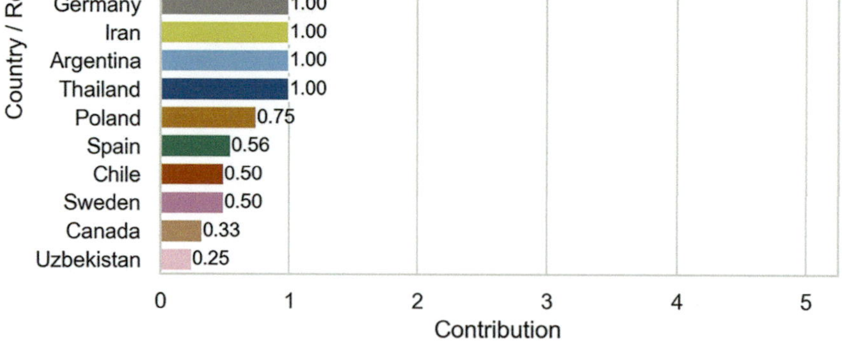

Figure 7.5 Distribution of author contributions by country.

2.1.3.2 ANSWERING RESEARCH QUESTIONS

2.1.3.2.1 RQ1: What technologies and tools are commonly employed in developing virtual museums? **Game Engines as the Dominant Platform:** Since the 2000s, game engines such as Unreal Engine and Unity have dominated the field of video game production and have also become widely used platforms for cultural heritage preservation and virtual museum development (Chia, 2022). In addition to the role of game engines in developing virtual museums, a significant portion of the literature highlights the role of digital photogrammetry (also known as captured reality) and computer-based 3D modelling tools as foundational steps in virtual museum creation. Typically, after acquiring a 3D model of cultural artifacts or heritage sites, these models are integrated into a virtual museum environment using a game engine (Aiello, Fai, & Santagati, 2019; Kersten et al., 2017; Loaiza Carvajal et al., 2020; Lombardo et al., 2023; Maddy & Narman, 2023; Skamantzari & Georgopoulos, 2016; Wu et al., 2021).

While these technologies have significantly advanced virtual museum development, the manual process of integrating models, arranging exhibits, and designing layouts within game engines can still be time-consuming and resource-intensive. To address these challenges and streamline the development process, Goto et al. (2018) proposed an automated system that uses SuperSQL queries to arrange objects from a relational database and render scenes directly in Unity. This approach significantly reduces the effort required to create virtual museums, particularly for large-scale exhibitions or frequently changing layouts. By automating the rendering process, the system alleviates many challenges associated with manual museum creation and ongoing content management.

Beyond the use of high-end game engines, the cost-effectiveness of the development process is a significant consideration in virtual museum creation. Some studies focus on low-code or no-code solutions, which offer simplified implementation while reducing the need for programming expertise. For example, Pirbazari and Tabrizi (2022) developed a low-cost online virtual museum for Iran's cultural heritage using 3DVista, software that does not require coding and is accessible across devices. In their implementation, 3D models were uploaded to Sketchfab and embedded into virtual tours. However, a limitation of 3DVista is that it restricts users to fixed panoramic views, limiting free navigation within the environment.

Immersive Virtual Museums (VR and AR Solutions): The integration of VR and AR has been a key trend in enhancing the immersive experience of virtual museums. VR-based solutions dominate the literature, but cost and accessibility remain critical factors in their adoption. Schofield et al. (2018) highlighted the high cost of professional VR equipment and opted to use Google Cardboard as a low-cost alternative. Their approach balanced cost-effectiveness and performance by designing a mask inspired by a Viking bone mount and utilizing a Samsung Galaxy S6 as the display device. However, the cost of VR headsets is no longer a major concern due to the emergence of low-cost devices such as the Meta Quest 3 (Grande et al., 2024).

While most virtual museum implementations focus on VR, some studies explore AR-based museum experiences. Li et al. (2021) introduced CubeMuseum, an AR prototype that allows users to interact with 3D museum objects displayed on a wooden cube. The AR application was developed using Unity and Vuforia 5 AR SDK, demonstrating how AR can provide interactive engagement with museum artifacts. Shirai et al. (2015) also integrated AR, but through a different approach by incorporating gaming elements. Their study involved the creation of a virtual field museum in Sagamihara, Japan, utilizing Ingress, an AR-based mobile game. Although their implementation did not achieve a fully immersive virtual museum, the project successfully increased public engagement, as evidenced by higher web traffic and participation in cultural events. This study underscores the potential of smartphone and web technologies in enhancing interaction with local history and heritage.

Web-Based Virtual Museum Platforms: While many studies focus on game-engine-based or AR/VR solutions, some researchers have explored web-based approaches to improve accessibility and content management. Kiourt et al. (2016) developed DynaMus, a web-based virtual museum framework that functions as a 3D content management system (3D-CMS). DynaMus provides front-end and back-end functionalities, allowing users to manage and curate digital exhibits efficiently.

Similarly, Zidianakis et al. (2021) introduced a web platform to assist museum curators and novice users in creating personalized museum experiences. Their system provides tools that enable curators to develop digital exhibit collections, design virtual exhibitions, encourage participatory content creation, provide personalized visitor experiences, and visualize museums as immersive VR environments. This approach facilitates greater audience engagement by allowing museums to tailor content to different user preferences.

Analysis of Technologies Employed: The key technologies and tools employed in virtual museum development include game engines, 3D modelling software, web-based solutions, and VR/AR applications. Game engines such as Unity and Unreal Engine dominate the field, providing advanced rendering capabilities and interactive environments. Additionally, 3D modelling software, including 3ds Max and ZBrush, is frequently used to create realistic digital representations of artifacts and exhibition spaces.

For those seeking a more accessible and low-code approach, tools like 3DVista and Sketchfab have emerged as viable solutions, enabling users to create and publish virtual museums with minimal technical expertise. Meanwhile, web-based solutions such as DynaMus and SuperSQL-powered platforms offer greater accessibility by providing integrated content management systems, allowing curators and museum professionals to organize exhibits efficiently.

Figure 7.6a presents a word cloud in which larger words represent the most commonly used tools for developing museum applications. Figure 7.6b illustrates the classification of technologies employed in the development of various aspects of virtual museums, including digital photogrammetry, 3D modelling applications, rendering engines, database technologies, and more.

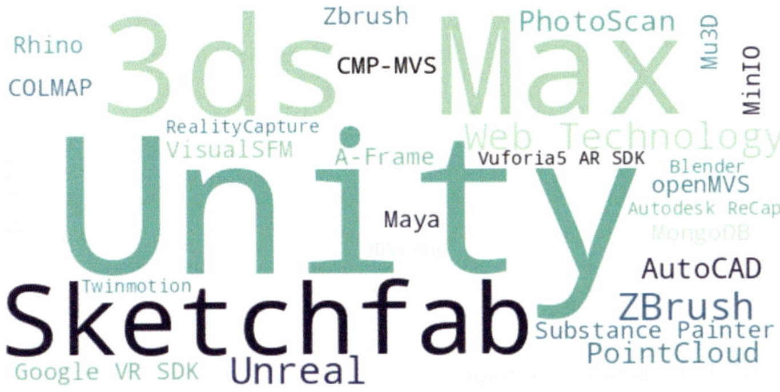

Figure 7.6a Word cloud of the most used tools for developing virtual museums.

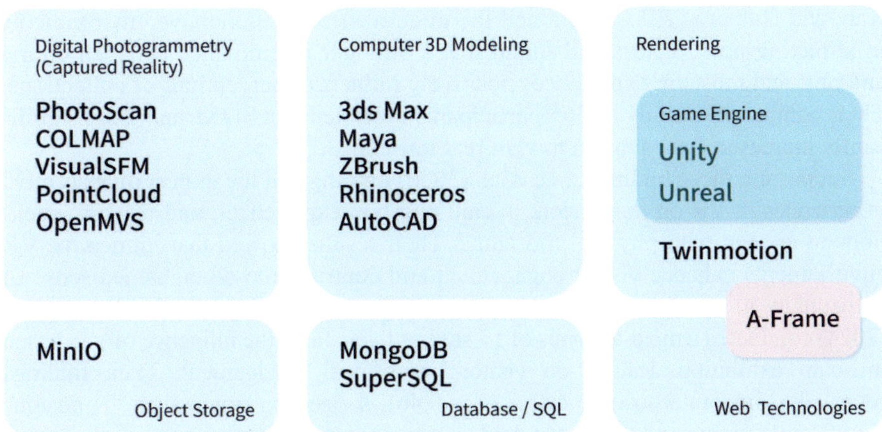

Figure 7.6b Classification of technologies employed in the selected publications.

2.1.3.2.2 RQ2: What are the potential benefits of virtual museums? **Educational Improvement:** Immersive technologies in virtual museums have demonstrated significant potential for enhancing educational outcomes. For example, Kang, Choi, and Nam (2022) introduced the Virtual Art Museum Education Program (VAMEP), utilizing Roblox Studio to facilitate students' collaborative creation of a virtual art museum. Pre- and post-activity questionnaires revealed substantial improvements in students' knowledge acquisition about museums and artworks, highlighting the effectiveness of metaverse technology in fostering participatory learning.

Furthermore, Katz and Halpern (2015) found that 3D virtual museums can significantly enhance users' cognitive involvement and sense of telepresence. Their research suggests that 3D virtual tours engage learners more effectively by providing richer perceptual cues, increasing interest in cultural content. Extending this focus on engagement, Aiello, Fai, and Santagati (2019) underscored the importance of integrating playful and educational elements in VR environments, demonstrating how immersive storytelling can enhance cultural education and users' engagement.

Building upon these studies, Vasic et al. (2024) introduced a novel use of generative artificial intelligence (AI) within a virtual museum context. By integrating GPT-4o into a web-based 3D panoramic tour of the Civic Art Gallery of Ascoli, they enabled visitors to interact with the museum using natural language. This AI-driven guide not only offered a more personalized and adaptive experience but also enhanced visitors' understanding of the museum and its artifacts. Their findings highlight the educational value of intelligent virtual guides in overcoming the limitations of static or generic tours, reinforcing the role of immersive and interactive technologies in deepening cultural engagement.

Enhanced Visitor Engagement and Emotional Experiences: Virtual museums also contribute to greater visitor engagement and enriched emotional experiences. Katz and Halpern (2015) examined the effectiveness of immersive environments in attracting new visitors and found that a stronger resemblance between online and physical museum experiences positively influences perceptions of collections. Their comparative study of 565 participants indicated that 3D virtual tours significantly increased the intention to visit real museums.

Supporting these findings, Lee et al. (2020) investigated the impact of absorptive experiences in VR on immersion, overall museum experiences, and visitors' intentions to engage with physical museums. Their results showed that immersive VR environments enhance visitor engagement and contribute to a heightened sense of authenticity in museum experiences. Regarding emotional experiences, Dong et al. (2024) conducted a meta-analysis of 17 studies to evaluate the influence of VR-based museum exhibition designs on visitors' emotional engagement. Their findings revealed a moderate positive effect ($g = 0.46$), suggesting immersive VR designs significantly enhance enjoyment and emotional engagement, particularly in art museums.

Increased Accessibility: Virtual museums play a crucial role in improving accessibility, particularly for individuals with disabilities and those in remote or isolated areas. Also, VR technology can bridge educational gaps by providing inclusive cultural experiences to users who might otherwise be unable to access traditional museums.

Żyła et al. (2020) found that while respondents appreciated VR's potential for accessibility and cultural preservation, they also expressed concerns regarding its inability to replicate a physical museum's authentic atmosphere fully. However, their findings suggest that VR can complement rather than replace traditional museum visits by increasing interest in physical exhibitions and enabling broader engagement, particularly for individuals with disabilities.

Additionally, Aiello, Fai, and Santagati (2019) developed the Timeless Museum, a virtual museum prototype that combines art masterpieces from various physically distant institutions into a single digital space. Their research highlights VR's potential to bridge geographical barriers, granting users access to cultural experiences that would otherwise be unavailable to them.

Figure 7.7 illustrates the distribution of topics in the selected HCI publications, with education, users' engagement, and accessibility emerging as the most frequently discussed themes. These findings underscore the growing recognition of virtual museums as powerful tools for enhancing learning, engagement, emotional experiences, and accessibility within cultural institutions.

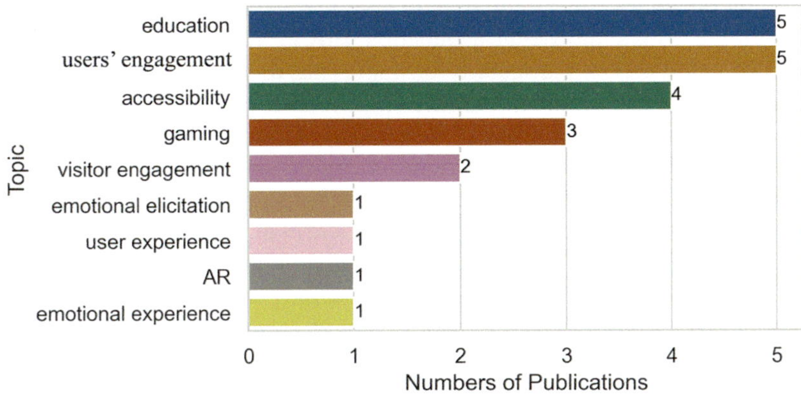

Figure 7.7 Topic distribution of the selected HCI publications.

2.1.3.2.3 RQ3: What techniques should be used to design an efficient virtual museum? **Accessibility features:** Ensuring accessibility in virtual museums is crucial for providing an inclusive and user-friendly experience, allowing all users—regardless of their technological background or physical limitations—to interact easily with these applications (Darejeh & Mashayekh, 2023). Zhang et al. (2024) investigated the challenges and preferences of older adults in VR museum visits, recommending several accessibility-oriented approaches. These include voice- and gesture-based navigation, AI-driven personalized interpretive content, multi-sensory experiences, visualization of imagined content, and varied exploration modes that support both individual and social interactions.

Campitiello et al. (2022) further emphasized accessibility enhancements in virtual museums, highlighting features such as the "vignette effect" and adjustable avatar height to improve comfort and reduce motion sickness. Their study also noted the importance of customizable brightness and color intensity to support visually impaired individuals, as well as font variety to improve readability for dyslexic users. These findings reinforce the necessity of designing VR museums with user-centered adaptations to accommodate diverse audiences.

Serious Games and Gamification: Gamification and serious games have emerged as effective strategies for enhancing engagement in virtual museums (Darejeh, 2023). *Gamification* involves incorporating game-like elements—such as points, levels, or rewards—into non-game contexts to motivate users, while *serious games* are fully developed games designed for purposes beyond entertainment, such as education or cultural engagement (Darejeh & Salim, 2016). Aiello, Fai, and Santagati

(2019) conducted a UX evaluation, offering insights that support the further development of virtual museums through the integration of serious game elements, making them more game-like to encourage users to explore the entire museum environment.

Chernbumroong et al. (2024) compared gamified and non-gamified virtual museum experiences, examining their impact on user engagement and satisfaction. While gamification was found to enhance motivation and interaction, its impact on learning outcomes was minimal, with both gamified and non-gamified groups demonstrating similar knowledge acquisition. However, the study suggests that the benefits of gamification, particularly in improving user satisfaction and engagement, justify its implementation in museum contexts.

Multi-Sensory Interactions: The integration of multi-sensory elements—such as smell, touch, 3D audio, and temperature changes—into virtual museums significantly enhances realism, immersion, and visitor engagement. Unexpected sensory elements can further deepen immersion; Schofield et al. (2018) noted that visitors' feedback often mentioned incidental sensory cues—such as the wooden scent of a VR headset—highlighting how even subtle sensory inputs contribute to a more immersive and memorable experience. Zhang et al. (2024) observed that while older adults engage less frequently with interactive exhibits, they express enthusiasm comparable to younger users when experiencing rich sensory environments.

Extending beyond ambient sensory inputs, more structured tactile interactions have also been explored. Kim et al. (2023) developed virtual museum content that integrates 3D-printed replicas of artifacts scanned with high-precision technology. By connecting these physical replicas to sensors, users can touch and manipulate virtual artifacts, enhancing their sense of realism and spatial understanding within the museum environment.

Beyond physical and tactile dimensions, VR has also proven effective for emotional engagement. Marín-Morales et al. (2019) compared emotional responses elicited in real and virtual art museum environments using electroencephalographic and electrocardiographic data. Their findings suggest that immersive virtual environments can evoke emotional reactions comparable to those experienced during physical museum visits, reinforcing the role of three-dimensional immersive virtual environments (3D IVEs) in fostering emotional connection to cultural content.

Further advancing interactive engagement, Xu et al. (2023) proposed an interactive VR installation that utilizes movement and facial expression tracking. Using lightweight, unobtrusive motion-tracking technologies, their system aims to create a multi-sensory and immersive experience. Although challenges related to tracking accuracy and participant interaction were identified, their approach envisions VR as a complementary extension to physical museums, enriching cultural experiences through dynamic storytelling and integrated sensory feedback.

2.2 *Exploring Virtual Museums Based on Existing Applications on Steam Platform*

Steam, a game distribution and social entertainment platform developed by Valve (Valve Corporation, n.d.), has profoundly shaped the gaming industry since its launch in 2003, amassing millions of active users (Werning, 2019). In 2015, Valve introduced the OpenVR software development kit (SDK) alongside the SteamVR application. OpenVR significantly impacted VR application development by offering developers greater flexibility (Wawro, 2015). Instead of creating applications separately for each manufacturer's VR device, developers could utilize the OpenVR Application Programming Interface (API) to ensure compatibility across a range of headsets, including Meta, MR, VIVE, and other brands (Casey et al., 2021; OpenVR API Documentation, n.d.; Werning, 2019). This approach streamlined the development process, fostering a more unified VR ecosystem and facilitating broader accessibility and innovation in VR content creation (Grolleman, 2024; Valve, 2024).

With the growing prominence of virtual museums within the metaverse, their presence on platforms like Steam warrants closer examination. However, virtual museums have yet to be studied from a scientific perspective in the same structured manner as in the previous section of this chapter. This section aims to explore the landscape of virtual museums on Steam, focusing on the technologies and features that define these applications. Through quantitative and qualitative analyses, we will assess the distribution of virtual museums by release year features and examine the technologies employed in their development.

To guide this systematic review, we established three key research questions:

- RQ1: What technologies and tools are commonly employed in developing virtual museums?
- RQ2: What are the key features and challenges of existing virtual museums?
- RQ3: What are the different types of virtual museums?

This structured approach will provide a comprehensive understanding of virtual museums on Steam, highlighting their technological foundations, functional attributes, and categorization within the broader metaverse ecosystem.

2.2.1 Inclusion Criteria for Studies

To address these research questions, we established the following criteria for selecting relevant literature and determining whether an application should be retained on the list:

1 Application Nature

- a The application must not be a casual game or general entertainment software.
- b The application must be a virtual museum, an exhibition platform, or contain museum-like elements, such as historical residences, government buildings, or art galleries.
- c The application may include immersive experiences designed for educational purposes, such as introductions to historical events or cultural heritage.

2 Educational Nature

- a The application should have a clear educational or informational function.

3 Exhibit Requirements

a With the exception of virtual artworks or installations, the exhibits featured in the virtual museum must originate from the real world.
b If the application displays virtual artworks or installations, it should not function solely as an independent art installation.

2.2.2 Search Method

To identify relevant virtual museum applications on Steam, we initially conducted a search using "Museum" as the primary keyword, setting the "VR Supported" filter to "Yes" while leaving the "Narrow by Language" option unspecified. This search yielded a total of 68 results.

We then manually examined the Store Page descriptions to exclude applications clearly categorized as entertainment or casual games, as well as those labelled as "coming soon" and not yet released. In addition, we reviewed the "More Like This" recommendations associated with relevant applications to identify further titles related to the museum field that did not explicitly include the term "Museum" in their titles or descriptions.

To complement the initial search, a second query was conducted using "Virtual Museum" as the keyword, this time without applying the "VR Supported" filter, while maintaining all other parameters. This extended search produced 75 results. After applying consistent exclusion criteria across both search sets, a total of 63 applications were retained for further evaluation, as detailed in Appendix A.

For each selected application, we collected key metadata, including release dates, categories, and genres, through a combination of Steam API queries and manual documentation. We also analyzed the technologies employed by examining publicly available resources such as file directories, application icons, and compiled logs. The collected data underwent quantitative analysis to identify patterns and trends in virtual museum development on the Steam platform.

2.2.3 Testing Environment and Equipment

After compiling the preliminary list, each application was downloaded and evaluated on a Windows 11 computer equipped with an NVIDIA RTX 3090 GPU, Intel Core i9–11900KF CPU, and 32 GB RAM. The evaluation was conducted without administrator privileges to replicate a standard UX.

A VIVE Pro 2 headset and HTC VIVE Controllers were used for testing. The VIVE Pro 2 features 5K resolution, a 120° field of view, and a 120 Hz refresh rate, specifications that are considered advanced for current VR technology (HTC, n.d.). These high-performance specifications ensured a high-quality visual experience, which was crucial for accurately assessing each application's effectiveness.

This setup enabled a comprehensive evaluation of the immersive environment, providing valuable insights into VR technology's potential benefits and limitations in practical museum applications.

2.2.4 Results and Findings

2.2.4.1 RESULTS OF THE SEARCH

The systematic search analysis shows that the use of virtual environments and the metaverse in the context of heritage and museums has garnered consistent attention over the past seven years, particularly since 2017. As shown in Figure 7.8, the distribution of virtual museums by release year reveals notable trends. In particular, the number of applications launched in 2019, 2020, and 2023 is significantly higher than in previous years, indicating a dramatic increase in the development of virtual museums on the Steam platform.

The surge in 2019 and 2020 can be attributed to the impact of the COVID-19 pandemic, which accelerated the adoption of virtual technologies across various industries, including museums and cultural institutions. Meanwhile, the increase in 2023 likely reflects the growing popularity of the metaverse concept and a rising interest in VR technologies. As the metaverse continues to evolve, more developers and institutions are investing in immersive museum experiences, further driving the expansion of virtual museum applications. These trends underscore an overall upward trajectory in the development of virtual museums, particularly during periods influenced by external factors, such as a global shift toward virtual engagement and technological advancements in VR and digital culture.

Figure 7.8 Distribution of virtual museums applications by release year.

To gain deeper insights into the characteristics and themes associated with virtual museums, we analyzed keywords from the descriptions, categories, and genres sections of all selected applications on Steam. This information was extracted from the Steam webpage of each application. Based on this analysis, we generated three word clouds (Figures 7.9a, 7.9b, and 7.9c).

As illustrated in Figure 7.9a, the word cloud highlights the most frequently used terminology to describe virtual museum applications in the description sections of these applications. Based on this word cloud, the terms “virtual”, “VR”, and “museum” are prominently featured at the center, with words like “art” and “digital painting” reflecting a strong emphasis on historical and artistic themes. The frequent appearance of terms such as “experience,” “explore,” and “immersive” underscores the focus on creating engaging narratives that transport users into virtual environments that feel both familiar and transformative.

Figure 7.9a Word cloud showing the frequency of keywords extracted from the short descriptions of museum applications on Steam platform.

In the word cloud related to genres (Figure 7.9b), words like “simulation,” “casual,” “adventure,” and “education” indicate the diverse experiences available in these virtual museums, catering to a wide range of audiences.

Figure 7.9b Word clouds of genres of virtual museums on Steam platform.

Finally, in the word cloud related to categories (Figure 9c), while “single-player” remains the dominant mode, the growing presence of terms like “multiplayer” and “online co-op” suggests a shift toward collaborative exploration in virtual museum settings. This trend enhances social interaction, enabling users to share and experience cultural content collectively, aligning with the metaverse’s social characteristics.

Figure 7.9c Word clouds of categories of virtual museums on Steam platform.

2.2.4.2 ANSWERING RESEARCH QUESTIONS

2.2.4.2.1 RQ1: What technologies and tools are commonly employed in developing virtual museums? An analysis of technological choices in virtual museums on Steam reveals a strong preference among developers for two major game engines. Figure 7.10 illustrates the distribution of virtual museums by game engine type and release date. The bar chart demonstrates that Unity is the most commonly used game engine, appearing in 41 instances, which accounts for 65.1% of the total technologies utilized.

Following Unity, Unreal Engine is employed in 21 instances, representing 33.3% of the total. This trend aligns with Holfeld (2024), who found that Unity leads the game engine market on Steam, with Unreal Engine ranking second. Additionally, only one instance of the GoDot Engine was identified, illustrating that while Unity and Unreal Engine dominate, other engines are occasionally used in virtual museum development.

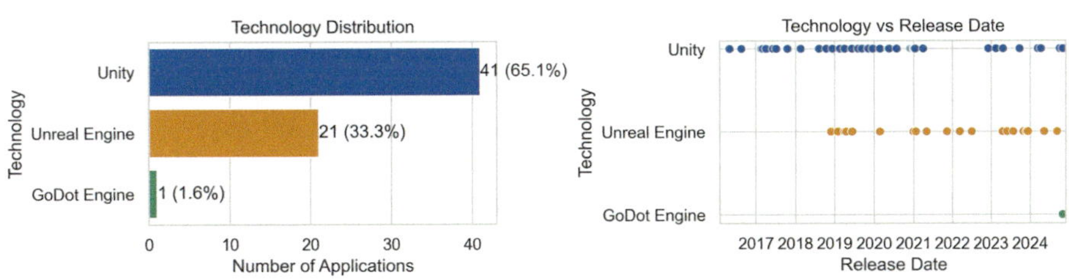

Figure 7.10 Distribution of the three main tools used for developing virtual museums. Left: The number of applications developed using each tool. Right: The number of applications developed each year using each game engine.

The scatter plot illustrating game engine usage by release date further highlights the timeline of game engine adoption in virtual museum applications. Before mid-2018, all virtual museum applications examined were developed using Unity.

Since then, both Unity and Unreal Engine have been used more evenly, though Unity still maintains a slight edge. The single application developed with the GoDot Engine, released in 2024, suggests that alternative game engines have gained some traction in recent years, particularly between 2022 and 2023, reflecting a shift in the competitive landscape. The open-source nature and flexible licensing of GoDot may have contributed to its adoption (Holfeld, 2024).

The continued dominance of Unity suggests that developers benefit from its extensive resources and community support, making it a preferred choice for virtual museum development. Meanwhile, Unreal Engine, known for its powerful rendering capabilities, aids in enhancing environmental realism (Silić et al., 2024). At the same time, the GoDot Engine, with its open-source model and flexible licensing, offers an alternative path for developers, hinting at the possibility of increased diversification in-game engine usage within virtual museums.

2.2.4.2.2 RQ2: What are the key features and challenges of existing virtual museums? The second research question examines the challenges encountered in existing virtual museum applications, which is important for ensuring that future projects can avoid these issues. While some challenges are technical and likely to be resolved over time by technology advancements and developers, others refer to the design of immersive environments, influencing user perception and engagement.

During the evaluation of 63 applications, 7 crashed on launch, and 12 could not be assessed due to various issues. Some VR applications lacked compatibility with certain controllers, limiting interaction and underscoring the need for broader hardware support. Additionally, network-related issues, such as offline back-end servers, prevented access to online features, disrupting the UX.

Some applications, such as London Museum of Water & Steam (ThinQ Digital Media Limited, 2018) and Roald Amundsen's House (Glitch Studios, 2020), attempt to replicate real-world museum visits faithfully. However, virtual spaces feel constricted or experience glitches where they pass through walls, which detracts from immersion. While these applications include teleportation-based navigation, their implementation is restrictive, allowing users to move only to specific marked locations. This constraint requires precise aiming, making navigation feel cumbersome and unnatural, ultimately reducing the spontaneity and ease of movement expected in a virtual space.

In addition to these technical and navigation-related issues, performance limitations were also evident. Applications such as *Parliament* (DPT, 2020) and *LUX SINE* (Suber et al., 2020) demonstrated high latency and poor frame rates during evaluation. Users frequently reported experiencing dizziness and discomfort within seconds of use, discouraging prolonged engagement. These challenges highlight the importance of optimization to ensure smooth, immersive experiences that sustain user interest without compromising comfort.

Beyond technical performance, content design itself presents another challenge to maintaining user engagement. Among the 44 successfully evaluated applications, most primarily featured simple 3D models of museum venues accompanied by a selection of artifacts. While these applications serve as virtual representations of physical spaces, the overwhelming prevalence of similar designs can lead to aesthetic fatigue among users after repeated interactions. This repetitiveness diminishes the initial excitement and curiosity that virtual museums are intended to evoke.

Social Interaction in Virtual Museums: Regarding social interaction, a key feature of the metaverse, many multi-user applications failed to foster meaningful user engagement. Although a number of virtual museums offered multiplayer features, a common sentiment among users was a persistent sense of isolation, as they often navigated these virtual spaces alone due to the lack of concurrent online participants. This absence of interaction diminishes the communal atmosphere that virtual museums aim to cultivate, emphasizing the need for effective marketing strategies to attract and engage users simultaneously.

Internationalization and Localization Challenges: Internationalization efforts vary significantly across applications. While some developers have taken essential steps to accommodate a diverse audience, the localization quality is often inconsistent. Except for well-funded projects backed by major institutions—such as Mona Lisa: Beyond The Glass (Emissive & HTC VIVE, 2019)—many applications suffer from poor translations, often relying on machine-generated text that results in confusing phrasing and placeholders. This lack of attention to detail can alienate non-native speakers and detract from the immersive experience that virtual museums strive to provide. Improving localization practices is essential for enhancing accessibility and inclusivity, ensuring that users from different linguistic backgrounds can fully appreciate the content.

2.2.4.2.3 RQ3: What are the different types of virtual museums? Based on the purpose, functionality, features, and design of virtual museums on the Steam platform, we categorize them into three distinct categories:

A **Replicating Real-Life:** This category of museum applications recreates physical museums or exhibitions in online or virtual environments. The primary focus of these applications is on designing objects, their surrounding environments, colors, and textures to closely resemble real-world events or artifacts with a high degree of realism.

B **Completely Virtual:** This category includes virtual museums that feature venues or exhibition halls that do not exist in the real world and **display** artifacts that either originate from real-life objects or consist entirely of digital content (e.g., digital artworks).

C **Immersive Experiences:** Immersive experiences prioritize immersion and engagement, often taking the form of guided tours, serious games, interactive simulations, or VR narratives. Unlike other categories that rely on replicating physical spaces like galleries, exhibition halls, or traditional museum layouts, these applications focus on dynamic storytelling, user participation, and sensory-rich environments to **captivate** users, offering flexible and creative alternatives to structured venues.

The categories above can be used as a framework for understanding the diverse virtual museum applications before a developer begins to design a new virtual museum. Many applications integrate features from multiple categories while incorporating functionalities such as interconnectivity, social engagement, interactivity, and VR capabilities. Consequently, a virtual museum can be conceptualized as combining these categories and features rather than fitting into a single rigid classification.

Based on the categories outlined above, three of the most comprehensive virtual museum applications available on Steam are chosen for further investigation. These are listed as follows:

- Japanese Cultural Property VR Museum (Xeen Inc. & Sougogiken Co., Ltd, 2023)
- Museum of Other Realities (MOR Museum Inc., 2020)
- Mona Lisa: Beyond the Glass (Emissive & HTC VIVE, 2019)

These applications were chosen because they are among Steam's most feature-rich virtual museum experiences, covering all features of our classification framework. According to our framework, Figure 7.11 illustrates the categories and functionalities of the three selected virtual museums.

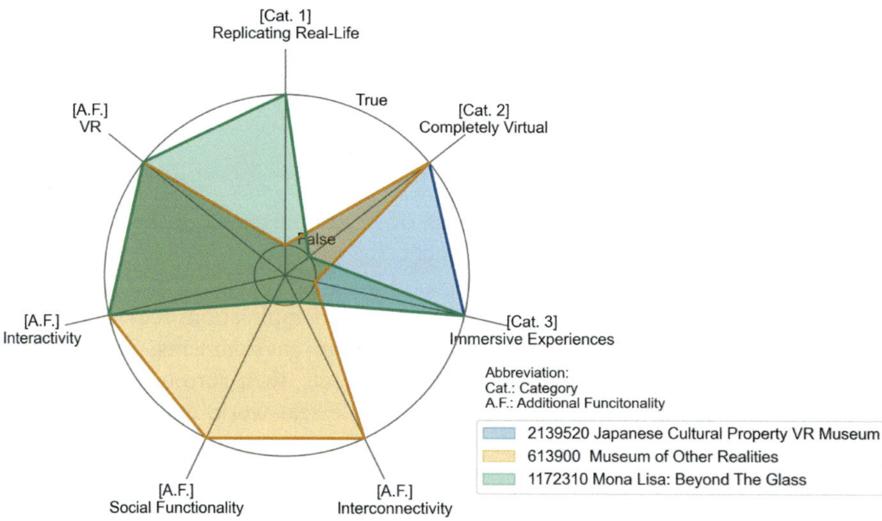

Figure 7.11 Categories and additional functionalities for selected virtual museums.

Japanese Cultural Property VR Museum ***(Xeen Inc. & Sougogiken Co., Ltd., 2023)***
This application falls under the "Completely Virtual" and "Immersive Experiences" categories, featuring independent virtual museums, some of which are modeled after physical locations but redesigned specifically for VR. The application offers rich educational content, including detailed descriptions, audio guides, and videos. However, it lacks a guided tour that takes users through different parts of the

museum while narrating the stories behind the exhibits and artifacts (Figure 7.12b). The interaction methods are well-designed, offering different modes tailored for novice and experienced users (Figure 7.12c).

Users can manipulate artifacts, examine details using magnifying glasses and flashlights, or engage in historical scenarios with candles in dimly lit settings (Figure 7.12a). The Iwabitsu Sanada Ninja section of the museum also features an immersive combat experience, enhancing user engagement (Figure 7.12d). Despite these strengths, the application remains single-player, with no multiplayer or social features, limiting the collaborative aspects of exploration within the virtual museum.

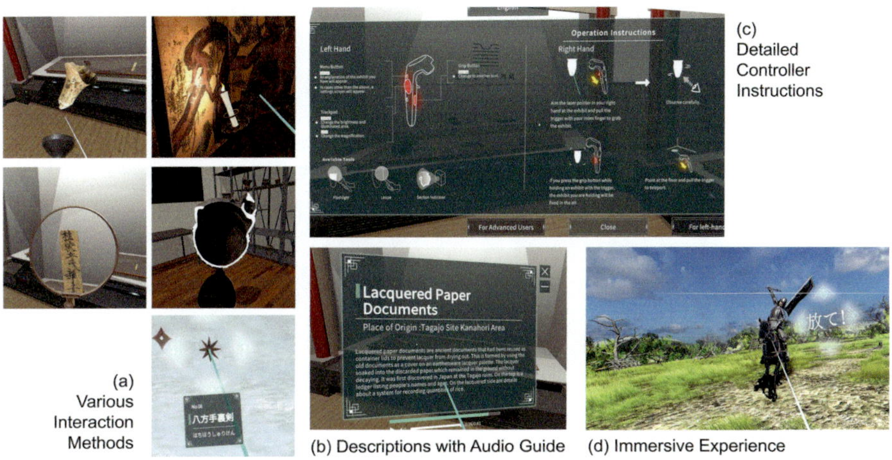

Figure 7.12 Screenshots of Japanese cultural property VR museum (Xeen Inc. & Sougogiken Co., Ltd., 2023).

Museum of Other Realities **(*MOR Museum Inc., 2020*)** Categorized as both a "Completely Virtual" and "Immersive Experiences" digital exhibition space, this application presents VR-based artworks, such as interactive installations, 3D sculptures, and animated creations. This virtual museum offers an immersive environment where users can explore, interact with, and even enter the artworks, redefining traditional art engagement. One of its key strengths is its integration of social features, aligning with the metaverse concept. The application supports multiplayer interactions and customizable avatars, fostering shared experiences. A unique interactive element, "turtles," allows users to stand on them while conversing and admiring the artwork, adding a playful and engaging dimension to the experience.

Mona Lisa: Beyond the Glass **(*Emissive & HTC VIVE, 2019*)** This application fits into the "Replicating Real-Life" and "Immersive Experiences" categories and is supported by the Louvre Museum. It faithfully recreates key locations, including the Louvre Pyramid and Grande Galerie (Figure 7.13a), while also incorporating an "Immersive Experience" that deconstructs the Mona Lisa, exploring its history, artistic techniques, and cultural significance (Figure 7.13b). Users can experience Da Vinci's world by interacting with the painting's backdrop and even flying in an aircraft he designed (Figure 7.13c). An interactive interface allows further exploration, including historical insights, hidden details through X-ray analysis, and the painting's artistic legacy (Figure 7.13d).

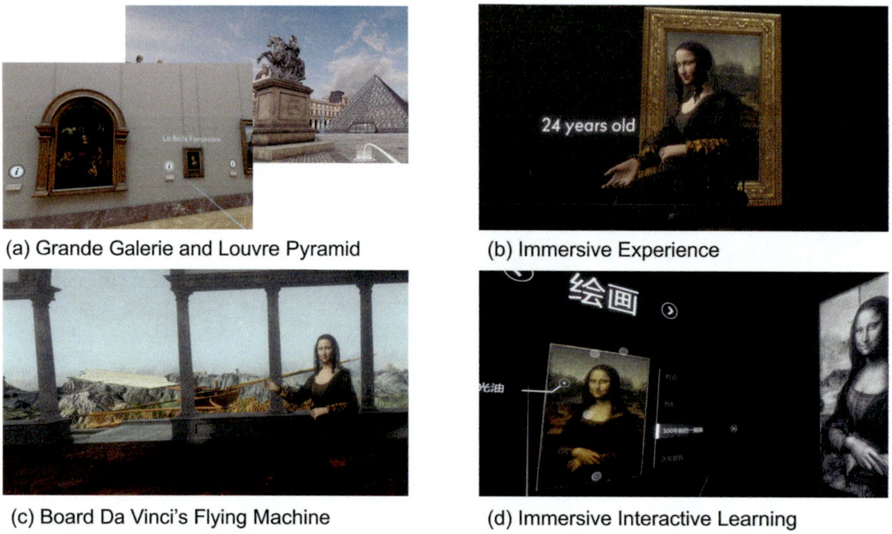

Figure 7.13 Screenshots of Mona Lisa: Beyond the glass (Emissive & HTC VIVE, 2019).

The analysis of existing literature and Steam-based applications revealed several gaps in current virtual museums, including limited interactive storytelling, restrictive navigation, and weak social engagement features. To address these challenges, the next section introduces our metaverse-based virtual museum, featuring integrated storytelling and enhanced user interaction. It outlines the application's key features and presents findings from a user study evaluating its effectiveness.

3 Empirical Study: A VR-Based Virtual Museum System for Cultural Heritage Preservation

This study explores the development of our Virtual Museum System, created as part of the UNSW Metaverse Platform. The system is designed to provide users with an interactive and immersive experience for engaging with historical artifacts and cultural heritage. By leveraging VR and interactive storytelling techniques, it enhances traditional museum experiences, allowing users to explore a 3D Earth environment, discover artifact histories, trace their origins, and learn about production methods.

The integration of dynamic virtual exhibitions with guided storytelling offers users an innovative approach to cultural education. Visitors can navigate historical sites, explore porcelain craftsmanship and international trends, and follow the historical journey of artifacts within a cohesive narrative framework. This system fosters deeper cultural understanding by bridging the gap between physical and digital museums, creating a truly engaging and educational virtual environment.

3.1 Application Development for the Study

This Virtual Museum System is developed using Unreal Engine 5 and deployed on HTC VIVE Pro 2 headsets, supported by high-performance computing systems. The system consists of three interconnected modules, each offering a unique way for users to engage with historical artifacts and events. In the first module, users enter the main museum hall, where they experience a narrative-driven exhibit featuring an animated panoramic artwork displayed along the walls. This artwork illustrates key aspects of a specific era and country, providing essential historical and cultural context.

As users walk along the exhibition, a voice narration automatically plays, explaining the historical significance of different scenes depicted in the artwork. This dynamic storytelling approach enhances user engagement by combining visual, auditory, and interactive elements. Additionally, users can access detailed information about significant artifacts and historical moments from that period, enriching their understanding of the explored era (Figure 7.14).

Figure 7.14 The main museum hall.

The second module takes users to a dedicated artifact exhibition hall, where they can explore and interact with various objects related to the selected historical era (Figure 7.15). When users grab an artifact, they have the option to transition to the third module, which offers a deeper exploration of the artifact's origins.

Figure 7.15 Artifacts and heritage objects displayed in the selected museum space.

In the third module (Figure 7.16), users embark on a historical journey through a 3D interactive map, travelling to where the artifact was discovered. There, they can learn about the artifact's history in its original setting before returning to the museum.

Figure 7.16 3D interactive map of a historical archaeology site in Parramatta, Sydney.

Finally, the multiplayer system allows users to wear VR headsets, create personalized avatars, and explore the museum together. They can interact with one another, discuss artifacts, and share insights, making the experience more collaborative and engaging.

3.2 Methodology for Conducting the Study

To evaluate the effectiveness of the Virtual Museum System, user testing was conducted with ten general users interested in cultural heritage. All participants were recruited in accordance with a human research ethics protocol approved by the University of New South Wales and sanctioned by its institutional review board. After signing a consent form, participants were given a brief introduction and then allowed 20 minutes to explore the museum environment. Following the exploration, data was collected through a post-experience survey and a semi-structured interview. The responses were analyzed using thematic analysis to identify trends in user satisfaction and areas for improvement.

The survey and interview questions focused on five key areas: user engagement, educational value, interactivity and navigation, storytelling impact, and technical performance. User engagement examined how immersive and engaging the experience was compared to traditional museum visits. Educational value assessed whether the system enhanced understanding of porcelain production and its global significance. Interactivity and navigation evaluated the intuitiveness of controls and ease of movement. Storytelling impact was analyzed how effectively the narrative guided the UX. Finally, technical performance assessed the visual quality and responsiveness of the system.

3.3 Results from the Study

Findings from the evaluation revealed several strengths and areas for improvement. High engagement levels were reported, as users found that exploring historical locations in a 3D Earth environment provided a sense of immersion and presence that traditional static museum exhibits lack. The dynamic transitions between locations, such as traveling through a wormhole to historical sites, were particularly praised for maintaining user engagement and fostering curiosity. Participants also highlighted that the ability to navigate freely, rather than being confined to pre-set viewpoints, created a sense of agency and exploration, which enhanced their overall experience.

Improved learning outcomes were observed, with users reporting that interactive elements significantly enhanced their understanding of porcelain production. The handscroll painting exhibit, which allowed users to view the entire porcelain-making process sequentially, was found to be more effective than traditional text-based learning methods. Participants noted that highlighting specific steps, accessing contextual descriptions, and visualizing intricate craftsmanship helped them retain information more effectively. Additionally, the exhibition hall's thematic organization, which grouped artifacts by historical period and cultural influence, provided a structured learning experience that aided knowledge retention.

The storytelling system was similarly effective in enhancing user engagement. Participants highlighted that the narrative structure in the main hall helped them connect more deeply with historical events and artifacts. The integration of animated timelines and artifact-specific narratives made historical content feel more accessible and engaging. Some users particularly appreciated the use of first-person storytelling (e.g., a Ming bowl narrating its journey), which created an emotional connection with the artifacts, making the experience both informative and memorable.

Multiplayer functionality was another highly valued feature, allowing users to explore the virtual museum collaboratively. Participants appreciated the ability to interact with peers, discuss artifacts in real-time, and share their interpretations, making the experience feel more social and engaging than a solitary virtual tour. Some users suggested that the presence of avatars and the ability to navigate together helped simulate the social aspects of visiting a real museum, enhancing their overall enjoyment.

However, several challenges and usability issues were identified. Navigation difficulties were frequently reported, as some users found the movement controls unintuitive and difficult to master. Suggested improvements included wayfinding markers, tutorial guidance, and alternative control options for smoother navigation. Limited object interactivity was another major concern, as users expressed interest in features such as zooming into fine details, rotating objects freely, and accessing more in-depth historical annotations. Some participants also noted technical inconsistencies, such as delayed responsiveness in interactive elements and graphical glitches when transitioning between different museum sections. Finally, a few users mentioned that certain exhibits' lighting and visual contrasts made text and details harder to discern, suggesting the need for improved accessibility features such as adjustable brightness and enhanced text legibility.

3.4 Insights from the Study

Traditional museum experiences rely on static exhibits, often leading to cognitive overload or disengagement due to excessive textual information. Our proposed virtual museum addresses these challenges through a narrative-driven approach and spatial exploration, integrating visual storytelling and experiential learning in ways that align with constructivist educational theories. However, this also presents a challenge: does immersion always lead to better knowledge retention? While all users in the study reported increased engagement, the depth of learning depends on how effectively information is integrated into the interactive experience rather than just its presence. Future iterations of the system should examine whether interactive storytelling alone enhances learning outcomes or if additional reinforcement mechanisms, such as quizzes or guided discussions, are needed.

Navigation challenges highlight the trade-off between freedom of movement and usability. While allowing users to freely explore the environment was a valuable feature, it also introduced usability issues such as disorientation and difficulty locating specific exhibits. While open-world exploration fosters agency, it also increases cognitive load, making navigation difficult for novice users.

Implementing a hybrid approach—offering both structured paths and free exploration—could improve accessibility without compromising engagement.

Another critical issue is artifact interactivity. While the system allows users to explore exhibits by grabbing and rotating objects, there is a growing expectation for greater control, such as zooming, highlighting the demand for more hands-on digital engagement. This reflects a shift in how users interact with cultural heritage in virtual spaces—they seek more than passive observation. Future development should incorporate gesture-based interactions and haptic feedback to create a more immersive and tactile experience.

Finally, visual accessibility concerns, such as lighting inconsistencies and readability issues, underscore the importance of inclusive design. Unlike physical museums, where visitors adapt to lighting conditions naturally, digital spaces require deliberate optimization for clarity and comfort. Implementing adaptive visual settings—such as adjustable brightness and contrast—would enhance usability for diverse audiences.

4 Discussion

Integrating virtual museums within the metaverse offers new opportunities for presenting and preserving cultural heritage while delivering novel experiences to visitors. The empirical study in this chapter, and the application developed as part of it, sought to demonstrate the potential of virtual museums for future development into metaverse environments. This discussion synthesizes insights from the literature and findings from the study to identify key areas for improvement and future directions in virtual museum development.

One of the most critical limitations of current virtual museums is their reliance on static 3D replications of physical spaces. While photorealistic reconstructions contribute to a sense of presence, they often fail to foster meaningful engagement (Lepouras et al., 2004). Research emphasizes the importance of interactive storytelling, where dynamic narratives enhance cognitive retention by encouraging users to interact with historical contexts rather than passively viewing exhibits (Aiello, Fai, & Santagati, 2019; Darejeh, Marcus, & Sweller, 2021).

Findings from our study reinforce this point. The narrative-driven design of the Virtual Museum System—particularly through the handscroll painting exhibit and historical journey mode—was found to significantly improve users' understanding. Participants reported that sequential storytelling and contextualized artifact presentations were more effective than traditional text-based displays. These results support the argument that future virtual museums should move beyond static 3D modeling, incorporating multi-layered, interactive storytelling techniques to facilitate deeper engagement and learning experiences.

Interactivity is another key determinant of user engagement, yet many existing applications offer limited means of interaction beyond basic navigation and object manipulation (Kersten et al., 2017). Research suggests that integrating gesture-based controls, haptic feedback, and adaptive learning pathways can create more immersive and engaging experiences (Kim et al., 2023).

Our study findings strongly support this need. While users appreciated the ability to grab and rotate objects within the virtual museum, many found the lack of zoom functionality and detailed exploration limiting. Participants also emphasized the importance of multi-sensory engagement—such as spatial audio, scent diffusion, and dynamic lighting—to replicate the sensory attributes of physical museum visits. However, these immersive elements must be balanced carefully to avoid overwhelming users, as excessive stimulation may hinder rather than enhance learning (Darejeh, Marcus, & Sweller, 2022; Marín-Morales et al., 2019).

A further limitation of many virtual museums is the absence of robust social features. While some applications incorporate multiplayer functionalities, most remain isolated experiences that do not leverage the collaborative potential of the metaverse (Cheng, 2023). Research demonstrates that social interaction enhances engagement and knowledge retention, as peer discussions foster deeper understanding (Brown, Metz, & Campione, 1996).

To address this gap, successful metaverse-based virtual museums should integrate structured social mechanics, such as expert-led tours, collaborative annotations, and shared exploration experiences. Asynchronous features—such as recorded guided tours and community-curated exhibits—could address time-zone and connectivity challenges, enabling users to engage even when real-time interaction is not feasible. Additionally, cross-institutional collaborations could transform virtual museums from isolated digital replicas into dynamic cultural ecosystems, linking exhibitions across institutions to create unified, expansive experiences.

Also, to develop a successful metaverse-based virtual museum, accessibility challenges must be addressed. These include barriers to cultural access, as well as technical limitations such as high hardware requirements, inconsistent localization, and limited accessibility settings, all of which can exclude many users (Campitiello et al., 2022).

Our study findings align with these concerns. Several participants reported difficulties with complex control schemes, highlighting the need for more intuitive navigation, tutorial guidance, and alternative control options. Issues with text readability and limited localization support further reinforce the necessity for adaptive UI/UX features. Implementing customizable text sizes, alternative navigation modes, improved localization, and adjustable lighting and contrast settings would significantly enhance accessibility, ensuring a more inclusive and satisfying UX.

Although the application developed and presented in this study demonstrated a high degree of novelty, the limited number of participants may have affected the generalizability of the findings. Therefore, future work should test the application with a larger sample size and collect quantitative data to enhance the reliability and validity of the results. Future research should also explore AI-driven personalization and adaptive learning to further enhance virtual museum experiences (Darejeh et al., 2024c; Sargazi Moghadam et al., 2024). Natural Language Processing (NLP) techniques could enable AI-driven museum guides capable of responding to user queries and providing personalized recommendations based on their interactions (Saparamadu et al., 2025). These AI-driven avatars could serve as virtual museum tour guides, leading users through the museum while answering questions about different artifacts and tailoring the experience to individual visitor interests.

Another critical area is multi-sensory interaction and haptic feedback. Future studies should examine how touch-sensitive interfaces and environmental soundscapes affect user perception, allowing visitors to experience artifact textures or historical environments more realistically.

Additionally, longitudinal studies are needed to assess the educational impact of metaverse-integrated virtual museums over time. Current research focuses on short-term engagement, but little is known about long-term knowledge retention and historical understanding. Tracking users' learning progression over months or years would provide valuable insights into the cognitive and behavioral effects of immersive museum experiences.

5 Conclusion

This chapter has presented a comprehensive examination of metaverse-enabled virtual museums, combining a systematic literature review, an analysis of existing applications, and an empirical study to explore both the opportunities and challenges of immersive cultural heritage experiences. Unlike general discussions on the potential of the metaverse, our findings offer tangible insights grounded in empirical analysis and real-world implementation.

Through the systematic literature review, we identified the core technologies and tools used in virtual museum development, such as Unity and Unreal Engine, and outlined best practices in accessibility, engagement, and interactivity. Importantly, we highlighted a key gap: while current virtual museums excel in photorealistic 3D reconstructions, they often lack depth in narrative design, multi-sensory interaction, and social collaboration.

Our study addressed these gaps by introducing a novel virtual museum system developed as part of our UNSW metaverse platform. This system goes beyond static exhibits by incorporating three interactive modules that blend storytelling, spatial exploration, and collaborative learning. User testing revealed strong engagement and educational outcomes, particularly through features like dynamic artifact timelines, multi-user exploration, and contextual storytelling. These findings underscore the value of narrative structure and social presence in enhancing the cultural and cognitive impact of virtual museums.

Moreover, our application served as a testing ground for several lessons learned: the need for intuitive navigation to prevent disorientation, the importance of adaptive content delivery based on user behavior, and the role of accessibility features such as adjustable lighting and text visibility. These lessons are not abstract principles but are based on direct user feedback and thematic analysis.

Ultimately, this chapter contributes to the growing body of knowledge on immersive digital heritage by demonstrating that successful virtual museums must do more than replicate physical spaces—they must reimagine them. The path forward lies in merging adaptive storytelling, collaborative interaction, and inclusive design. As metaverse technologies evolve, virtual museums have the potential to become dynamic cultural ecosystems—bridging geographic divides, personalizing learning, and preserving heritage in ways that traditional institutions alone cannot achieve.

Appendix A

Virtual museums from the Steam platform that are included in this study:

In the Evaluated column, C indicates a crash upon launch, I denotes evaluation with issues, and Y signifies successful evaluation.

	Name	*Evaluated*	*Published Year*	*Link*
1	Vintage VR	C	2016	https://store.steampowered.com/app/466720
2	The VR Museum of Fine Art	Y	2016	https://store.steampowered.com/app/515020
3	Inception VR	I	2017	https://store.steampowered.com/app/592920
4	GIPHY Museum of GIF Art	C	2017	https://store.steampowered.com/app/593090
5	The Cable Center - Virtual Archive	C	2017	https://store.steampowered.com/app/612440
6	ObserVRtarium	Y	2017	https://store.steampowered.com/app/634060
7	Douarnenez VR	Y	2017	https://store.steampowered.com/app/656770
8	Museu do Círio de Nazaré em Realidade Virtual	Y	2017	https://store.steampowered.com/app/727030
9	Dreams of Dali	Y	2018	https://store.steampowered.com/app/591360
10	London Museum Of Water	Y	2018	https://store.steampowered.com/app/831040
11	Steam	Y	2018	https://store.steampowered.com/app/831040
12	Museum of Symmetry	Y	2018	https://store.steampowered.com/app/870890
13	Dinosaur Safari VR	I	2018	https://store.steampowered.com/app/953290
14	Experience: Colorblindness	I	2018	https://store.steampowered.com/app/979100
15	Cappasity Demo	Y	2018	https://store.steampowered.com/app/982830
16	WellTown	C	2019	https://store.steampowered.com/app/974150
17	The Virtual Reality Museum of Immersive Experiences	Y	2019	https://store.steampowered.com/app/1015650
18	ESKO	C	2019	https://store.steampowered.com/app/1045140
19	Infinite Art Museum	Y	2019	https://store.steampowered.com/app/1011000
20	Blanket Heavy With Nightmares	I	2019	https://store.steampowered.com/app/1045020
21	Raatihuone	I	2019	https://store.steampowered.com/app/1045130

(*Continued*)

	Name	*Evaluated*	*Published Year*	*Link*
22	The Homestead	Y	2019	https://store.steampowered.com/app/731950
23	Age-Old Cities VR	Y	2019	https://store.steampowered.com/app/1025500
24	Smithsonian American Art Museum “Beyond The Walls”	Y	2019	https://store.steampowered.com/app/1087320
25	Church Art Of Sweden	Y	2019	https://store.steampowered.com/app/1123150
26	Mona Lisa: Beyond The Glass	Y	2019	https://store.steampowered.com/app/1172310
27	Mannerheim Virtual Experience	Y	2019	https://store.steampowered.com/app/1179090
28	World of Guns: VR	I	2019	https://store.steampowered.com/app/1111760
29	Porsche Hall of Legends VR	Y	2020	https://store.steampowered.com/app/1231900
30	Museum of Other Realities	Y	2020	https://store.steampowered.com/app/613900
31	Parliament – Parlement	Y	2020	https://store.steampowered.com/app/1232920
32	Future Fragments - Command Z	Y	2020	https://store.steampowered.com/app/1295250
33	Gerrymander Madness	Y	2020	https://store.steampowered.com/app/1358180
34	Javols VR	Y	2020	https://store.steampowered.com/app/1460230
35	Roald Amundsen’s House	Y	2020	https://store.steampowered.com/app/1435700
36	Dört Model Müzesi	Y	2020	https://store.steampowered.com/app/1318520
37	LUX SINE	Y	2020	https://store.steampowered.com/app/1229740
38	AVROD	I	2021	https://store.steampowered.com/app/1495350
39	Great Paintings VR	Y	2021	https://store.steampowered.com/app/1511090
40	Beyond The Diorama: Caribou World	Y	2021	https://store.steampowered.com/app/1488050
41	Baalbek Reborn: Temples	Y	2021	https://store.steampowered.com/app/1370800
42	The OmniGallery	Y	2021	https://store.steampowered.com/app/1587200
43	Curatours	I	2021	https://store.steampowered.com/app/1532110
44	VRUSEUM	Y	2022	https://store.steampowered.com/app/1909230
45	Maria Blanchard Virtual Gallery	I	2022	https://store.steampowered.com/app/2026860

(*Continued*)

	Name	*Evaluated*	*Published Year*	*Link*
46	VR Archeology: Secrets of Kulikovo Field	I	2022	https://store.steampowered.com/app/2209990
47	Angels of Amsterdam	Y	2022	https://store.steampowered.com/app/2133100
48	Digital Pierrot Museum	C	2023	https://store.steampowered.com/app/2099240
49	VR Museum	Y	2023	https://store.steampowered.com/app/1373100
50	Japanese Cultural Property VR Museum	Y	2023	https://store.steampowered.com/app/2139520
51	Metaart	Y	2023	https://store.steampowered.com/app/2385390
52	VMVA - Virtual Museum for Virtual Art	Y	2023	https://store.steampowered.com/app/2236930
53	Testimony of HELP	Y	2023	https://store.steampowered.com/app/2595440
54	Yoshitaka Amano VR Museum	C	2023	https://store.steampowered.com/app/1853360
55	AGO BRISTOL 1775: From Warship to Prison Hulk	Y	2023	https://store.steampowered.com/app/2605210
56	MalovMetaArt Metaverse	Y	2023	https://store.steampowered.com/app/2224020
57	MalovModernArt Virtual Museum	Y	2023	https://store.steampowered.com/app/2680700
58	Kottabos VR	Y	2024	https://store.steampowered.com/app/2626890
59	Anne Frank House VR	I	2024	https://store.steampowered.com/app/2877690
60	Pronovost Art Gallery	Y	2024	https://store.steampowered.com/app/2761450
61	Twin It! Virtual Museum	Y	2024	https://store.steampowered.com/app/3150630
62	Civilización Caral Perú, la más antigua de América	I	2024	https://store.steampowered.com/app/3004270
63	D-Day VR Museum	Y	2024	https://store.steampowered.com/app/2512870

References

Aiello, D., Fai, S., & Santagati, C. (2019). Virtual museums as a means for promotion and enhancement of cultural heritage. *The International Archives of the Photogrammetry, Remote Sensing and Spatial Information Sciences*, *XLII-2-W*15, 33–40. 27th CIPA International Symposium "Documenting the past for a better future" (Volume XLII-2/W15), 1–5 September 2019, Ávila, Spain. https://doi.org/10.5194/isprs-archives-XLII-2-W15-33-2019

Anwar, M. S., Yang, J., Frnda, J., Choi, A., Baghaei, N., & Ali, M. (2025). Metaverse and XR for cultural heritage education: Applications, standards, architecture, and technological insights for enhanced immersive experience. *Virtual Reality*, *29*(2), 51.

Berros, G. P. (2025). The lockdown historic house museum: The subversion of the domestic ideal during the COVID-19 pandemic through Bo Burnham's inside. *Res Mobilis: Revista internacional de investigación en mobiliario y objetos decorativos*, *13*(19), 52–71.

Blanchard, C., Burgess, S., Harvill, Y., Lanier, J., Lasko, A., Oberman, M., & Teitel, M. (1990). Reality built for two: A virtual reality tool. *Proceedings of the 1990 Symposium on Interactive 3D Graphics*, 35–36. https://doi.org/10.1145/91385.91409

Brown, A. L., Metz, K. E., & Campione, J. C. (1996). Social interaction and individual understanding in a community of learners: The influence of Piaget and Vygotsky. In A. Tryphon & J. Vonèche (Eds.), *Piaget-Vygotsky: The Social Genesis of Thought* (pp. 145–170). Psychology Press.

Burke, V., Jørgensen, D., & Jørgensen, F. A. (2020). Museums at home: Digital initiatives in response to COVID-19. *Norsk Museumstidsskrift*, *6*(2), 117–123.\

Campitiello, L., Caldarelli, A., Todino, M. D., Tore, P. A. D., Tore, S. D., & Lecce, A. (2022). Maximizing accessibility in museum education through virtual reality: An inclusive perspective. *Italian Journal of Health Education, Sport and Inclusive Didactics*, *6*(4), 44–58. https://doi.org/10.32043/gsd.v6i4.731

Casey, P., Baggili, I., & Yarramreddy, A. (2021). Immersive virtual reality attacks and the human joystick. *IEEE Transactions on Dependable and Secure Computing*, *18*(2), 550–562. IEEE Transactions on Dependable and Secure Computing. https://doi.org/10.1109/TDSC.2019.2907942

Cheng, S. (2023). Metaverse. In S. Cheng (Ed.), *Metaverse: Concept, content and context* (pp. 1–23). Springer Nature Switzerland. https://doi.org/10.1007/978-3-031-24359-2_1

Chernbumroong, S., Ariya, P., Yolthasart, S., Wongwan, N., Intawong, K., & Puritat, K. (2024). Comparing the Impact of Non-Gamified and Gamified Virtual Reality in Digital Twin Virtual Museum Environments: A Case Study of Wieng Yong House Museum, Thailand. *Heritage*, *7*(4, 4), 1870–1892. https://doi.org/10.3390/heritage7040089

Chia, A. (2022). The metaverse, but not the way you think: Game engines and automation beyond game development. *Critical Studies in Media Communication*, *39*(3), 191–200. https://doi.org/10.1080/15295036.2022.2080850

Darejeh, A. (2023). Empowering education through eerp: A customizable educational vr escape room platform. In *2023 IEEE International Symposium on Mixed and Augmented Reality Adjunct (ISMAR-Adjunct)* (pp. 764–766). IEEE.

Darejeh, A., Huynh, L., Ji, X., Marcus, N., & Polly, P. (2024a). Assessing the efficacy of virtual reality-based laboratories versus real-life settings for PCR procedures, within medical science laboratories. In *2024 IEEE International Symposium on Mixed and Augmented Reality Adjunct (ISMAR-Adjunct)* (pp. 489–492). IEEE.

Darejeh, A., Marcus, N., & Sweller, J. (2022). Increasing learner interactions with E-learning systems can either decrease or increase cognitive load depending on the nature of the interaction. *L'Année psychologique*, *122*(3), 405–437.

Darejeh, A., Marcus, N., & Sweller, J. (2021). The effect of narrative-based E-learning systems on novice users' cognitive load while learning software applications. *Educational Technology Research and Development*, *69*(5), 2451–2473.

Darejeh, A., & Mashayekh, S. (2023). Mind-object control as a new form of interaction in metaverse for users with motor disability. *Journal of Biomedical Physics & Engineering*, *13*(1), 257.

Darejeh, A., Moghadam, T. S., Delaramifar, M., & Mashayekh, S. (2024c). A framework for AI-powered decision making in developing adaptive e-learning systems to impact learners' emotional responses. In *2024 11th International and the 17th National Conference on E-Learning and E-Teaching (ICeLeT)* (pp. 1–6). IEEE.

Darejeh, A., Mondal, M., Marcus, N., & Vassar, A. (2024b). Exploring the cognitive load effects of diverse virtual reality interaction methods in interactive educational platforms. In *2024 IEEE International Symposium on Mixed and Augmented Reality Adjunct (ISMAR-Adjunct)* (pp. 435–438). IEEE.

Darejeh, A., & Salim, S. S. (2016). Gamification solutions to enhance software user engagement—a systematic review. *International Journal of Human-Computer Interaction, 32*(8), 613–642.

Dong, H., Cai, Y., & Wei, Q. (2024). The impact of VR museum exhibition design on user emotional experience: A meta-analysis. *Proceedings of the 2024 International Conference on Cloud Computing and Big Data*, 144–149. https://doi.org/10.1145/3695080.3695105

DPT. (2020). *Parliament - Parlement on steam.* NFB - ONF; Library of Parliament - Bibliothèque du Parlement. Steam.https://store.steampowered.com/app/1232920

Emissive, & HTC VIVE. (2019). Mona Lisa: Beyond the glass. *Steam.* https://store.steampowered.com/app/1172310

Glitch Studios. (2020). Roald Amundsen's house. *Steam.* https://store.steampowered.com/app/1435700

Goto, K., Kotani, M., & Toyama, M. (2018). 3D Virtual museum generation by using SuperSQL. *International Journal of Pervasive Computing and Communications, 14*(2), 124–146. https://doi.org/10.1108/IJPCC-D-18-00026

Grande, R., Albusac, J., Vallejo, D., Glez-Morcillo, C., & Castro-Schez, J. J. (2024). Performance evaluation and optimization of 3D models from low-cost 3D scanning technologies for virtual reality and metaverse E-commerce. *Applied Sciences, 14*(14), 6037.

Grolleman, J. (2024). *Steam VR vs OpenXR: Which runtime is best?* Pimax. https://pimax.com/blogs/blogs/steam-vr-vs-openxr-which-runtime-is-best

Hazan, D. S., & Hamaagal, R. (2008). *Musing the metaverse.* In *Proceedings of the CIDOC 2008 Annual Conference.* International Council of Museums (ICOM).

Holfeld, J. (2024, January 10). *On the relevance of the Godot engine in the indie game development industry.* Preprint arXiv. https://doi.org/10.48550/arXiv.2401.01909

HTC. (n.d.). *VIVE Pro 2 specs.* HTC VIVE. https://www.vive.com/us/product/vive-pro2/specs/

Kang, D., Choi, H., & Nam, S. (2022). Learning cultural spaces: A collaborative creation of a virtual art museum using Roblox. *International Journal of Emerging Technologies in Learning (Online), 17*(22), 232–245. https://doi.org/10.3991/ijet.v17i22.33023

Katz, J. E., & Halpern, D. (2015). Can virtual museums motivate students? Toward a constructivist learning approach. *Journal of Science Education and Technology, 24*(6), 776–788. https://doi.org/10.1007/s10956-015-9563-7

Kersten, T., Tschirschwitz, F., & Deggim, S. (2017). *Development of a virtual museum including a 4D presentation of building history in virtual reality. The International Archives of the Photogrammetry, Remote Sensing and Spatial Information Sciences, 42*(2/W3), 361–367. International Society for Photogrammetry and Remote Sensing (ISPRS). https://doi.org/10.5194/isprs-archives-XLII-2-W3-361-2017

Kim, K., Kwon, O., & Yu, J. (2023). Evaluation of an HMD-based multi-sensory virtual museum experience for enhancing sense of presence. *IEEE Access, 11*, 100295–100308. IEEE Access. https://doi.org/10.1109/ACCESS.2023.3311135

Kiourt, C., Koutsoudis, A., & Pavlidis, G. (2016). DynaMus: A fully dynamic 3D virtual museum framework. *Journal of Cultural Heritage, 22*, 984–991. https://doi.org/10.1016/j.culher.2016.06.007

Lee, H., Jung, T. H., tom Dieck, M. C., & Chung, N. (2020). Experiencing immersive virtual reality in museums. *Information & Management*, *57*(5), 103229. https://doi.org/10.1016/j.im.2019.103229

Lepouras, G., Katifori, A., Vassilakis, C., & Charitos, D. (2004). Real exhibitions in a virtual museum. *Virtual Reality*, *7*(2), 120–128. https://doi.org/10.1007/s10055-004-0121-5

Li, Y., Yu, L., & Liang, H.-N. (2021). CubeMuseum: An augmented reality prototype of embodied virtual museum. *2021 IEEE International Symposium on Mixed and Augmented Reality Adjunct (ISMAR-Adjunct)*, 13–17. https://doi.org/10.1109/ISMAR-Adjunct54149.2021.00014

Loaiza Carvajal, D. A., Morita, M. M., & Bilmes, G. M. (2020). Virtual museums. Captured reality and 3D modeling. *Journal of Cultural Heritage*, *45*, 234–239. https://doi.org/10.1016/j.culher.2020.04.013

Lombardo, V., Lauro, V., Murtas, V., & Goud, S. (2023). Merging archaeological site recreation and museum exhibition. In L. Holloway-Attaway & J. T. Murray (Eds.), *Interactive storytelling* (pp. 68–84). Springer Nature Switzerland. https://doi.org/10.1007/978-3-031-47658-7_6

Lu, J., Xiao, X., Xu, Z., Wang, C., Zhang, M., & Zhou, Y. (2022). The potential of virtual tourism in the recovery of tourism industry during the COVID-19 pandemic. *Current Issues in Tourism*, *25*(3), 441–457. https://doi.org/10.1080/13683500.2021.1959526

Maddy, J., & Narman, H. S. (2023). Virtual Reality Museum Application for the Arts. *2023 IEEE Integrated STEM Education Conference (ISEC)*, 286–292. https://doi.org/10.1109/ISEC57711.2023.10402270

Marín-Morales, J., Higuera-Trujillo, J. L., Greco, A., Guixeres, J., Llinares, C., Gentili, C., Scilingo, E. P., Alcañiz, M., & Valenza, G. (2019). Real vs. Immersive-virtual emotional experience: Analysis of psycho-physiological patterns in a free exploration of an art museum. *PLOS ONE*, *14*(10), e0223881. https://doi.org/10.1371/journal.pone.0223881

MOR Museum Inc. (2020). Museum of other realities. *Steam*. https://store.steampowered.com/app/613900

Mozumder, M. A. I., Sheeraz, M. M., Athar, A., Aich, S., & Kim, H.-C. (2022). Overview: technology roadmap of the future trend of metaverse based on IoT, Blockchain, AI Technique, and Medical Domain Metaverse Activity. *2022 24th International Conference on Advanced Communication Technology (ICACT)*, 256–261. https://doi.org/10.23919/ICACT53585.2022.9728808

Mystakidis, S. (2022). Metaverse. *Encyclopedia*, *2*(1, 1), 486–497. https://doi.org/10.3390/encyclopedia2010031

OpenVR API Documentation. (n.d.). *GitHub*. Retrieved November 21, 2024, from https://github.com/ValveSoftware/openvr/wiki/API-Documentation

Pirbazari, A. G., & Tabrizi, S. K. (2022). RecorDIM of Iran's cultural heritage using an online virtual museum, considering the coronavirus pandemic. *Journal on Computing and Cultural Heritage*, *15*(2), 37:1–37:14. https://doi.org/10.1145/3500925

Ritterbusch, G. D., & Teichmann, M. R. (2023). Defining the metaverse: A systematic literature review. *IEEE Access*, *11*, 12368–12377. IEEE Access. https://doi.org/10.1109/ACCESS.2023.3241809

Saparamadu, P. V. I. N., Sepasgozar, S., Guruge, R. N. D., Jayasena, H. S., Darejeh, A., Ebrahimzadeh, S. M., & Eranga, B. A. I. (2025). Optimising contract interpretations with large language models: A comparative evaluation of a vector database-powered Chatbot vs. ChatGPT. *Buildings*, *15*(7), 1144.

Sargazi Moghadam, T., Darejeh, A., Delaramifar, M., & Mashayekh, S. (2024). Toward an artificial intelligence-based decision framework for developing adaptive e-learning systems to impact learners' emotions. *Interactive Learning Environments*, *32*(7), 3665–3685.

Schofield, G., Beale, G., Beale, N., Fell, M., Hadley, D., Hook, J., Murphy, D., Richards, J., & Thresh, L. (2018). Viking VR: Designing a virtual reality experience for a museum. *Proceedings of the 2018 Designing Interactive Systems Conference*, 805–815. https://doi.org/10.1145/3196709.3196714

Shirai, A., Kose, Y., Minobe, K., & Kimura, T. (2015). Gamification and construction of virtual field museum by using augmented reality game "Ingress". *Proceedings of the 2015 Virtual Reality International Conference*, 1–4. https://doi.org/10.1145/2806173.2806182

Silić, M., Džaja, B., Rogulj, R., & Truić, H. (2024, June 13–15). *Optimizing game ready asset creation pipeline: From concept to implementation in unreal engine 5. 6th International Scientific and Professional Conference Contemporary Issues in Economy and Technology* (pp. 732–744). University of Split, University Department of Professional Studies. https://www.croris.hr/crosbi/publikacija/prilog-skup/835073

Skamantzari, M., & Georgopoulos, A. (2016). 3D Visualization for virtual museum development. *The International Archives of the Photogrammetry, Remote Sensing and Spatial Information Sciences*, *XLI-B5*, 961–968. XXIII ISPRS Congress, Commission V (Volume XLI-B5) -12–19 July 2016, Prague, Czech Republic. https://doi.org/10.5194/isprs-archives-XLI-B5-961-2016

Spyrou, O., Hurst, W., & Krampe, C. (2025). A reference architecture for virtual human integration in the metaverse: Enhancing the Galleries, Libraries, Archives, and Museums (GLAM) sector with AI-driven experiences. *Future Internet*, *17*(1), 36.

Styliani, S., Fotis, L., Kostas, K., & Petros, P. (2009). Virtual museums, a survey and some issues for consideration. *Journal of Cultural Heritage*, *10*(4), 520–528. https://doi.org/10.1016/j.culher.2009.03.003

Suber, A., Turley, N., & Fingerhut, M. (2020). LUX SINE. *Steam*. https://store.steampowered.com/app/1229740

ThinQ Digital Media Limited. (2018). *London museum Of water & steam. Steam*. https://store.steampowered.com/app/831040

Tsichritzis, D., & Gibbs, S. (1991). Virtual Museums and Virtual Realities. *Hypermedia and Interactivity in Museums*. First International Conference on Hypermedia & Interactivity in Museums (ICHIM '91) (pp. 17–25). Archives & Museum Informatics. https://www.archimuse.com/publishing/hypermedia/hypermedia.Ch3.pdf

Valve Corporation. (n.d.). *Steam store*. https://store.steampowered.com/

Valve. (2024). *SteamVR on steam*. (n.d.). Retrieved November 21, 2024, from https://store.steampowered.com/app/250820/SteamVR/

Van der Land, S., Schouten, A., & Feldberg, F. (2011). Modeling the metaverse: A theoretical model of effective team collaboration in 3D virtual environments. *Journal of Virtual Worlds Research*, *4*(3), 1–16.

Varvello, M., Ferrari, S., Biersack, E., & Diot, C. (2011). Exploring Second Life. *IEEE/ACM Transactions on Networking*, *19*(1), 80–91. IEEE/ACM Transactions on Networking. https://doi.org/10.1109/TNET.2010.2060351

Vasic, I., Fill, H.-G., Quattrini, R., & Pierdicca, R. (2024). LLM-aided museum guide: Personalized tours based on user preferences. In L. T. De Paolis, P. Arpaia & M. Sacco (Eds.), *Extended reality* (pp. 249–262). Springer Nature Switzerland. https://doi.org/10.1007/978-3-031-71710-9_18

Wawro, A. (2015, April 30). Valve launches new OpenVR SDK to expand SteamVR development. *Game Developer*. https://www.gamedeveloper.com/vr/valve-launches-new-openvr-sdk-to-expand-steamvr-development

Werning, S. (2019). Disrupting video game distribution: A diachronic affordance analysis of steam's platformization strategy. *Nordic Journal of Media Studies*, *1*(1), 103–124. https://doi.org/10.2478/njms-2019-0007

Wu, L., Su, W., Ye, S., & Yu, R. (2021). Digital museum for traditional culture showcase and interactive experience based on virtual reality. *2021 IEEE International Conference on Advances in Electrical Engineering and Computer Applications (AEECA)*, 218–223. https://doi.org/10.1109/AEECA52519.2021.9574421

Xeen Inc. & Sougogiken Co., Ltd. (2023). Japanese cultural property VR museum [Video Game Software]. *Steam*. https://store.steampowered.com/app/2139520

Xu, J., Yang, L., Guo, M., Gao, F., & Yin, Y. (2023). Immersive museum: Design and develop an interactive virtual museum experience. In P.- L. P. Rau (Ed.), *Cross-cultural design* (pp. 152–161). Springer Nature Switzerland. https://doi.org/10.1007/978-3-031-35946-0_13

Zhang, J., Wei, Q., Wei, X., & Fan, M. (2024, October 19). *Exploring the design of virtual reality museums to support remote visitation with older adults*. Preprint arXiv. https://doi.org/10.48550/arXiv.2410.15092

Zidianakis, E., Partarakis, N., Ntoa, S., Dimopoulos, A., Kopidaki, S., Ntagianta, A., Ntafotis, E., Xhako, A., Pervolarakis, Z., Kontaki, E., Zidianaki, I., Michelakis, A., Foukarakis, M., & Stephanidis, C. (2021). The invisible museum: A user-centric platform for creating virtual 3D exhibitions with VR support. *Electronics*, *10*(3, 3), 363. https://doi.org/10.3390/electronics10030363

Żyła, K., Montusiewicz, J., Skulimowski, S., & Kayumov, R. (2020). VR technologies as an extension to the museum exhibition: A case study of the silk road museums in Samarkand. *Muzeológia a Kultúrne Dedičstvo*, *8*(4), 73–93. https://doi.org/10.46284/mkd.2020.8.4.6

8 Framework for Real-Time Robotic Manipulation in the Metaverse Using Digital Twins and Reinforcement Learning

Arkapravo Nandi, Ali Darejeh, Mahdi Bamdad, Angel Ayala, Francisco Cruz, and Samad M. E. Sepasgozar

1 Introduction

The convergence of robotics and immersive digital technologies has introduced a transformative paradigm in automation, simulation, and human–robot interaction. As digital ecosystems evolve, the integration of robotics into the metaverse presents a groundbreaking approach to bridging virtual and physical environments. This chapter explores how robotics can be embedded within the metaverse to establish a direct connection between virtual spaces and real-world robotic systems, with a focus on synchronization, control, and human interaction.

By leveraging robotics and digital twin-based metaverse frameworks, users can remotely operate physical robots through their digital counterparts, enabling seamless synchronization and real-time interaction. The advancement of artificial intelligence (AI), extended reality (XR), and reinforcement learning (RL) has further expanded the potential of metaverse-driven robotic control, supporting intuitive, adaptive, and collaborative modes of human-robot engagement. Understanding how humans interact with robots, and identifying the challenges and opportunities presented by AI and immersive technologies, remains central to advancing this emerging field.

The integration of machine learning, cloud computing, and digital twin technologies has significantly enhanced the remote management and optimization of robotic systems. Within the metaverse, these technologies converge to create scalable, interactive environments where users can manipulate robotic digital twins, issue control commands, and receive real-time feedback—effectively bridging the gap between virtual design and physical implementation. These environments also offer a natural and accessible platform for training in both foundational and advanced robotic skills, particularly for the digital-native generation (Kaarlela et al., 2023).

Unlike traditional robotic simulations confined to isolated systems, digital twins in the metaverse operate within interconnected, multi-user ecosystems that support live data sharing and iterative development. Platforms such as NVIDIA's Omniverse provide high-fidelity virtual settings where robotic agents can be trained and tested safely, significantly accelerating the development of complex behaviors and improving deployment outcomes (Ahmed et al., 2024).

DOI: 10.1201/9781003557555-8

AI-driven predictive analytics further augment these capabilities by forecasting robotic failures, optimizing motion planning, and reducing latency in teleoperation scenarios. RL, in particular, benefits from the simulated nature of the metaverse, where robots can learn through trial and error without real-world consequences. For instance, NVIDIA's Isaac Lab, built on the Isaac Sim engine and powered by Omniverse, streamlines robot training workflows by supporting both reinforcement and imitation learning (NVIDIA Developer, n.d.). By integrating advanced RL algorithms such as Twin Delayed Deep Deterministic Policy Gradient (TD3) and Soft Actor-Critic (SAC), robotic systems in the metaverse can dynamically refine control strategies based on real-time feedback, thereby improving motion efficiency, adaptability, and decision-making (Korayem, Bamdad, & Akbareh, 2010).

The convergence of motion planning, AI, and RL within metaverse frameworks is poised to transform the future of robotics—enabling more sophisticated, flexible, and autonomous systems. This synergy not only shortens development cycles but also unlocks new possibilities for human-robot collaboration across industrial and collaborative domains.

This chapter provides a comprehensive overview of the role of the metaverse in remote robotic control, highlighting key advancements in digital twins, AI-driven analytics, and immersive human–robot interaction. It concludes by presenting an empirical study that proposes a framework for controlling a real-world robot within a metaverse environment.

2 Literature Review

This literature review synthesizes key developments that underpin the design and implementation of intelligent, real-time robotic systems within virtual environments, including those enabled by the metaverse. It begins by outlining the metaverse-robotics framework, which serves as the technological foundation for achieving interoperability, real-time control, and seamless data exchange between physical robots and their virtual counterparts. Building on this, the review explores real-time robotic control mechanisms, focusing on motion planning, latency mitigation, and AI-enhanced decision-making.

The next section investigates user interaction paradigms, particularly the role of virtual reality (VR)-based and avatar-driven interfaces in enabling intuitive control. This is followed by an analysis of real-world applications, highlighting how these technologies are transforming industrial automation and remote training. Finally, the review addresses human-centered design principles, with an emphasis on trust, transparency, and user perception—critical factors for the broader adoption and ethical deployment of metaverse robotics.

2.1 Metaverse-Robotics Frameworks

Metaverse-Robotics Frameworks form the technological backbone of metaverse-integrated robotic systems by enabling seamless interaction between physical robots and immersive virtual environments. They support key functionalities such as real-time control, bidirectional data exchange, simulation, and remote operation—capabilities that are particularly valuable in domains like industrial automation, medical robotics, and hazardous environment interventions, where direct access is limited or unsafe.

To function effectively, metaverse-robotics frameworks must ensure high interoperability, low-latency communication, and adaptive learning capabilities. At their core are two foundational pillars: middleware platforms—such as the Robot Operating System (ROS)—and distributed computing infrastructures, including cloud-based and edge computing architectures. ROS provides the modular, sensor-integrated middleware necessary for synchronizing digital twins with physical robot systems, while cloud and edge computing deliver scalable, responsive computational resources to support real-time interactions across geographically distributed networks.

Together, these components form a dynamic and extensible ecosystem that allows multiple users to collaboratively visualize, operate, and refine robotic behaviors in real time. The following subsections examine the technical foundations of these frameworks, beginning with ROS and web-based integration, followed by an exploration of the cloud and edge computing infrastructures that enable scalable deployment.

2.1.1 ROS and Web-Based Integration

The ROS plays a foundational role in enabling metaverse-driven robotics by acting as middleware for integrating robotic simulations with immersive virtual environments. ROS facilitates modularity, real-time communication, and sensor-based control, which are essential for representing physical robotic systems as digital twins within the metaverse. It allows for the abstraction and management of robotic components—such as actuators, sensors, and controllers—across distributed systems (Koubaa, 2017; Macenski et al., 2022).

One of the key advantages of ROS in the metaverse context is its compatibility with 3D visualization and simulation tools. For example, ROS JavaScript libraries (e.g., roslibjs and ros3djs) enable real-time visualization and manipulation of robotic digital twins through web-based interfaces (Fresnillo et al., 2024). You et al. (2024) introduced a framework that utilizes URDF (Unified Robot Description Format) link/joint models to simulate robotic arm kinematics and joint constraints in real time within a metaverse environment. Such frameworks provide high-fidelity simulations for training, experimentation, and remote manipulation, allowing users to interact with virtual representations of physical robots in a responsive and immersive manner.

Modern web-based robotic frameworks build on these capabilities by incorporating a range of web technologies such as WebSockets, WebRTC, and WebGL (Koubaa, 2015; Krūmiņš et al., 2024). In particular, WebRTC-based streaming plays a critical role in supporting low-latency, bidirectional communication between users and robotic systems. This integration ensures smoother feedback loops for command execution and sensor monitoring, which are essential for real-time control and immersive user experiences. These tools collectively enable the development of browser-based robotic interaction platforms that eliminate the need for dedicated software installations, significantly improving accessibility and scalability (Toris et al., 2015). Researchers and practitioners can deploy robotic control systems directly over the internet, enabling experimentation and operation in metaverse environments—even from devices with limited local computational resources.

Together, ROS and web-based technologies provide a flexible and accessible foundation for metaverse robotics, supporting distributed, real-time, and visually immersive interaction with robotic systems in both research and applied contexts.

2.1.2 Cloud-Based and Edge Computing

Cloud and edge computing infrastructures complement middleware platforms like ROS by addressing the intensive computational demands of real-time robotic control in metaverse environments. These infrastructures enable dynamic scalability, distributed processing, and enhanced responsiveness, all of which are essential for supporting immersive and interactive robotic systems (Groshev et al., 2023; Zhang, 2021).

Cloud-based robotics in the metaverse facilitates centralized data processing, large-scale simulations, and global collaboration. Organizations leveraging cloud computing can deploy and manage robotic systems remotely, significantly reducing hardware dependency and computational overhead at the user level. Chen et al. (2024) demonstrated that cloud-integrated digital twins can enhance predictive control models, resulting in smoother interactions between metaverse platforms and physical robots.

A notable development in this area is the integration of ROS2 with cloud platforms. Robot Operating System 2 (ROS2) is the modern, re-architected version of ROS, specifically designed for secure, real-time, and distributed robotic applications. Unlike its predecessor ROS1, ROS2 offers cross-platform support, enhanced real-time communication through the Data Distribution Service (DDS) protocol, and better scalability—features that make it particularly well-suited for cloud-native and metaverse-based robotic systems (Bonci et al., 2023; Macenski et al., 2022). Cloud-based implementations of ROS2 allow for distributed robotic control through cloud networks while maintaining synchronization with their virtual digital twins. Kaarlela et al. (2023) proposed a hybrid ROS2-cloud integration framework in which real-time sensor data from physical robots is mirrored in the metaverse. This approach enables predictive modeling, remote automation, and collaborative access by multiple users from geographically dispersed locations—critical features for scalable and intelligent robotic systems in virtual environments.

Despite these advantages, cloud-based solutions face challenges such as network latency, data privacy, and bandwidth limitations, which can affect time-sensitive robotic tasks. To address these limitations, edge computing has emerged as a complementary technology that processes data closer to the robot's physical environment.

Unlike centralized cloud systems, edge nodes reduce transmission delays and enable faster system responses—especially critical in domains like robotic-assisted surgery, autonomous vehicles, and industrial automation, where even milliseconds of delay can compromise performance and safety (Zhang et al., 2022).

Recent advances in hybrid cloud-edge architectures offer an optimized balance between scalability and low-latency performance. These systems dynamically distribute tasks between edge nodes and centralized cloud servers, improving fault tolerance, load balancing, and system redundancy across large-scale networks. Furthermore, researchers are exploring federated learning approaches to support decentralized AI training without centralizing sensitive data, enhancing privacy and efficiency in metaverse-integrated robotic applications.

2.2 Real-Time Robotic Control

Building on the technological foundations outlined in the previous section, this part of the review shifts focus from infrastructure to functionality—specifically, how the components of metaverse-robotics frameworks enable real-time robotic control. While middleware platforms and cloud-edge computing architectures provide the infrastructure for synchronization and data exchange, their true value emerges through applications that require precise, adaptive, and low-latency interaction between physical robots and their digital twins.

Achieving such capabilities requires the convergence of advanced technologies that support real-time interaction, predictive decision-making, and intelligent control. This section explores four interrelated pillars that underpin real-time robotic control in metaverse environments: motion planning, latency mitigation, AI-driven predictive control, and sensor fusion. Together, these components ensure seamless synchronization between virtual environments and physical robotic systems, providing the responsiveness, adaptability, and precision required for high-performance operation.

2.2.1 Motion Planning

Motion planning in the metaverse leverages immersive virtual environments as dynamic simulation platforms where industrial robots can be safely developed and tested for precise trajectory generation and optimization (Korayem, Bamdad, & Najafi, 2010). These virtual settings allow developers to refine algorithms and validate robotic behaviors in a controlled, risk-free environment, thereby reducing development costs and deployment risks. Through RL and AI-driven simulations, robots can autonomously acquire motion planning strategies, accelerating innovation and adaptation across various industrial applications. However, the effectiveness of these simulations depends on the continued advancement of motion planning techniques and the establishment of clearly defined operational strategies that ensure seamless integration between virtual models and physical robotic systems (Korayem, Bamdad, & Bayat, 2009).

To enable such integration in real-world scenarios, robust control mechanisms and responsive feedback loops are essential—particularly in domains where precise motion and trajectory planning are critical. This need is addressed through technologies like ROS, which continuously streams various operational parameters, including joint positions, velocities, torques, temperature data, motion commands, and timestamps. This continuous synchronization of robotic status and trajectory data allows robots to operate dynamically and interact effectively within complex virtual environments (You et al., 2024).

A compelling example of these integration and control strategies in action is found in Construction Industry 4.0, where the metaverse functions as a virtual interface layer that connects robotic arms and autonomous systems to enable real-time interaction, control, and optimization across Architecture, Engineering, and Construction (AEC) applications (Sepasgozar, 2021a). Enabled by digital twins—synchronized via real-time data, IoT connectivity, and AI analytics—this setup allows for tasks such as 3D concrete printing to be remotely managed within immersive environments. Operators can visualize, monitor, and adjust operations through virtual site replicas, enhancing precision, safety, and collaboration. These capabilities also support applications in automated manufacturing, human–robot collaboration, and training environments for AEC professionals.

2.2.2 Addressing Latency Issues

One of the biggest challenges in metaverse-based robotic control is latency, which affects real-time interaction between operators and robotic systems. Wireless transmission delays, network congestion, and computational overhead are primary sources of latency that can hinder effective robotic manipulation in virtual environments (Chen et al., 2024; Hatami et al., 2024).

To mitigate these issues, researchers have proposed predictive modelling approaches such as Auto-Regressive Moving Average (ARMA) models to anticipate robotic movements and reduce lag. Chen et al. (2024) demonstrated how ARMA could improve teleoperation accuracy by predicting control signals before they are executed. An alternative approach to latency reduction is the use of 5G networks and low-latency communication protocols to improve data transmission speeds between the metaverse and real-world robots. Recent advancements in network slicing allow different robotic applications to be assigned priority bandwidth, ensuring mission-critical operations (such as industrial safety monitoring) are executed without network interference.

Further advancements in real-time rendering optimization, including edge-based GPU acceleration, can provide low-latency robotic visualization within metaverse applications. This enables more dynamic robotic simulations with higher visual fidelity, reducing perceptual lags for remote operators.

2.2.3 AI-Driven Predictive Control – Deep Reinforcement Learning

AI is playing an increasingly vital role in predictive control models for metaverse-driven robotics. AI-powered predictive analytics and machine learning models can anticipate user intentions, optimize robotic movements, and reduce redundant computations to enhance system efficiency. For instance, AI models trained on historical user interaction data can forecast the next likely robotic action, allowing systems to preload commands and reduce response latency. Lu et al. (2024) demonstrated that such AI-assisted predictive control systems achieved up to a 40% improvement in reaction speed compared to traditional teleoperation frameworks.

A particularly effective subset of these AI-driven approaches involves deep reinforcement learning (DRL), which fine-tunes robotic decision-making through iterative interactions with dynamic environments. DRL models such as TD3 and SAC have been successfully applied to robotic control systems, significantly improving responsiveness and adaptability. When trained on digital twin simulations within the metaverse prior to real-world deployment, these models reduce training time, minimize operational errors, and optimize motion planning. Villani et al. (2023) showed that DRL-enhanced predictive control led to notable improvements in robotic path optimization and obstacle avoidance, making it especially valuable for industrial automation and logistics scenarios.

2.2.4 Sensor Fusion and Multi-Modal Feedback Systems

Sensor fusion is another essential component in achieving real-time robotic interaction in the metaverse. Integrating data from multiple sensors—such as LiDAR, depth cameras, inertial measurement units (IMUs), and haptic feedback devices—allows robotic systems to generate a comprehensive situational awareness model, leading to more precise control and interaction (Uhlenberg, Derungs, & Amft, 2023).

Multi-modal feedback systems, which combine haptic feedback, audio cues, and real-time visual overlays, further enhance operator control, enabling users to receive tactile and auditory feedback that helps them refine robotic manipulation. This approach has been particularly useful in surgical robotics and remote robotic-assisted repair, where high precision and real-time adjustments are required (Caccianiga et al., 2021; Zhongda, 2023).

2.3 VR-Based and Avatar-Based Robotic Control in the Metaverse

This section explores how immersive VR platforms and embodied avatar representations enable intuitive, real-time interaction with robotic systems in virtual environments. The following subsections examine key components of this paradigm, including the advantages of VR-based systems for robotic manipulation, the role of avatar-based control mechanisms in enhancing embodiment and precision, and the importance of human-in-the-loop (HIL) teleoperation for collaborative and responsive decision-making.

2.3.1 VR-Based Systems for Robotic Manipulation

VR-based systems enable users to operate robotic arms, drones, and autonomous systems within immersive simulated environments that closely replicate real-world conditions. These environments leverage intuitive and adaptive input methods—including motion tracking, haptic feedback, and real-time visualization—to deliver a more natural and responsive control experience compared to traditional keyboard or remote-control interfaces (Gani et al., 2022). They also support accessibility features, including hands-free control options, making them adaptable to diverse user needs (Bamdad et al., 2022; Darejeh & Mashayekh, 2023).

One of the key advantages of VR-based teleoperation is its ability to enhance operational precision by allowing engineers and operators to practice robotic manipulation in a simulated space before issuing commands to physical systems. This preparatory layer reduces the likelihood of errors during real-world execution and improves overall task performance (Darejeh et al., 2024a; Pizzagalli et al., 2023). Moreover, VR-based robotic interfaces improve situational awareness through real-time, multi-perspective visualization of robotic tasks, enabling more accurate and informed decision-making. Gesture-based command frameworks further streamline interaction by allowing users to control robotic systems through natural hand movements, thereby minimizing reliance on manual controllers or complex input mechanisms (Buckingham, 2021; Kober et al., 2022).

2.3.2 Embodied Avatar Control for Collaborative Robotic Systems

In addition to VR, avatar-based control introduces a human-centered approach to robotic manipulation, allowing users to operate robotic systems through digital avatars that mirror their physical movements. This method enhances the efficiency of remote operations and supports more natural collaboration in shared virtual environments (Park et al., 2025). Avatar-based control mechanisms provide several advantages, including multi-user interaction, where multiple operators can collaborate in virtual environments to coordinate robotic functions (Dewez, 2021). This is particularly useful in complex industrial operations, emergency response scenarios, and telemedicine applications, where precision and team coordination are crucial (Park et al., 2024).

For example, Villani et al. (2023) introduced the Avatarm system, which allows multiple users to control robots within a shared metaverse environment. This avatar-based teleoperation system enables seamless integration of robotic control into daily industrial and research applications.

Recent developments in full-body motion capture technology have further enhanced avatar-based robotic control by allowing multi-degree-of-freedom (DOF) robotic arms and humanoid robots to be controlled using an operator's entire body movements. This facilitates more natural and immersive robotic interactions, increasing adaptability and reducing the cognitive load associated with teleoperation (Ha et al., 2022).

2.3.3 HIL Teleoperation and Adaptive Oversight

HIL systems play a critical role in metaverse-based robotic control by providing real-time feedback to human operators and enabling them to guide robotic functions through VR and avatar-based interfaces. These systems blend AI-driven automation with continuous human oversight, enhancing the safety, adaptability, and precision of robotic execution (Nertinger et al., 2024). HIL frameworks are particularly valuable in high-stakes environments such as space exploration, disaster response, and remote surgery, where fine-grained control and rapid human judgment are essential (Kumar et al., 2024a, 2024b).

One of the key advantages of HIL teleoperation lies in its error correction capabilities: when AI algorithms face uncertainty or unpredictable scenarios, human operators can intervene to adjust or override system actions. This hybrid model leverages the speed and consistency of machine learning with the contextual reasoning and decision-making strengths of human cognition, resulting in more robust and reliable performance across complex industrial and medical tasks (Zhang, Fernandez-Ayala, & Dimarogonas, 2024).

2.4 Applications of Metaverse Robotics in Industrial Automation

This section highlights how immersive technologies and digital twins are transforming traditional industrial workflows. It examines key use cases, including robotic systems in manufacturing, the integration of AI and DRL for intelligent automation, and the use of virtual environments for remote maintenance and industrial training.

2.4.1 Smart Manufacturing With Digital Twins and Collaborative Robots

Metaverse robotics revolutionizes industrial automation by enabling real-time monitoring, simulation, and training. Wang, Wang, and Zhang (2022) demonstrated that digital twins of robotic arms could be used to optimize industrial assembly line workflows, improving efficiency and reducing operational errors. These digital twins provide predictive insights into system performance, allowing manufacturers to simulate potential bottlenecks before they occur in physical production lines.

Virtual factory simulations powered by metaverse robotics also support remote monitoring and control, enabling experts to oversee production lines from any location. These systems facilitate the implementation of predictive maintenance strategies, allowing issues to be identified and addressed before they lead to failures. Such integrations enable automated fault detection, reduce unplanned downtime, and enhance the overall resilience and efficiency of manufacturing operations (Dolgui & Ivanov, 2023; Xie et al., 2024a, 2024b).

Additionally, metaverse-driven collaborative robots (cobots) enhance human–robot interaction in industrial settings by operating alongside human workers within shared virtual environments. In these simulated settings, cobots dynamically adjust their movements in response to real-time changes in the virtual environment, thereby improving safety protocols and coordination strategies before physical deployment. For example, in virtual assembly lines, cobots can detect a worker's proximity or gestures and adapt their path or pace accordingly to prevent collisions and streamline task sequencing—helping to reduce downtime and optimize collaborative workflows during planning or training phases (Li et al., 2024; Liu et al., 2024; Upadhyay et al., 2024).

2.4.2 *Advantages of AI and DRL in Industrial Robotics*

The integration of AI with metaverse robotics enables intelligent, adaptive task execution across virtual industrial environments. For instance, Fujimoto et al. (2018) demonstrated the use of TD3 to refine robotic decision-making, resulting in significantly improved accuracy during simulated task execution. By incorporating DRL algorithms such as TD3 and SAC, robotic systems within industrial metaverse platforms gain the ability to autonomously optimize virtual grasping, material handling, and path planning—enhancing adaptability to dynamic, simulated production conditions. These self-learning capabilities allow robots to iteratively improve their performance based on past outcomes within digital twin environments (Al Zami et al., 2025).

Additionally, AI-powered computer vision systems embedded in virtual metaverse robotics can detect defects in digital product models in real time, contributing to improved quality assurance and reduced rework during pre-deployment testing (The et al., 2023). For example, in a smart factory simulation, a DRL-trained virtual robotic arm can adjust its grip strength and movement trajectory in response to the simulated properties of digital objects during packaging tasks. At the same time, virtual computer vision systems inspect product surfaces for simulated defects and notify remote operators or trigger automated corrections—streamlining the virtual workflow before physical implementation and minimising potential disruptions in real-world production.

2.4.3 *Virtual Training and Remote Maintenance in Industrial Settings*

The metaverse has transformed industrial maintenance and workforce training by offering virtualized training environments where workers can interact with robotic systems without requiring direct access to physical hardware. Augmented reality (AR) and VR simulations allow technicians to practice troubleshooting and repair procedures in digital twins before applying them to real-world machinery (Bordegoni & Ferrise, 2023).

Research by Daling et al. (2023) highlights that VR-based training modules significantly improve retention rates and reduce errors in complex assembly tasks compared to traditional on-site training. By implementing industrial training within the metaverse, workers gain hands-on experience in risk-free, immersive environments, leading to improved efficiency in real-world applications.

Additionally, remote robotic maintenance is becoming more feasible due to metaverse integration. Engineers can use digital twins of industrial equipment to diagnose and resolve mechanical failures remotely, reducing the need for on-site visits and minimizing production downtime (Oppermann et al., 2023). AI-powered predictive maintenance models further enhance this capability by forecasting potential malfunctions before they occur.

2.5 Human-Centered Design Approaches in Robotic Interaction

As robotic systems become increasingly integrated into virtual environments, human-centered design has emerged as a critical consideration in shaping effective and ethical human–robot interaction. This section explores key factors that influence user acceptance and engagement, including the role of anthropomorphic design in building trust, the importance of social presence in collaborative virtual spaces, ethical concerns around transparency and explainability, and how user perceptions vary across industrial applications.

2.5.1 Trust, Anthropomorphism, and User Perception in Metaverse Robotics

Trust is a critical factor influencing the adoption of robotic systems, particularly in immersive environments such as the metaverse. Human-centered design plays a key role in fostering this trust by making robotic interactions more intuitive, engaging, and relatable. One widely used strategy is anthropomorphic design, which involves integrating human-like attributes—such as facial expressions, voice modulation, and natural movement patterns—into robotic systems (Roesler, Manzey, & Onnasch, 2021). Research has shown that such features positively influence user perceptions. For example, Wittmann (2024) found that robots exhibiting human-like gestures and facial cues were perceived as more reliable and easier to interact with. Similarly, Paliga (2023) observed that users are more inclined to collaborate with robots that display predictable, natural movements, as these reduce cognitive load and enhance interaction fluency.

However, anthropomorphism must be applied with caution. The Uncanny Valley Hypothesis suggests that robots that appear almost—but not entirely—human can evoke discomfort or unease among users. Oksanen and Savolainen (2025) stress the importance of balancing familiarity with functional clarity when designing metaverse-integrated robotic agents to ensure trust is cultivated without triggering negative emotional responses. Achieving this balance is essential for developing robotic systems that are both trustworthy and psychologically acceptable in immersive, user-centered environments.

In industrial applications, where robots are often perceived as purely functional tools, user trust can be more difficult to establish. These systems typically lack emotional intelligence, which can make human–robot collaboration challenging (Kumar et al., 2024a, 2024b). Studies have shown that integrating anthropomorphic traits such as voice modulation and adaptive motion planning not only enhances user engagement but also reduces operator fatigue (Nderitu, 2023). This is particularly important in high-stakes domains such as healthcare, manufacturing, and logistics, where safety and task efficiency are paramount.

To further support trust and adoption in these contexts, adaptive trust-building mechanisms should be implemented to facilitate a smoother transition from traditional industrial automation to metaverse-based robotic operations. These mechanisms work by gradually enhancing user familiarity and confidence, which are crucial for effective engagement with advanced robotic systems (Chang & Hasanzadeh,

2024; Halasz & Buhnova, 2024). Examples include gamified training modules that simulate real-world tasks, interactive onboarding experiences that progressively introduce system capabilities, and personalized robotic learning curves tailored to users' experience levels and cognitive preferences. By reinforcing the benefits of anthropomorphic design, such strategies help bridge the gap between functional efficiency and human-centered interaction while addressing the complex demands of industrial environments.

2.5.2 Enhancing Social Presence in Human–Robot Collaboration

Social presence—the sense of "being with" another entity—is a key dimension of human-centered design in metaverse-based robotics, particularly in applications requiring ongoing human–robot collaboration (Almeida, Menezes, & Dias, 2022; Kim, So, & Wirtz, 2022). This concept becomes especially important in domains such as telemedicine, remote industrial operations, and assistive robotics for elderly care, where the perception of shared space and mutual responsiveness directly influences user trust, comfort, and task engagement.

Villani et al. (2023) examined how metaverse-driven robotic interfaces shape user perceptions of social presence. Their findings suggest that enhanced sensory feedback—such as haptic responses, auditory cues, and adaptive AI behaviors—can strengthen this sense of co-presence by simulating physical interaction and emotional connectedness. These features make virtual collaboration feel more immersive and intuitive, helping to bridge the psychological gap between users and robotic agents.

Building on this, AI-driven emotional recognition and response models further personalize the interaction (Darejeh et al., 2024c). By dynamically adjusting tone, movement, and behavior in response to real-time user feedback, robotic systems can exhibit context-aware empathy and adapt to users' emotional states. Such mechanisms are especially valuable in emotionally sensitive domains like education, healthcare, and customer service, where emotional intelligence plays a pivotal role in shaping user satisfaction, communication quality, and overall task performance (Lu, Huang, & Lee, 2024; Vemuri & Thaneeru, 2023).

Taken together, these developments underscore the importance of designing for emotional and social presence in metaverse-based robotics. By combining sensory feedback with emotionally responsive AI, designers can foster deeper engagement, trust, and collaboration—paving the way for more meaningful human–robot interaction in virtual environments.

2.5.3 Ethical Safeguards for Trust and Social Presence in Metaverse Robotics

While anthropomorphic and emotionally responsive design strategies can enhance user trust and foster meaningful social presence, they also raise complex ethical concerns—particularly around transparency, informed consent, and the potential for overtrust (Corti, Stefano, & Bertolaso, 2023). As robotic agents become increasingly human-like in both appearance and behavior, users may incorrectly assume that these systems possess human-level reasoning, emotional understanding, or moral judgment. This misperception, driven by anthropomorphic cues and socially intelligent responses, can lead to misplaced trust and unrealistic expectations, especially in high-stakes or emotionally sensitive environments (Boada, Maestre, & Genís, 2021).

From an ethical standpoint, designers must carefully balance the benefits of human-like interaction with the need to maintain clarity about a robot's actual capabilities and limitations. In social contexts such as healthcare or education, where emotionally adaptive robots are deployed to support vulnerable populations, it is essential to ensure that users are aware they are engaging with algorithmic agents—not sentient beings. Without clear communication, users may attribute empathy or intentionality to robotic systems that merely simulate these traits, potentially undermining autonomy, accountability, and psychological well-being.

To address these risks, explainable AI (XAI) frameworks are increasingly integrated into metaverse robotics to improve transparency. These systems provide users with interpretable insights into how decisions are made—such as the agent's data sources, confidence levels, and reasoning pathways—thereby supporting more accurate mental models and calibrated trust (Thalpage, 2023). For example, Chen et al. (2024) propose a trust-calibration model where robotic agents disclose real-time decision-making parameters and signal their reliability through clear interface cues. Such mechanisms are critical in avoiding over-reliance, particularly when social presence and anthropomorphic design have made the agent seem more competent or emotionally attuned than it actually is.

Ultimately, the ethical design of AI-driven robotic systems in the metaverse must prioritize transparency, accountability, and user education. By embedding ethical safeguards alongside features that enhance trust and social presence, designers can foster interactions that are not only engaging and intuitive but also honest, safe, and aligned with human values.

In the following section, we present our research project, which aims to develop a real-time metaverse framework that supports a closed-loop interactive system for controlling a robotic arm's end-effector.

3 Empirical Study: A Real-Time Metaverse Framework for Robotic Arm Control

This section presents a real-time metaverse framework we propose, which enables a closed-loop interactive system for virtual control of a robotic arm's end-effector and its guidance toward target positions using RL models. By integrating virtual and physical robotic environments, the framework enhances human–robot interaction, optimizes motion planning, and offers a scalable solution for training and automation in both industrial and educational contexts.

Several related studies have explored similar approaches. Pizzagalli et al. (2023) introduced a VR simulation interface that integrates ROS with Unity, allowing users to virtually control a robotic end-effector and transfer learned trajectories to physical systems. However, their system faces limitations during the planning phase, such as collisions with virtual objects. These require manual replanning, demanding prior knowledge of system kinematics and spatial awareness.

Wang, Wang, and Zhang (2022) proposed a framework based on Deep Deterministic Policy Gradient (DDPG) to optimize robotic motion planning in the metaverse. However, their model relies heavily on sensor data to update the robot's state, without dynamically refining trajectories through learning mechanisms. Bahrpeyma, Sunilkumar, and Reichelt (2023) used Omniverse Isaac Sim to simulate multi-step robotic inspection tasks using the UR10 arm. Their approach optimized positioning in dynamic manufacturing contexts, demonstrating how high-fidelity simulations with realistic physics enable effective testing under real-world constraints such as time pressure and moving targets.

Building upon the foundations laid by previous studies, this research proposes an intelligent execution framework that leverages RL for adaptive and context-aware robotic motion within a real-time metaverse environment. Like Bahrpeyma, Sunilkumar, and Reichelt (2023), who utilized Omniverse Isaac Sim to simulate robotic inspection tasks with the UR10 arm under dynamic manufacturing conditions, our approach applies learning-based strategies to optimize motion planning. However, our framework extends this by tightly integrating both physical and virtual robotic environments to enable real-time interaction and closed-loop control.

We also draw conceptual inspiration from the Avatarm framework introduced by Villani et al. (2023), which enables teleoperation of physical systems via virtual manipulations. Unlike their approach—which enhances immersion by concealing the real-time video feed—our system directly addresses key functional aspects, including trajectory generation, collision avoidance, and motion execution under environmental constraints.

The proposed solution integrates a policy-learning control architecture within a 3D virtual environment developed using Three.js. It features collision-checking mechanisms and angular variation constraints to ensure smooth and safe manipulation of the robotic arm across both simulated and real-world domains. By using RL for real-time decision-making, the system supports continuous adaptation to changing conditions, enabling safer and more efficient execution of tasks.

3.1 *Environment Design*

3.1.1 *Digital Twin and Simulation Setup*

The UR5 robotic arm, as shown in Figure 8.1 (a), serves as the primary system for this study. A digital twin of the robot was developed using Three.js and placed on a virtual table, with a blue object representing the target for the arm's reaching task (Figure 8.1 (b)). The use of a digital twin enables real-time validation of robotic movements and allows for safe experimentation prior to deployment on the physical system. The state representation of the UR5 robotic arm at a given timestep t is expressed as $s_t = (\theta_1, \theta_2, \theta_3, \theta_4, \theta_5, \theta_6, d)$, where θ_1 to θ_6 represent the six joint angles of the arm and *d* represents the Euclidean distance from the 'wrist 3 joint' to the object. Robotic control is governed by the system's ability to perceive and respond to changes in its environment. In this context, the Euclidean distance *d* serves as a key metric for motion planning, allowing the arm to continuously adjust its trajectory toward the target.

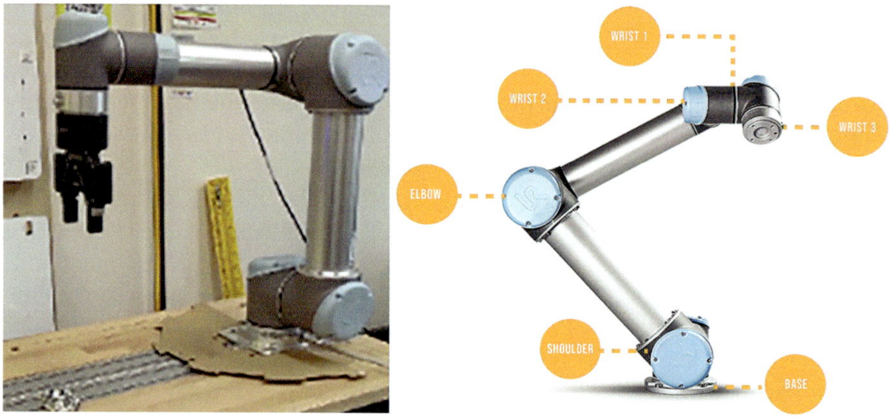

Figure 8.1(a) Robot state representation -UR5 with six joint states.

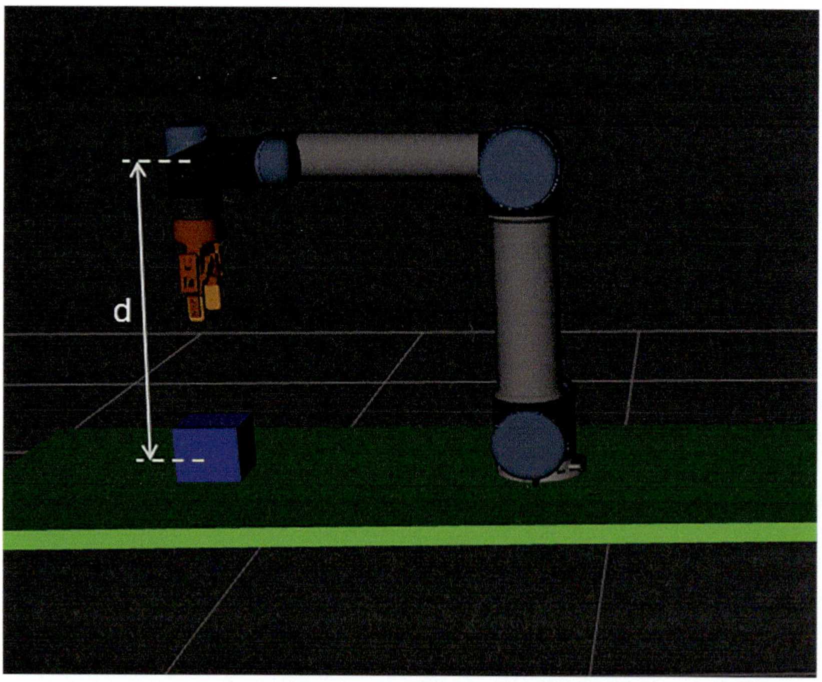

Figure 8.1(b) Digital twin and the object distance in the simulation environment.

3.1.2 State and Action Space Representations

The simulation environment is designed to replicate real-world physics and robotic constraints, enabling the RL agent to train under conditions that closely mirror real-world deployment. This high-fidelity simulation supports the development of robust control policies that can generalize across both virtual and physical contexts (Sepasgozar, 2021b). By integrating virtual representations with RL, digital twins play a central role in enabling the transfer of learned behaviors from simulated environments to real-world applications, including automation, manufacturing, 3D printing, and human–robot collaboration.

Within this simulation-driven framework, the RL model operates over a well-defined state and action space. The state space captures both the robot's physical configuration (e.g., joint angles) and the relative position of the target object, providing the agent with a comprehensive view of its environment. Correspondingly, the action space is defined as $a_t = (\Delta\theta_1, \Delta\theta_2, \Delta\theta_3, \Delta\theta_4, \Delta\theta_5, \Delta\theta_6)$, where $\Delta\theta_i$ denotes the angular displacement applied to each joint, influencing the movement trajectory toward the target object. The RL agent iteratively adjusts joint angles to minimize the distance to the target while avoiding obstacles and self-collisions.

3.1.3 Reward Function

The RL model optimizes robotic movement based on a well-defined reward function that encourages efficient and collision-free reaching behavior. This reward structure encourages the robotic arm to minimize its distance to the target while penalizing inefficient or erratic movements. The RL model progressively refines its trajectory by maximizing the cumulative reward over multiple training episodes.

The reward function at each timestep *t* using the Heaviside step function (*x*) can be defined as $r_t = -d_t + 30H(0.1 - \Delta d_t)H(\mathrm{T} - \mathrm{t})$ where dt is the Euclidean distance at timestep t and a positive reward of 30 is added when d_t is less than 0.1 m.

3.1.4 Execution Function for Robotic Arm Control

Building upon the reward function outlined in the preceding section, this section introduces the step function, a core component of the RL training loop. While the reward function provides feedback on the effectiveness of each action, the step function executes those actions, updates the robot's joint angles, and generates the next state and reward. It governs how the robotic arm interacts with its environment at each training step by ensuring that predicted actions are safely translated into physical movements, enforcing joint limits, incorporating motion planning and collision avoidance, and evaluating termination conditions.

The Step Function Algorithm for Trajectory Execution is detailed below:

Input:

- Action a_t (predicted from Actor Network), object position, current state s_t.

Output:

- Updated state s_{t+1}, reward r_t, and termination flags.

Process:

1. Clipping Action Values: The action a_t is clipped within [−0.2, 0.2] radians to maintain safe joint movements and prevent abrupt transitions.
2. Computing New Joint Angles: The new joint angles are calculated by summing clipped with previous joint angles.
3. Ensuring Joint Limits: The computed angles are constrained within predefined limits to prevent mechanical failures.
4. Executing Motion Planning: The Rapidly exploring Random Tree (RRT) planner algorithm in Move It generates a feasible and collision-free trajectory.
5. Collision Checking: A collision detection framework verifies if the new joint positions result in self-collision or collision with external objects.
6. Computing Next State: The updated state s_{t+1} is derived from the adjusted joint angles and computed trajectory.
7. Evaluating Termination Conditions:
   - If the wrist 3 joint is within 0.1 meters of the object, the episode ends.
   - If the step count reaches 100, the episode is truncated.

Returning Results: The algorithm outputs the new state, reward value, and termination conditions.

This step function was developed and validated by the authors as part of the proposed framework. Its effectiveness in enabling the RL model to perform precise and controlled movements was empirically demonstrated through the simulation results shown in Figures 8.2–8.8, where smooth and efficient trajectory execution was consistently observed.

3.2 RL Models

This section outlines the RL algorithms implemented in the proposed metaverse framework to train the UR5 robotic arm for adaptive motion control. Two widely adopted off-policy algorithms—SAC and TD3—are employed due to their effectiveness in high-dimensional, continuous action spaces. These models enable the robotic arm to learn stable and efficient policies in both deterministic and randomized simulation environments. Importantly, both algorithms are well-suited to the demands of real-time decision-making and generalization across virtual and physical domains, making them ideal for integration within a digital twin-based robotic metaverse framework.

3.2.1 SAC

SAC is an off-policy RL algorithm that balances exploration and exploitation through entropy regularization, promoting stable learning in dynamic environments

(Haarnoja et al., 2018). Its stochastic policy formulation prevents premature convergence and supports continuous adaptation to environmental changes. Within the robotic metaverse framework, SAC allows the UR5 arm to explore multiple action strategies while remaining robust to unexpected changes in the virtual environment. This is especially beneficial in domain-randomized simulations, where the robot must generalize control policies across varying object positions and scene configurations. The algorithm's entropy-driven mechanism encourages broad exploration during training, leading to flexible and transferable behavior during deployment.

3.2.2 TD3

TD3 is an advanced off-policy RL algorithm designed to improve the stability and reliability of policy learning in continuous action spaces (Fujimoto et al., 2018). It addresses the overestimation bias common in Q-value predictions by introducing twin critic networks and target smoothing techniques, which enhance robustness during training and improve generalization across tasks. In the context of the robotic metaverse, TD3 ensures the UR5 arm can develop precise and consistent manipulation strategies even under noisy or variable training conditions. Its improved value estimation supports more accurate motion planning, enabling the robotic arm to perform reaching tasks with greater reliability.

While DDPG models have also been applied in robotic control, they typically require a large number of training samples, making them computationally expensive. In response to this, Xie et al. (2024b) proposed the Context Recognition Network to predict the likelihood of action failure and improve grasping safety. Compared to DDPG-based baselines, TD3 not only achieves higher sample efficiency but also maintains more stable policy learning, making it highly compatible with safety-critical and performance-sensitive tasks in immersive robotic environments.

3.3 Domain Randomization Techniques for Robust Learning

Domain randomization introduces controlled variations in the simulation environment, enabling the RL model to generalize beyond deterministic training conditions (Josifovski et al., 2024). By exposing the model to different disturbances, domain randomization enhances the robustness of learned policies and improves real-world transferability.

3.3.1 Latency Randomization

Latency randomization simulates real-world communication delays that may arise when controlling the robotic arm remotely. Instead of assuming instant feedback, latency randomization introduces random time delays between action execution and state updates. This variation prevents the model from overfitting to an idealized simulation and ensures that it can handle variable response times in real-world applications.

3.3.2 Noise Randomization

Noise randomization introduces sensor inaccuracies into the simulated environment, reflecting real-world uncertainties in robotic perception. Gaussian noise is applied to joint angle readings, target object locations, and collision detection feedback. By training under noisy conditions, the RL model learns to operate effectively despite minor inaccuracies in sensory inputs, ensuring robust real-world performance.

The combination of latency and noise randomization enables the RL model to develop a more resilient policy that accounts for imperfections in robotic control, making it highly adaptable when controlling the robot in a simulated environment and preparing it for reliable real-world deployment.

3.4 Evaluation of RL Algorithms in Simulation

To validate the effectiveness of the RL models, we conducted the experiments in the simulated environment. Figures 8.2 and 8.3 present the movement of the UR5 under different RL strategies. Figure 8.2(a) illustrates the trajectory executed by the UR5 trained using the SAC RL algorithm. The arm smoothly transitions from its initial position to the target, avoiding collisions and self-interference. Figure 8.2(b) demonstrates the trajectory obtained using the TD3 algorithm, which exhibits a more direct path but with occasional deviations due to overestimated Q-values.

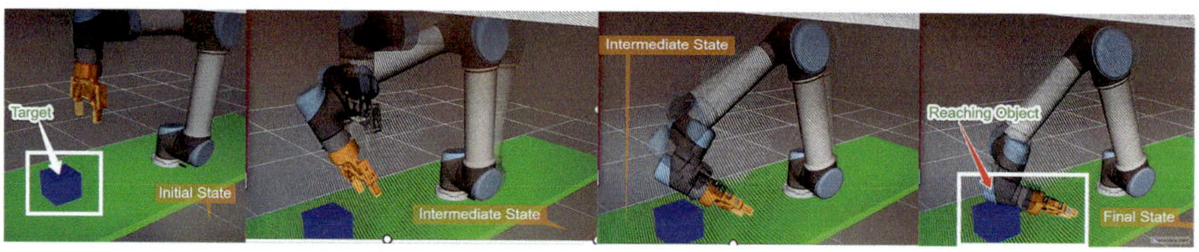

Figure 8.2(a) Motion Planning Using the SAC Algorithm.

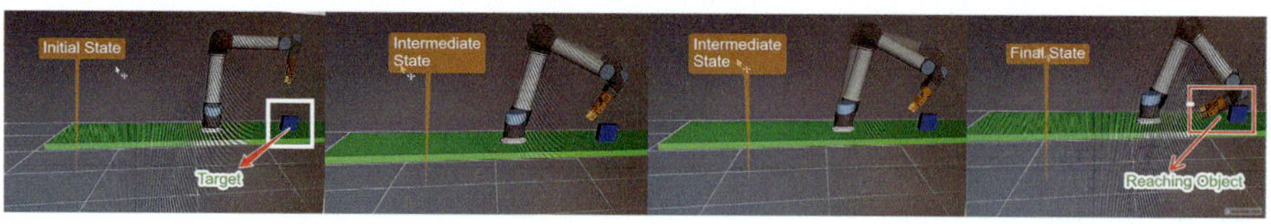

Figure 8.2(b) Motion Planning Using the TD3 Algorithm.

Figures 8.3(a) and 8.3(b) introduce domain randomization effects into the training process. When domain randomization is applied, SAC produces more stable trajectories, while TD3 exhibits reduced variation in execution paths. These results confirm that domain randomization enhances robustness and ensures that the learned policies generalize effectively to real-world conditions.

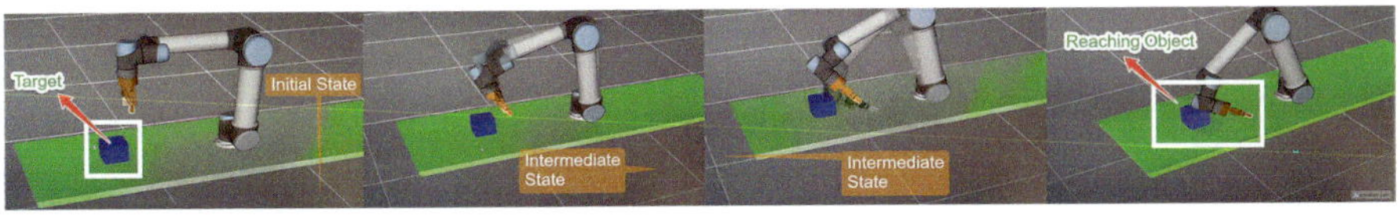

Figure 8.3(a) Application of Domain Randomization to the State Space Using the SAC Algorithm.

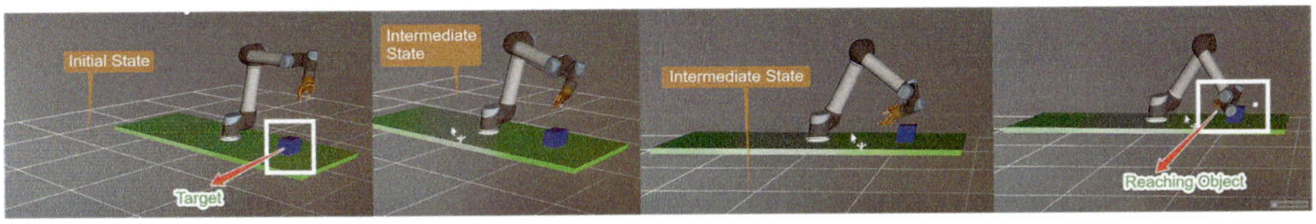

Figure 8.3(b) Application of Domain Randomization to the State Space Using the TD3 Algorithm.

3.4.1 Evaluation Metrics

To quantitatively assess model performance, we employed several key evaluation metrics. These metrics help analyze the effectiveness of RL models in controlling the UR5 under various conditions, including domain randomization. The following subsections discuss the evaluation metrics in detail and provide visual representations through Figures 8.4 and 8.9.

3.4.1.1 REACHING DISTANCE

The reaching distance measures the final Euclidean distance between the robotic arm's 'wrist 3 joint' and the target object. This metric evaluates how precisely the RL model can guide the arm to the desired position. Lower reaching distances indicate better performance, as they reflect the robot's ability to reach the object efficiently.

Figures 8.4 and 8.5 present the trajectory executed by the robotic arm in each episode of evaluation. The plots display the executed path, where the red point in the bottom left of the 3D plots denotes the goal position, and the green point marks the initial or starting coordinates of the wrist 3 joint. The pink point represents the final position of the wrist 3 joint reached by the arm at the end of the simulation. As shown in Figure 8.4, the SAC algorithm results in more diverse transition paths across different episodes, visualized by the presence of multiple distinct trajectories. In contrast, the TD3 algorithm tends to follow a more repetitive and consistent path in each episode. When domain randomization is introduced—illustrated in Figure 8.5—both algorithms demonstrate improved stability in controlling the robotic arm, producing more refined and repeatable trajectories across episodes.

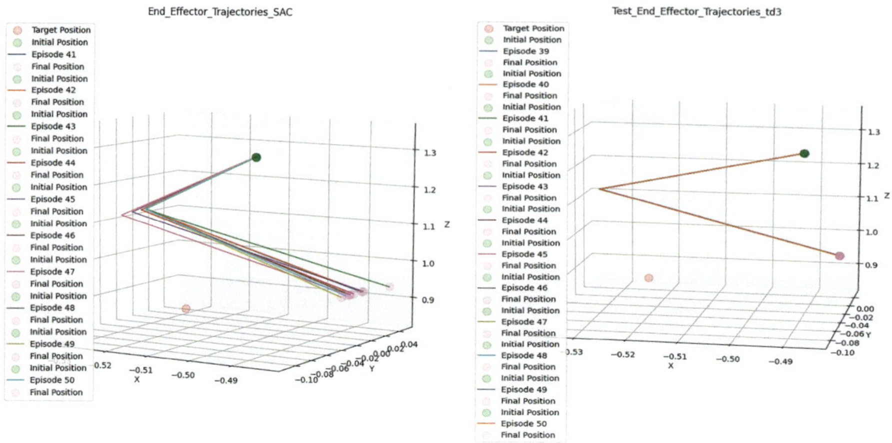

Figure 8.4 Trajectory Planning Guided by the SAC Algorithm (left) and the TD3 Algorithm (right).

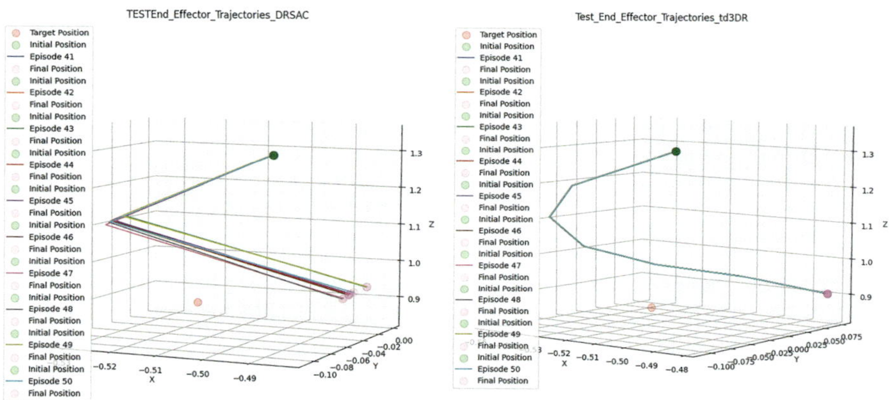

Figure 8.5 Trajectory Planning in a Randomized State Space: Comparison of SAC (left) and TD3 (right) Algorithms.

Finally, Figure 8.6 illustrates the reaching distances achieved by the robotic arm under various learning conditions. The plot compares SAC and TD3 RL models, both with and without domain randomization. The results demonstrate that domain randomization improves reaching accuracy, reducing the variability in final distances. The white circles in the violin plot indicate the median distance values, showing that SAC achieves a more stable performance compared to TD3. The presence of outliers in the TD3 model without domain randomization suggests that some executions resulted in suboptimal trajectories, reinforcing the necessity of domain adaptation techniques.

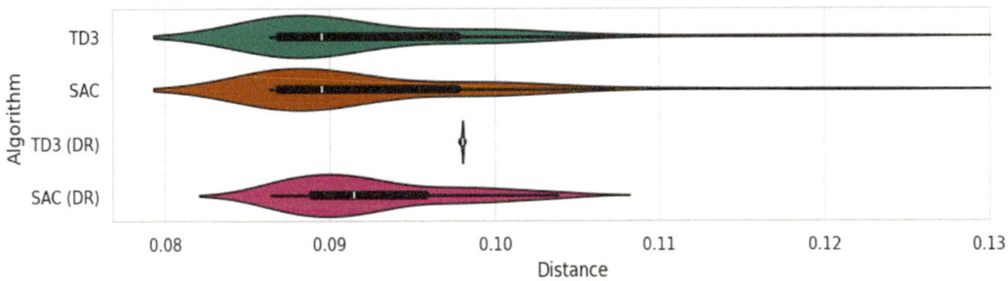

Figure 8.6 Variation in Reaching Distance (in meters) Measured from the 'Wrist 3' Joint.

3.4.1.2 CONTINUITY COST

Continuity cost (in the range between 0 and 1) measures the smoothness of robotic movements by calculating the difference between consecutive joint actions. A lower continuity cost indicates smoother and more natural transitions between movements, reducing the risk of abrupt or erratic behavior. This metric is particularly relevant in applications where jerk-free motion is critical, such as industrial automation and surgical robotics.

Figure 8.7 shows the continuity cost for SAC and TD3 across different training conditions. SAC consistently achieves lower continuity costs, meaning that its actions produce smoother trajectories. TD3, while effective, exhibits slightly more variability in its movement transitions. The introduction of domain randomization helps mitigate high discontinuity, ensuring that RL models generate stable and human-like motions. The results indicate that incorporating randomization techniques significantly improves motion consistency, which is essential in scenarios requiring precise control, such as robotic assembly lines or autonomous navigation in cluttered environments.

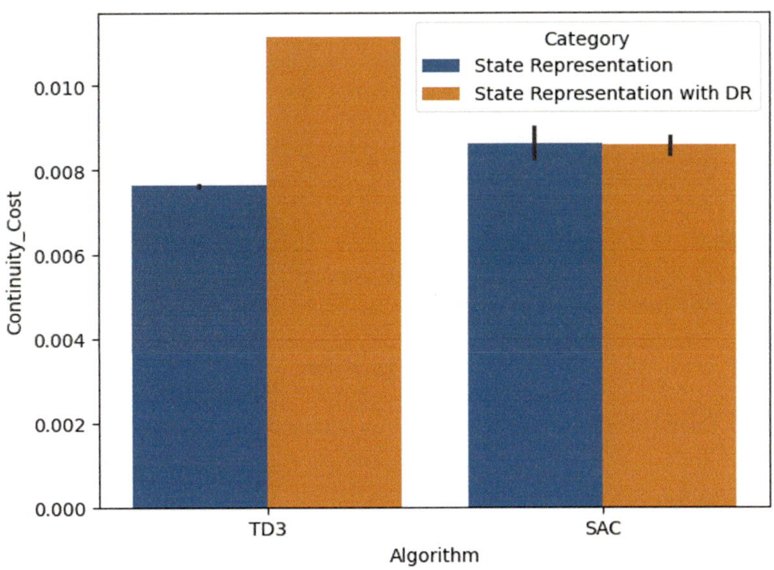

Figure 8.7 Continuity Cost Across Different State Representations.

3.4.1.3 CUMULATIVE EPISODIC REWARD

Cumulative episodic reward is a key metric in RL that sums the total reward collected by the agent over an episode. Higher rewards correspond to more optimal robotic behaviors, where the model efficiently moves the robotic arm toward the target while avoiding collisions and unnecessary detours. This metric helps assess the long-term efficiency of different learning strategies and determines whether models are learning the desired behaviors effectively.

Figure 8.8 presents the cumulative episodic reward achieved by SAC and TD3 across different state representations. The results highlight that models trained with domain randomization consistently achieve higher cumulative rewards, suggesting that these models are better at generalizing to new and unseen scenarios. The graph also shows that SAC tends to achieve more stable and higher rewards compared to TD3, reinforcing its suitability for robotic control tasks. The variance in TD3 reward accumulation without domain randomization indicates that its performance is highly dependent on environmental conditions, making it less reliable in real-world applications.

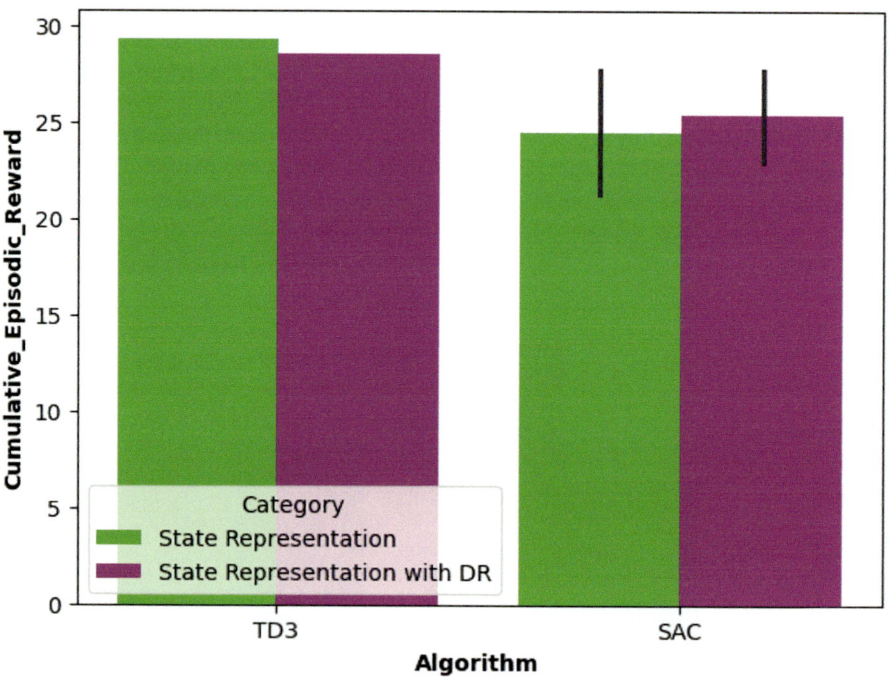

Figure 8.8 Cumulative Episodic Reward Across Different State Representations.

3.4.1.4 EXECUTION TIME AND TRAINING STABILITY

Beyond reward-based metrics, execution time and training stability play crucial roles in assessing model efficiency. Execution time measures the duration required for the robotic arm to complete a reaching task, while training stability evaluates the convergence of RL algorithms over multiple training episodes. Models that converge faster with minimal variance are preferred in real-world applications where computational efficiency and predictability are essential.

Figure 8.9 illustrates the average execution time required for SAC and TD3 models to complete the reaching task across different state representations. The results indicate that SAC generally achieves faster execution times with less variation, while TD3 exhibits slightly longer and more inconsistent execution durations. The presence of execution delays in TD3 without domain randomization suggests that these models occasionally take inefficient paths, requiring additional corrective movements.

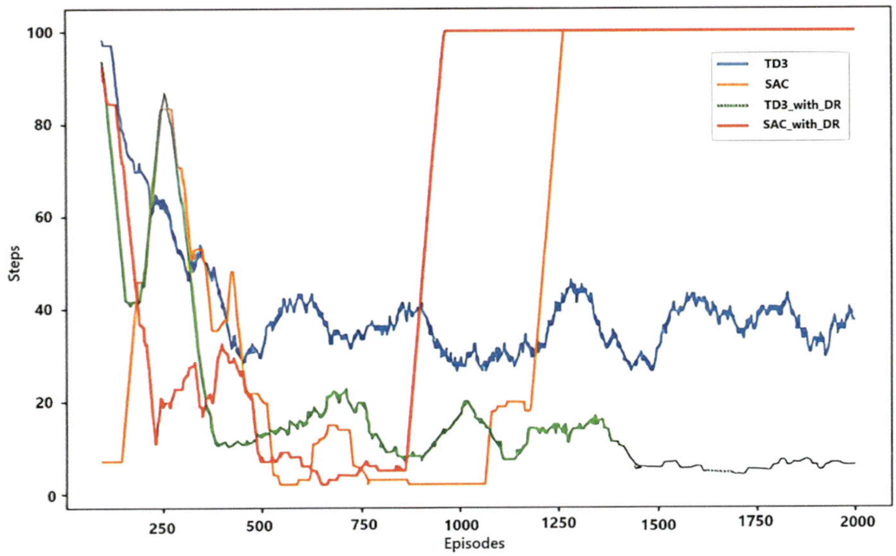

Figure 8.9 Training Episodes with and without domain randomization (DR).

3.4.2 Insight from the Findings

A critical comparison of SAC and TD3 within the metaverse-based robotic framework highlights clear performance differences. While both algorithms are capable of training effective control policies, SAC demonstrates superior stability, adaptability, and precision—particularly when trained with domain randomization.

SAC's entropy-based exploration enables it to maintain robust policy behavior under dynamic and uncertain conditions. Its stochastic formulation leads to smoother and more reliable trajectories, making it better suited for tasks that require continuous adaptation. In contrast, TD3, though architecturally enhanced with twin critic networks and delayed updates, tends to follow more conservative learning patterns and exhibits higher sensitivity to state variability. Domain randomization plays a crucial role in amplifying these differences. SAC shows strong resilience to noise and latency, adapting well to randomized environments and generalizing effectively to new conditions. TD3 benefits from domain randomization to some extent but struggles to internalize variability during training, resulting in less flexible performance during deployment.

SAC also demonstrates better execution efficiency, requiring fewer corrective actions and maintaining trajectory smoothness. This leads to more consistent performance, particularly in high-dimensional control spaces. TD3, while stable in structured settings, shows increased variance under noisy or delayed conditions. Overall, SAC proves to be more reliable and generalizable for complex robotic tasks in immersive, unpredictable environments, making it the preferred choice for real-world metaverse deployments.

3.4.3 A Framework for Integration with VR and the Metaverse

In this section, we propose a real-time robotic control framework informed by the empirical findings presented in the preceding sections. The comparative evaluation of SAC and TD3 under varying state representations and domain randomization strategies revealed key design priorities—namely, smooth motion execution, policy stability, and adaptability to environmental uncertainty. These insights directly informed the architectural decisions of the system.

Although both algorithms were implemented and assessed, the final framework adopts SAC as the primary RL model due to its superior performance in dynamic and noisy environments. SAC demonstrated greater consistency in trajectory control and reward optimization, while TD3, though effective in controlled settings, proved more sensitive to latency and sensor variability—limiting its reliability for real-world deployment in immersive systems.

To operationalize the SAC-trained policies, the model was integrated into a VR-enabled control interface. This environment allows users to interact with the UR5 robotic arm in real time via hand-tracking controllers and a VR headset. Movements performed in the virtual space are translated into joint commands for the robot's digital twin, which are transmitted to the physical system using ROS and WebSocket protocols. The platform also provides live feedback on robot status, trajectory, and execution metrics, enabling intuitive teleoperation and simulation-based validation.

As illustrated in Figure 8.10, the modular structure of the proposed system integrates simulation, RL, middleware, and real-world execution components. Real-time communication across these four core modules ensures synchronization between virtual and physical environments, enabling seamless integration of real-time robotic control with immersive technologies. The details of each module are presented below:

1 **Simulation and Visualization Module (Three.js):** Provides a 3D interface for interacting with the UR5 digital twin. Developed using *Three.js*, it renders object dynamics, detects collisions, and supports user input for both training and deployment.
2 **Reinforcement Learning Engine (SAC):** Hosts the SAC model responsible for generating action policies under domain-randomized training. This engine serves as the core planner and decision-maker for robotic behavior.

3 **Middleware and Communication Layer (ROS + WebSockets):** Facilitates low-latency, bi-directional communication between simulation and the physical UR5 robot. ROS manages control and sensing, while WebSockets support seamless data exchange with the web-based interface.
4 **Digital Twin Synchronization Layer:** Ensures real-time alignment between the virtual and physical robot by continuously updating joint states and performance metrics. This supports safe testing, remote execution, and VR immersion.

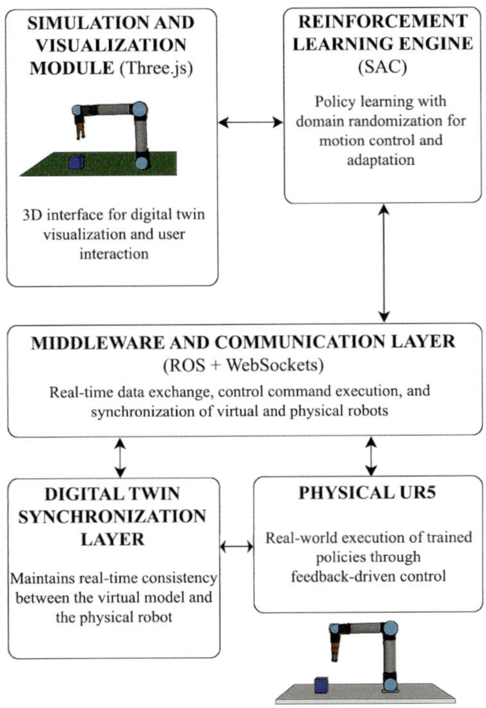

Figure 8.10 System Architecture of the Robotic Control Framework.

Figure 8.11 further illustrates the proposed framework, highlighting the real-time feedback loop between the physical UR5 robotic arm and the VR interface, with ROS and MoveIt managing motion planning and WebXR enabling immersive rendering.

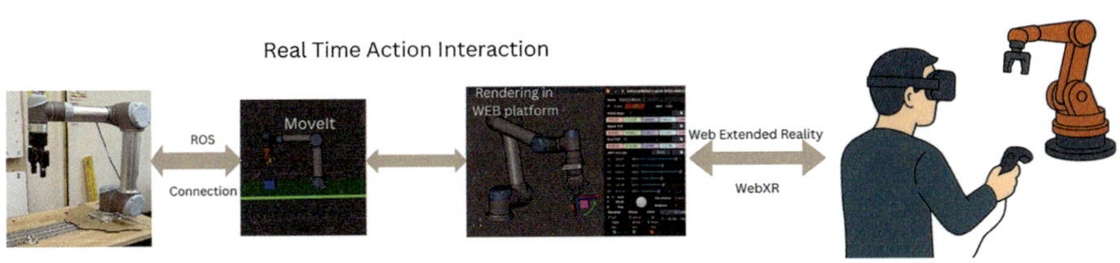

Figure 8.11 Real-Time VR Interaction with the UR5 Robotic Arm.

To enhance scalability and accessibility, the framework is further extended into a metaverse environment that supports multi-user interaction. In this setting, users—represented as avatars—can collaboratively engage in training, trajectory evaluation, or remote supervision. This extension enables immersive, distributed, and socially interactive robotic learning for both educational and industrial applications.

4 Discussion

The integration of robotics into virtual environments, particularly the metaverse, marks a paradigm shift in how autonomous systems are trained, deployed, and managed. Our Real-Time Human–Robotic Arm Interaction framework exemplifies this convergence by implementing a closed-loop metaverse system in which RL, digital twins, and immersive interfaces work cohesively to enable real-time robotic control. This synthesis of capabilities offers new possibilities for education, industrial automation, and human–robot interaction.

RL within metaverse platforms builds upon a growing body of research advocating for safe, scalable, and adaptable training environments. As demonstrated in NVIDIA's Isaac Sim and Isaac Lab (Ahmed et al., 2024; NVIDIA Developer, n.d.), RL algorithms such as SAC and TD3 benefit from the high-fidelity simulation of dynamic environments. Our comparative evaluation affirms that SAC outperforms TD3 in terms of reaching precision, trajectory smoothness, and execution time. These findings align with earlier studies that emphasize SAC's entropy-based exploration strategy as more suitable for high-dimensional robotic tasks, particularly in uncertain environments (Fujimoto et al., 2018; Haarnoja et al., 2018). By introducing domain randomization—including noise and latency—the trained SAC policy achieved stronger generalization to real-world deployment, consistent with insights from Josifovski et al. (2024) regarding the importance of variability during training.

These capabilities are grounded in the implementation of digital twins, which serve not merely as visual representations but as interactive control layers enabling safe experimentation and real-time feedback. Our framework's integration of angular constraints and collision detection within the virtual model allowed RL agents to train under realistic mechanical limitations. This dual-validation approach reflects the evolution of digital twins from static simulators to dynamic, adaptive systems—an advancement supported by work from You et al. (2024) and Kaarlela et al. (2023). In scenarios where safety and accuracy are critical, such as industrial manufacturing or surgical robotics, the ability to simulate and test before execution will be indispensable.

To support these capabilities in real-time settings, the framework employs a modular architecture designed for responsiveness, scalability, and real-time interaction. It leverages Three.js for 3D simulation and cloud-integrated ROS for robotic control. While our current implementation relies on centralized computation, the architecture is designed to support future integration of edge computing to mitigate latency in distributed or bandwidth-constrained environments.

This forward-compatible design reflects emerging trends in hybrid cloud-edge robotic systems (Bonci et al., 2023; Chen et al., 2024; Macenski et al., 2022) and enables flexible collaboration across virtual and physical domains.

Building on this technical foundation, the system introduces immersive interaction capabilities through a VR interface that enables intuitive robot teleoperation. The use of hand tracking significantly improves usability, situational awareness, and training efficiency. Compared to conventional control interfaces, this approach supports more natural, gesture-based manipulation and reduces cognitive overhead—findings that align with prior work by Bamdad et al. (2022), Darejeh and Mashayekh (2023), and Pizzagalli et al. (2023). Our implementation also demonstrated the importance of tight integration between the VR layer and the ROS/WebSocket communication infrastructure, which enables real-time synchronization between the digital twin and the physical UR5 robot.

In extending this interaction model, the framework introduces multi-user support within the metaverse, allowing for distributed collaboration and shared control scenarios. This concept echoes avatar-based control systems such as Avatarm (Villani et al., 2023), where operators interact with robotic systems through full-body avatars. Our framework embeds RL and trajectory planning into this avatar interaction loop, enabling dynamic error correction and adaptive execution. This supports not only efficient task performance but also user trust and safety—critical aspects of human-centered robotic interaction (Darejeh, Marcus, & Sweller, 2022; Paliga, 2023; Wittmann, 2024).

The applicability of our framework spans industrial automation and educational innovation. In manufacturing, the ability to simulate workflows and execute robotic actions remotely reduces physical risk, enables predictive maintenance, and improves process optimization. These advantages are consistent with recent applications of digital twins in smart factories (Dolgui & Ivanov, 2023; Wang, Wang, & Zhang, 2022; Xie et al., 2024a, 2024b). Meanwhile, in education, the framework provides immersive, remote-access learning environments where students can interact with robotic systems safely and meaningfully. This is especially valuable in domains like 3D printing, construction robotics, or engineering education, where proximity to machinery is often hazardous. As shown in studies by Daling et al. (2023), Darejeh (2023), and Darejeh et al. (2024a, 2024b), such environments significantly improve skill retention, reduce learner anxiety, and expand access to hands-on training.

Despite its strengths, the system also faces limitations. Latency and network instability remain critical challenges, particularly in multi-user deployments or remote execution scenarios. While our implementation of latency randomization improved model robustness, future enhancements could benefit from 5G integration, network slicing, or federated learning to further mitigate these constraints (Hatami et al., 2024). Additionally, the current framework is primarily designed for reaching tasks with a robotic arm. Expanding it to support manipulation, collaborative lifting, or pathfinding in cluttered environments would improve its generalizability and application scope.

Looking ahead, advancing the framework with capabilities such as user intent recognition, emotional state modelling, and semantic understanding could significantly enhance the quality of interaction—particularly in educational and assistive applications. When combined with RL, as suggested by Sargazi Moghadam et al. (2024) and Lu et al. (2024), these features can enable emotionally intelligent robotic agents that dynamically adapt to users' goals, behaviors, and affective cues. To complement this adaptivity, integrating XAI and trust-calibration mechanisms—as proposed by Thalpage (2023) and Chen et al. (2024)—will be crucial for ensuring transparency, fostering trust, and preventing overreliance on autonomous agents. As immersive metaverse systems continue to evolve and become embedded in real-world training, healthcare, and automation settings, these innovations will be key to developing interactions that are not only intelligent but also ethically grounded, user-centric, and socially meaningful.

5 Conclusion

This chapter has outlined a forward-looking approach to integrating RL, digital twins, and immersive technologies into a cohesive framework for robotic manipulation. Rather than viewing virtual environments as mere simulation tools, the proposed system positions them as dynamic, interactive platforms for real-time control, experimentation, and collaboration.

Beyond the technical implementation, the framework reflects a broader shift toward scalable, decentralized, and human-centered robotics. By aligning virtual and physical interactions through a closed-loop architecture, it opens up new opportunities for remote training, education, and industrial automation—particularly in scenarios where safety, accessibility, or cost are critical factors.

Looking ahead, this framework offers a strong foundation for advancing multi-agent collaboration within the metaverse, enabling adaptive shared autonomy, and integrating emerging modalities such as brain–computer interfaces and generative AI. As the boundaries between physical and virtual spaces continue to blur, architectures like the one presented here will be central to enabling the next generation of responsive, intelligent, and immersive robotic systems.

References

Ahmed, N., Afyouni, I., Dabool, H., & Al Aghbari, Z. (2024). A systemic survey of the Omniverse platform and its applications in data generation, simulation and metaverse. *Frontiers in Computer Science*, *6*, 1423129.

Al Zami, M. B., Shaon, S., Quy, V. K., & Nguyen, D. C. (2025). *Digital twin in industries: A comprehensive survey*. IEEE Access.

Almeida, L., Menezes, P., & Dias, J. (2022). Telepresence social robotics towards co-presence: A review. *Applied Sciences*, *12*(11), 5557.

Bahrpeyma, F., Sunilkumar, A., & Reichelt, D. (2023, July). Application of reinforcement learning to ur10 positioning for prioritized multi-step inspection in nvidia omniverse. In *2023 IEEE Symposium on Industrial Electronics & Applications (ISIEA)* (pp. 1–6). IEEE.

Bamdad, M., Kirtas, O., Mohammadi, M., & Struijk, L. N. A. (2022, November). Towards the space design for user interface elements of inductive tongue-computer interface. In *2022 E-Health and Bioengineering Conference (EHB)* (pp. 1–4). IEEE.

Boada, J. P., Maestre, B. R., & Genís, C. T. (2021). The ethical issues of social assistive robotics: A critical literature review. *Technology in Society*, *67*, 101726.

Bonci, A., Gaudeni, F., Giannini, M. C., & Longhi, S. (2023). Robot operating system 2 (ros2)-based frameworks for increasing robot autonomy: A survey. *Applied Sciences*, *13*(23), 12796.

Bordegoni, M., & Ferrise, F. (2023). Exploring the intersection of metaverse, digital twins, and artificial intelligence in training and maintenance. *Journal of Computing and Information Science in Engineering*, *23*(6), 060806.

Buckingham, G. (2021). Hand tracking for immersive virtual reality: Opportunities and challenges. *Frontiers in Virtual Reality*, *2*, 728461.

Caccianiga, G., Mariani, A., de Paratesi, C. G., Menciassi, A., & De Momi, E. (2021). Multi-sensory guidance and feedback for simulation-based training in robot assisted surgery: A preliminary comparison of visual, haptic, and visuo-haptic. *IEEE Robotics and Automation Letters*, *6*(2), 3801–3808.

Chang, W. C., & Hasanzadeh, S. (2024). Toward a framework for trust building between humans and robots in the construction industry: A systematic review of current research and future directions. *Journal of Computing in Civil Engineering*, *38*(3), 03124001.

Chen, K., Meng, Z., Xu, X., She, C., & Zhao, P. G. (2024, October). Real-time interactions between human controllers and remote devices in the metaverse. In *2024 IEEE International Conference on Metrology for eXtended Reality, Artificial Intelligence and Neural Engineering (MetroXRAINE)* (pp. 353–358). IEEE.

Corti, L., Stefano, N. D., & Bertolaso, M. (2023). Artificial emotions: Toward a human-centric ethics. *International Journal of Social Robotics*, *15*(12), 2039–2053.

Daling, L. M., Tenbrock, M., Isenhardt, I., & Schlittmeier, S. J. (2023). Assemble it like this!–Is AR-or VR-based training an effective alternative to video-based training in manual assembly? *Applied Ergonomics, 110*, 104021.

Darejeh, A. (2023). Empowering education through eerp: A customizable educational vr escape room platform. In *2023 IEEE International Symposium on Mixed and Augmented Reality Adjunct (ISMAR-Adjunct)* (pp. 764–766). IEEE.

Darejeh, A., & Mashayekh, S. (2023). Mind-object control as a new form of interaction in metaverse for users with motor disability. *Journal of Biomedical Physics & Engineering*, *13*(1), 257.

Darejeh, A., Huynh, L., Ji, X., Marcus, N., & Polly, P. (2024a). Assessing the efficacy of virtual reality-based laboratories versus real-life settings for PCR procedures, within medical science laboratories. In *2024 IEEE International Symposium on Mixed and Augmented Reality Adjunct (ISMAR-Adjunct)* (pp. 489–492). IEEE.

Darejeh, A., Marcus, N., & Sweller, J. (2022). Increasing learner interactions with E-learning systems can either decrease or increase cognitive load depending on the nature of the interaction. *L'Année Psychologique*, *122*(3), 405–437.

Darejeh, A., Moghadam, T. S., Delaramifar, M., & Mashayekh, S. (2024c). A framework for AI-powered decision making in developing adaptive e-learning systems to impact learners' emotional responses. In *2024 11th International and the 17th National Conference on E-Learning and E-Teaching (ICeLeT)* (pp. 1–6). IEEE.

Darejeh, A., Mondal, M., Marcus, N., & Vassar, A. (2024b). Exploring the cognitive load effects of diverse virtual reality interaction methods in interactive educational platforms.

In *2024 IEEE International Symposium on Mixed and Augmented Reality Adjunct (ISMAR-Adjunct)* (pp. 435–438). IEEE.

Dewez, D. (2021). *Towards avatar-based interaction in virtual reality* (Doctoral dissertation, INSA de Rennes).

Dolgui, A., & Ivanov, D. (2023). Metaverse supply chain and operations management. *International Journal of Production Research, 61*(23), 8179–8191.

Fresnillo, P. M., Vasudevan, S., Garcia, J. A. P., & Lastra, J. L. M. (2024). *An open and reconfigurable user interface to manage complex ROS-based robotic systems*. IEEE Access.

Fujimoto, S., Hoof, H., & Meger, D. (2018, July). Addressing function approximation error in actor-critic methods. In *Proceedings of the International Conference on Machine Learning* (pp. 1587–1596). PMLR.

Gani, A., Pickering, O., Ellis, C., Sabri, O., & Pucher, P. (2022). Impact of haptic feedback on surgical training outcomes: A randomised controlled trial of haptic versus non-haptic immersive virtual reality training. *Annals of Medicine and Surgery, 83*, 104734.

Groshev, M., Baldoni, G., Cominardi, L., de la Oliva, A., & Gazda, R. (2023). Edge robotics: Are we ready? An experimental evaluation of current vision and future directions. *Digital Communications and Networks, 9*(1), 166–174.

Ha, E., Byeon, G., & Yu, S. (2022). Full-body motion capture-based virtual reality multi-remote collaboration system. *Applied Sciences, 12*(12), 5862.

Haarnoja, T., Zhou, A., Hartikainen, K., Tucker, G., Ha, S., Tan, J., … & Levine, S. (2018). Soft actor-critic algorithms and applications. *arXiv preprint arXiv:1812.05905*. https://arxiv.org/abs/1812.05905

Halasz, D., & Buhnova, B. (2024, June). Achieving adaptive safety via trust building in autonomous ecosystems. In *Doctoral Conference on Computing, Electrical and Industrial Systems* (pp. 181–191). Springer Nature Switzerland.

Hatami, M., Qu, Q., Chen, Y., Kholidy, H., Blasch, E., & Ardiles-Cruz, E. (2024). A survey of the real-time metaverse: Challenges and opportunities. *Future Internet, 16*(10), 379.

Josifovski, J., Auddy, S., Malmir, M., Piater, J., Knoll, A., & Navarro-Guerrero, N. (2024). Continual domain randomization. In *2024 IEEE/RSJ International Conference on Intelligent Robots and Systems (IROS)* (pp. 4965–4972). IEEE.

Kaarlela, T., Pitkäaho, T., Pieskä, S., Padrão, P., Bobadilla, L., Tikanmäki, M., Haavisto, T., Blanco Bataller, V., Laivuori, N., & Luimula, M. (2023, May). Towards metaverse: Utilizing extended reality and digital twins to control robotic systems. *Actuators, 12*(6), 219. MDPI. https://doi.org/10.3390/act12060219

Kim, H., So, K. K. F., & Wirtz, J. (2022). Service robots: Applying social exchange theory to better understand human–robot interactions. *Tourism Management, 92*, 104537.

Kober, S. E., Settgast, V., Brunnhofer, M., Augsdörfer, U., & Wood, G. (2022). Move your virtual body: differences and similarities in brain activation patterns during hand movements in real world and virtual reality. *Virtual Reality, 26*, 501–511. https://doi.org/10.1007/s10055-021-00588-1

Korayem, M. H., Bamdad, M., & Akbareh, A. (2010). Trajectory optimization of cable parallel manipulators in point-to-point motion. *Journal of Industrial Engineering, 5*, 29–34.

Korayem, M. H., Bamdad, M., & Bayat, S. (2009). Optimal trajectory planning with maximum load carrying capacity for cable suspended robots. *2009 6th International Symposium on Mechatronics and Its Applications (ISMA)*, 1–6. https://doi.org/10.1109/ISMA.2009.5164817.

Korayem, M. H., Bamdad, M., & Najafi, K. (2010). Synthesis of cable driven robots' dynamic motion with maximum load carrying capacities: Iterative linear programming approach. *Scientia Iranica, 17*(3), 229–239.

Koubaa, A. (2015). ROS as a service: Web services for robot operating system. *Journal of Software Engineering for Robotics*, *6*(1), 1–14.

Koubaa, A. (Ed.). (2017). *Robot Operating System (ROS)* (Vol. 1, pp. 112–156). Springer.

Krūmiņš, D., Schumann, S., Vunder, V., Põlluäär, R., Laht, K., Raudmäe, R., …& Kruusamäe, K. (2024). *Open remote web lab for learning robotics and ROS with physical and simulated robots in an authentic developer environment*. IEEE Transactions on Learning Technologies.

Kumar, N., Pal, S. K., Agarwal, P., Rosak-Szyrocka, J., & Jain, V. (Eds.). (2024a). *Human-machine collaboration and emotional intelligence in industry 5.0*. IGI Global.

Kumar, S., Datta, S., Singh, V., Datta, D., Singh, S. K., & Sharma, R. (2024b). *Applications, challenges, and future directions of human-in-the-loop learning*. IEEE Access.

Li, S., Xie, H. L., Zheng, P., & Wang, L. (2024). Industrial metaverse: A proactive human-robot collaboration perspective. *Journal of Manufacturing Systems*, *76*, 314–319.

Liu, L., Guo, F., Zou, Z., & Duffy, V. G. (2024). Application, development and future opportunities of collaborative robots (cobots) in manufacturing: A literature review. *International Journal of Human–Computer Interaction*, *40*(4), 915–932.

Lu, I. C., Huang, J. Y., & Lee, W. P. (2024). An emotion-driven and topic-aware dialogue framework for human–robot interaction. *Advanced Robotics*, *38*(4), 267–281.

Macenski, S., Foote, T., Gerkey, B., Lalancette, C., & Woodall, W. (2022). Robot operating system 2: Design, architecture, and uses in the wild. *Science Robotics*, *7*(66), eabm6074.

Nderitu, J. H. (2023). Mental state adaptive interfaces as a remedy to the issue of long-term continuous human machine interaction. *Journal of Robotics Spectrum*, *1*, 078–089.

Nertinger, S., Kirschner, R. J., Naceri, A., & Haddadin, S. (2024). Acceptance of remote assistive robots with and without human-in-the-loop for healthcare applications. *International Journal of Social Robotics*, *16*(6), 1131–1150.

NVIDIA Developer. (n.d.). *Isaac Sim*. Retrieved from https://developer.nvidia.com/isaac/sim

Oksanen, A., & Savolainen, I. (2025). Trust in emerging technologies and artificial intelligence. In K. J. Rotenberg, S. Petrocchi, A. Levante, & F. Lecciso (Eds.), *Handbook of Trust and Social Psychology* (pp. 184–205). Edward Elgar Publishing Limited. https://doi.org/10.4337/9781803929415.00023

Oppermann, L., Uzun, Y., Buchholz, F., Riedlinger, U., Fuchs, S., Stenzel, H., …& Bau, M. (2023, June). Industrial metaverse? Human-centred design for collaborative remote maintenance and training using XR-technologies. In *International XR Conference* (pp. 195–209). Springer Nature Switzerland.

Paliga, M. (2023). The relationships of human-cobot interaction fluency with job performance and job satisfaction among cobot operators—the moderating role of workload. *International Journal of Environmental Research and Public Health*, *20*(6), 5111.

Park, M., Cho, Y., Na, G., & Kim, J. (2024). Application of virtual avatar using motion capture in immersive virtual environment. *International Journal of Human–Computer Interaction*, *40*(20), 6344–6358.

Park, M., Lee, J., Yang, H., & Kim, J. (2025). *Designing and analyzing virtual avatar based on rigid-body tracking in immersive virtual environments*. IEEE Access.

Pizzagalli, S. L., Bondarenko, Y., Kuts, V., O'Connell, E., Murray, N., & Otto, T. (2023, September). ROS-based augmented and virtual reality path planning interface for industrial robotic arms: A preliminary assessment. In *Proceedings of the 2023 6th International Conference on Robot Systems and Applications* (pp. 38–48). ACM.

Roesler, E., Manzey, D., & Onnasch, L. (2021). A meta-analysis on the effectiveness of anthropomorphism in human-robot interaction. *Science Robotics*, *6*(58), eabj5425.

Sargazi Moghadam, T., Darejeh, A., Delaramifar, M., & Mashayekh, S. (2024). Toward an artificial intelligence-based decision framework for developing adaptive e-learning systems to impact learners' emotions. *Interactive Learning Environments*, *32*(7), 3665–3685.

Sepasgozar, S. M. (2021a). Digital technology utilisation decisions for facilitating the implementation of Industry 4.0 technologies. *Construction Innovation*, *21*(3), 476–489.

Sepasgozar, S. M. (2021b). Differentiating digital twin from digital shadow: Elucidating a paradigm shift to expedite a smart, sustainable built environment. *Buildings*, *11*(4), 151.

Thalpage, N. (2023). Unlocking the black box: Explainable artificial intelligence (XAI) for trust and transparency in ai systems. *Journal of Digital Art Humanities*, *4*(1), 31–36.

The, T. H., Pham, Q. V., Pham, X. Q., Do-Duy, T., & Reddy Gadekallu, T. (2023). AI and computer vision technologies for metaverse. *Metaverse Communication and Computing Networks: Applications, Technologies, and Approaches*, 85–124.

Toris, R., Kammerl, J., Lu, D. V., Lee, J., Jenkins, O. C., Osentoski, S., …& Chernova, S. (2015, September). Robot web tools: Efficient messaging for cloud robotics. In *2015 IEEE/RSJ International Conference on Intelligent Robots and Systems (IROS)* (pp. 4530–4537). IEEE.

Uhlenberg, L., Derungs, A., & Amft, O. (2023). Co-simulation of human digital twins and wearable inertial sensors to analyse gait event estimation. *Frontiers in Bioengineering and Biotechnology*, *11*, 1104000.

Upadhyay, P., Mittal, V., Kriplani, P., & Kautish, S. (2024). Metaverse in healthcare integrated with explainable AI and Blockchain: Enabling immersiveness, ensuring trust, and providing patient data security. In *Metaverse Driven Intelligent Information Systems: Emerging Trends and Future Directions* (pp. 177–194). Springer Nature Switzerland.

Vemuri, N. N., & Thaneeru, N. (2023). Enhancing human-robot collaboration in industry 4.0 with AI-driven HRI. *Power System Technology*, *47*(4), 341–358.

Villani, A., Cortigiani, G., Brogi, B., D'Aurizio, N., Baldi, T. L., & Prattichizzo, D. (2023, May). AVATARM: An avatar with manipulation capabilities for the physical metaverse. In *2023 IEEE International Conference on Robotics and Automation (ICRA)* (pp. 11626–11632). IEEE.

Wang, Y., Wang, C., & Zhang, H. (2022, November). Industrial robotic intelligence simulation in metaverse scenes. In *2022 China Automation Congress (CAC)* (pp. 1196–1201). IEEE. https://doi.org/10.1109/CAC55689.2022.9984552

Wittmann, M. (2024, June). Exploring the effect of anthropomorphic design on trust in industrial robots: Insights from a metaverse cobot experiment. In *2024 21st International Conference on Ubiquitous Robots (UR)* (pp. 118–124). IEEE.

Xie, J., Liu, Y., Wang, X., Fang, S., & Liu, S. (2024a). A new XR-based human-robot collaboration assembly system based on industrial metaverse. *Journal of Manufacturing Systems*, *74*, 949–964.

Xie, S., Meng, Z., Bamdad, M., & Cruz, F. (2024b). Contextual recognition network: Combining DDPG and contextual affordances for robotic safe grasping. *Companion of the 2024 on ACM International Joint Conference on Pervasive and Ubiquitous Computing (UbiComp'24)*, 41–45. https://doi.org/10.1145/3675094.3677581

You, J., Wu, Z., Wei, W., Li, N., & Yang, Y. (2024). Evolution of industrial robots from the perspective of the metaverse: Integration of virtual and physical realities and human–robot collaboration. *Applied Sciences, 14*(14), 6369. https://doi.org/10.3390/app14146369

Zhang, J., Keramat, F., Yu, X., Hernández, D. M., Queralta, J. P., & Westerlund, T. (2022, December). Distributed robotic systems in the edge-cloud continuum with ros 2: A review on novel architectures and technology readiness. In *2022 Seventh International Conference on Fog and Mobile Edge Computing (FMEC)* (pp. 1–8). IEEE.

Zhang, N. (2021). A cloud-based platform for big data-driven cps modeling of robots. *IEEE Access*, *9*, 34667–34680.

Zhang, Y., Fernandez-Ayala, V. N., & Dimarogonas, D. V. (2024, May). Multi-robot human-in-the-loop control under spatiotemporal specifications. In *2024 IEEE International Conference on Robotics and Automation (ICRA)* (pp. 4841–4847). IEEE.

Zhongda, S. (2023). *Ai-enhanced human-machine interfaces using integrated multi-modal sensing and haptic-augmented functions for digital twin and metaverse* (Doctoral dissertation, National University of Singapore (Singapore)).

9 Privacy and Ethics in Educational Metaverse Platforms

Risks, Regulations, and Design Strategies

Rachid Hamadi and Ali Darejeh

1 Introduction

The metaverse stands at the vanguard of the digital revolution (Yang et al., 2024), representing a sophisticated convergence of physical and digital realities that creates immersive virtual environments where users can interact, socialize, learn, and conduct various activities (Li & Yu, 2023). It integrates multiple technologies, including blockchain, virtual reality (VR), augmented reality (AR), and mixed reality (MR) (Yinhui et al., 2024), all contributing to high levels of immersion, engagement, and interactivity in user experiences (Darejeh et al., 2024a, 2024b; Xianing, 2024). The metaverse presents transformative opportunities across various sectors, including education, healthcare, and commerce (Darejeh, 2023; De Felice et al., 2023). However, as metaverse platforms grow in complexity and scale, privacy concerns become increasingly critical and may significantly undermine these systems' efficiency, trust, and widespread adoption.

A major contributor to these concerns is the inherently immersive and data-intensive nature of metaverse interactions. Creating personalized and engaging virtual experiences often requires collecting, processing, and sharing vast amounts of sensitive user data—including biometric identifiers, behavioral patterns, and personal preferences (Alnuaimi & Alawida, 2024; Darejeh & Mashayekh, 2023; Özdemir et al., 2024). While such data is instrumental in enhancing interactivity and tailoring user experiences, it also introduces substantial risks related to privacy, security, and ethical use. Therefore, ensuring robust data protection frameworks and strict adherence to privacy regulations is essential not only for safeguarding users but also for sustaining trust and long-term participation in these immersive environments (Özdemir et al., 2024).

This chapter examines privacy preservation and awareness in educational metaverse platforms. It reviews key literature, identifies critical privacy concerns—such as data collection, user consent, and regulatory compliance—and highlights major ethical challenges. It also presents scenario-based analyses to illustrate common privacy issues and potential solutions through practical examples. Finally, the chapter provides actionable recommendations for developers, policymakers, and users to support the responsible development and use of educational metaverse technologies, with a focus on ethical considerations and data privacy.

DOI: 10.1201/9781003557555-9

2 Literature Review

This section provides a comprehensive review of the existing literature, highlighting key studies and findings that have shaped our understanding of privacy issues in the metaverse. By examining historical contexts and contemporary research, we aim to identify emerging themes and critical insights that inform the development of privacy-conscious metaverse platforms.

2.1 Evolving Privacy Challenges in Digital and Metaverse Environments

The concept of privacy in digital environments has evolved significantly over the past few decades. Privacy concerns were primarily focused on data protection in web services and social media platforms (Hamadi, Paik, & Benatallah, 2007). As the internet became more integrated into daily life, the scope of privacy issues expanded to include the protection of personal information shared online. Early studies, such as those by Westin (1968) on privacy and freedom, laid the groundwork for understanding the importance of privacy in digital interactions. The advent of social media further complicated privacy concerns, as platforms such as Facebook and Twitter (known as X as of July 2023) began collecting vast amounts of user data, leading to increased scrutiny and the development of privacy regulations such as the General Data Protection Regulation (GDPR) in the European Union (Wang et al., 2022).

With the rise of the metaverse, privacy concerns have expanded significantly due to these environments' immersive and interactive characteristics, which involve collecting sensitive data such as biometric information, behavioral patterns, and personal preferences. This shift has sparked a growing body of research to identify and address the unique privacy risks associated with virtual worlds. For instance, Hector, Mendana-Cuervo, and Juan Luis (2024) conducted a systematic literature review outlining major privacy and information security threats in the metaverse and corresponding mitigation strategies. In a similar vein, Alkaeed et al. (2024) explored privacy preservation in artificial intelligence (AI) and extended reality (XR) metaverses, focusing on protecting user data, ensuring secure information handling, and safeguarding user interactions. Their work emphasized the application of advanced technical solutions such as differential privacy, homomorphic encryption, and federated learning.

In addition to these contemporary investigations, foundational work has contributed to our understanding of privacy in digitally mediated environments. Hamadi, Paik, and Benatallah (2007) proposed a conceptual model for privacy-aware web service protocols, highlighting the necessity of incorporating privacy considerations—such as data collection, disclosure, and retention—into service design. More recently, Parlar (2023) addressed the broader challenges of data privacy and security within metaverse applications, advocating for stricter limitations on the types and volume of personal data collected and for transparency in third-party data-sharing practices. From a legal and regulatory perspective, Gulati and Pal (2024) examined the governance of privacy rights in the Indian context, illustrating how privacy, intellectual property, and information technology converge in metaverse environments and reinforcing the need for robust and adaptable legal frameworks.

Emerging technologies further complicate the privacy landscape in the metaverse. Bhattacharya et al. (2021) investigated the intersection of 6G networks and blockchain in AR/VR applications, acknowledging the potential of 6G to enhance immersive experiences through ultra-low latency, high bandwidth, and large-scale connectivity. However, they cautioned that these advancements also heighten privacy risks, as vast amounts of data may be transmitted over untrusted channels. Their study advocated for blockchain integration to enable secure user access control, intelligent resource management, and transparent auditing of transactions. Collectively, these studies underscore the multifaceted nature of privacy in the metaverse and the pressing need for interdisciplinary solutions that combine legal, technical, and ethical considerations.

2.2 *Privacy Themes in Educational Metaverse Platforms*

The preceding review of related work underscores that, while metaverse technologies offer unprecedented opportunities for immersive learning, they also present significant challenges concerning user privacy—particularly in the tracking and analysis of student behavior for educational purposes. Additionally, emotion recognition systems, which aim to address emotional isolation by providing educators with feedback on students' emotional states, raise significant ethical concerns regarding consent, privacy, and algorithmic bias (Darejeh et al., 2024c; Di Dario et al., 2024; Sargazi Moghadam et al., 2024).

Based on the related studies discussed in Section 2.1, six emerging themes concerning privacy in educational metaverse platforms have been identified:

1. Data minimization: Collecting only the necessary data and implementing robust data protection measures is important.
2. User consent and transparency: Ensuring users are informed about data collection practices and obtain explicit consent.
3. Data security: Metaverse platforms must implement strong encryption to protect data during transmission and storage.
4. Privacy by design: Integrating privacy considerations into developing metaverse services from the outset.
5. Privacy Control by Users: Metaverse platforms must respect user autonomy by allowing users to control their data and make informed decisions about their privacy.
6. Regulatory compliance: Metaverse platforms need to comply with evolving privacy regulations such as GDPR, the Family Educational Rights and Privacy Act (FERPA), and the Children's Online Privacy Protection Act (COPPA).

The following section provides an in-depth discussion of each theme.

2.2.1 Data Minimization

Educational metaverse platforms collect a wide range of data to support personalized and interactive learning experiences. This includes personal identifiers such as names, email addresses, and contact details used for user accounts and communication. They also gather educational records, including data on academic performance, course enrollment, and attendance. Additionally, behavioral data is tracked to monitor user interactions, navigation patterns, and engagement metrics within the virtual environment. More advanced platforms may collect biometric data through VR devices, capturing facial expressions, body movements, brain signals, and voice recordings. Furthermore, device and environmental data—such as IP addresses, geolocation, and hardware usage—are often recorded to optimize system performance and user experience (Kovacova, Horak, & Popescu, 2022).

It is essential to adhere to data minimization principles to protect user privacy within educational metaverse platforms. Data minimization begins with collecting only the strictly necessary information to support specific educational objectives. This targeted approach limits the exposure of sensitive information and reduces the potential for misuse or accidental disclosure. Avoiding the accumulation of excessive or irrelevant data is critical, as such information can introduce unnecessary privacy risks and complicate data governance. Also, metaverse platforms should establish clear data retention policies that specify how long collected data will be kept and the criteria for its deletion. Data should only be retained for as long as necessary to fulfill the purposes for which it was collected. Once the data is no longer needed, it should be securely deleted to minimize privacy risks (Trigo Kramcsak & Papakonstantinou, 2024; Pałczyńska, 2023).

To effectively operationalize data minimization principles, educational platforms should establish and maintain structured processes that support compliance. One of the most practical and impactful measures is conducting regular data audits. These audits review the types of data being collected, evaluate whether each data point aligns with the platform's stated purposes, and identify any practices that result in the collection or storage of superfluous or outdated data (Chen et al., 2024). Through this continuous review process, platforms can ensure compliance with data minimization policies and relevant legal standards and demonstrate a proactive commitment to protecting user privacy. Moreover, data audits create opportunities to refine platform design, adjust data workflows, and build a privacy-focused culture within development teams.

2.2.2 User Consent and Transparency

Metaverse platforms must clearly define the purposes for data collection and ensure that data is used solely for those explicitly stated objectives. This principle not only supports regulatory compliance but also helps sustain user trust by preventing unauthorized data repurposing. For instance, data collected to facilitate educational experiences should not be diverted to marketing or unrelated analytics without explicit user consent.

Building on this, transparency in data usage and disclosure is equally essential. Users should be clearly informed about how their data will be handled, including whether it will be shared with third-party partners. This can be accomplished through clear, accessible privacy policies that outline specific data practices. Moreover, maintaining ongoing transparency requires platforms to regularly update users on any changes to these practices and provide straightforward tools for users to access, modify, or manage their privacy settings (Sciolla, 2024).

Before sharing data with third-party partners, metaverse platforms must obtain explicit user consent and ensure that they comply with privacy standards and have appropriate data protection measures in place. Data-sharing agreements should specify the purposes for data use and include data security and user privacy provisions (Zhang et al., 2024). For instance, educational metaverse platforms might share data with external educational tools or content providers, but only under strict data protection agreements. Where possible, anonymize data before sharing it with third parties. Anonymization techniques can help protect user identities while allowing valuable data analysis and insights. This is particularly important when data is used for research or analytics purposes.

2.2.3 Data Security

Metaverse platforms should employ robust encryption techniques to safeguard data both in transit and at rest. This ensures that information remains inaccessible to unauthorized individuals, reducing the risk of data breaches and unauthorized access. Advanced Encryption Standards (AESs), such as AES-256, should be employed to secure sensitive information. Additionally, end-to-end encryption can be used to protect data as it moves between users and servers, ensuring that only the intended recipients can decrypt and access the information. Additionally, as quantum computing advances, current encryption techniques may become vulnerable, necessitating the development of quantum-resistant encryption techniques (Mohammed et al., 2024).

Role-based access controls (RBACs) are also essential for limiting data access to only authorized personnel. By assigning specific roles and permissions to users based on their job functions, RBAC helps prevent unauthorized access to sensitive data. Regularly reviewing and updating access permissions is crucial to maintaining data security. This includes conducting periodic audits to ensure access rights are appropriate and revoking access for users who no longer require it (Truong et al., 2023; Upadhyay et al., 2024).

In addition to encryption and access controls, metaverse platforms should implement comprehensive security measures to mitigate cyber threats. These measures include the deployment of firewalls, intrusion detection systems, and anti-malware software to protect the platform's infrastructure. Conducting regular security assessments and vulnerability testing is crucial for identifying and addressing potential weaknesses. Furthermore, implementing multi-factor authentication (MFA) enhances user authentication by requiring multiple forms of verification before granting access to sensitive data (Yang et al., 2023). By adopting a multi-layered security strategy, metaverse platforms can establish a robust defense against data breaches and other forms of cyberattacks, thereby ensuring a safer and more trustworthy virtual learning environment (VLE).

Another significant challenge in the metaverse is the risk of Distributed Denial-of-Service (DDoS) attacks, as highlighted by Alkaeed et al. (2024). As the metaverse expands and wearable devices become increasingly integrated, malicious actors may exploit their vulnerabilities to create botnets—such as the Mirai botnet—that can flood servers with traffic, resulting in network outages and service disruptions (Arya et al., 2022). In addition, metaverse platforms that integrate blockchain technologies—such as those supporting non-fungible tokens (NFTs)—may face DDoS threats, particularly when certain NFT functionalities operate off-chain due to blockchain limitations. These off-chain processes can become targets for attackers, potentially leading to system downtime and service unavailability. Thus, mitigating DDoS risks is critical to maintaining reliable access to data and services in virtual environments (Ali et al., 2023; Arya et al., 2022).

Overall, the key security challenges in the metaverse arise from the lack of strong user two-factor authentication (Yang et al., 2023), the absence of robust end-to-end data encryption (Mohammed et al., 2024), insufficient access control mechanisms (Truong et al., 2023), failure to ensure secure communication protocols (Kumar et al., 2023), inadequate threat detection systems (Novianti et al., 2024; Nkoro et al., 2024), non-compliance with privacy regulations (Gulati & Pal, 2024; Liu et al., 2023), and the lack of effective incident response procedures (Alkaeed et al., 2024; Pappachan & Bansal, 2024; Wodi, 2023).

2.2.4 Privacy by Design

Metaverse platforms must incorporate privacy considerations into the design and development of their services from the outset. Such a proactive approach enables developers to identify and address privacy risks early in the development process, ensuring that privacy is embedded as a fundamental component of the system rather than treated as an afterthought. This includes implementing data minimization techniques, ensuring user consent mechanisms are in place, and providing users with control over their data (Paneva et al., 2024).

Similar to the practices outlined in the Data Minimization and Data Security sections, regular privacy audits are essential for evaluating compliance with privacy policies and regulatory requirements. These audits involve systematically reviewing the platform's data handling practices, identifying potential privacy risks, and verifying that appropriate safeguards are in place. The insights gained from audit findings should inform continuous improvements in privacy management, addressing any gaps or vulnerabilities. By conducting regular audits, metaverse platforms can not only enhance their data protection strategies but also demonstrate a strong commitment to user privacy, thereby fostering greater user trust (Wang et al., 2022).

In addition to audits, metaverse platforms should engage their staff in ongoing privacy training and awareness programs. Educating employees about best practices for privacy and the importance of protecting user data can help create a culture of privacy within the organization. This includes training on data protection regulations, secure data handling procedures, and recognizing potential privacy threats.

2.2.5 Privacy Control by Users

Metaverse platforms should uphold user autonomy by giving individuals control over their data and enabling them to make informed choices regarding their privacy. This involves providing users with clear and accessible information about data collection practices and options to manage their privacy settings. Users should be able to easily access, modify, and delete their data, ensuring they control their personal information. Transparent communication about data practices helps build trust and empowers users to make informed choices (Amano et al., 2024; Falchuk, Loeb, & Neff, 2018).

2.2.6 Regulatory Compliance

Ensuring regulatory compliance is one of the most critical considerations for educational metaverse platforms, as it brings together all the previously discussed themes to effectively safeguard user privacy and fulfill legal obligations. A thorough understanding of relevant data protection regulations is essential for maintaining user trust, protecting sensitive information, and avoiding legal consequences. The following outlines key regulatory frameworks that govern privacy in educational contexts, including the GDPR, the FERPA, and the COPPA (Moerel, 2024; Raab, 2021; Skowronski, 2022).

GDPR sets strict requirements for data collection, processing, and storage for users within the European Union. Compliance with GDPR involves several key obligations (Frosio & Obafemi, 2025):

- Explicit user consent: Obtain clear and explicit consent from users before collecting and processing their personal data.
- Data portability: Ensure that users can easily transfer their data from one service provider to another.
- Right to be forgotten: Implement mechanisms that allow users to request the deletion of their personal data when it is no longer needed for the purposes for which it was collected.
- Data Protection Impact Assessments (DPIAs): Conduct DPIAs to identify and mitigate privacy risks associated with data processing activities.
- Breach notification: Notify relevant authorities and affected individuals promptly in the event of a data breach.

FERPA is a U.S. federal law that protects the privacy of students' education records. Educational metaverse platforms must comply with FERPA requirements, which include (Garon, 2023):

- Parental and student rights: Provide parents and eligible students with the right to access and review their education records.

- Consent for disclosure: Obtain written consent from parents or eligible students before disclosing personally identifiable information from education records, except in certain permitted situations.
- Recordkeeping: Maintain records of requests for and disclosures of student information to ensure transparency and accountability.

COPPA regulates the collection of personal information from children under 13 in the United States. Metaverse platforms targeting children must adhere to COPPA guidelines, which include (Guhathakurta, 2024):

- Parental consent: Obtain verifiable parental consent before collecting, using, or disclosing personal information from children.
- Privacy policy: Provide a clear and comprehensive privacy policy that explains data collection practices and how personal information is used and shared.
- Data minimization: Limit the collection of personal information to what is reasonably necessary for the child's participation in the online activity.
- Security measures: Implement robust security measures to protect the confidentiality, integrity, and security of children's personal information.

2.3 Challenges in Preserving Privacy in Educational Metaverse Environments

While the six privacy themes outlined above provide a conceptual framework for safeguarding user data in educational metaverse platforms, their practical implementation presents a range of complex challenges. These include technical limitations, regulatory constraints, and the evolving nature of user interaction in immersive environments. The following section explores these real-world challenges in greater detail, highlighting the barriers that must be addressed to ensure effective privacy preservation in educational metaverse settings.

2.3.1 Volume and Variety of Data

As outlined in the previous section, educational metaverse platforms collect unprecedented volumes of multimodal data, including behavioral patterns, biometric identifiers, and device-specific metadata. This diversity complicates data anonymization, as seemingly harmless datasets can reveal sensitive information when correlated across modalities (Kourtesis, 2024). For instance, VR headset motion data and voice recordings might expose users' emotional states or physical disabilities.

The integration of AI-driven analytics further amplifies risks. Machine learning models trained on aggregated datasets could infer protected attributes such as learning disabilities or cultural backgrounds, raising concerns about unintended profiling. Technical limitations in data minimization strategies exacerbate these challenges, as immersive educational experiences often require continuous and high-fidelity data streams to function effectively (Alja'Afreh et al., 2024).

2.3.2 Interoperability

Cross-platform interoperability introduces vulnerabilities through inconsistent privacy protocols. While blockchain-based decentralized identity systems (e.g., self-sovereign identity frameworks) promise secure credential sharing (Kim et al., 2023), implementation disparities between institutions create attack surfaces. For example, a university's metaverse classroom using Ethereum-based authentication might expose user metadata when interacting with corporate training platforms employing alternative blockchain architectures.

Standardization gaps also hinder compliance with regulations such as GDPR. For example, many metaverse ecosystems lack support for granular consent revocation across interconnected services, leaving users vulnerable to data retention loopholes when transitioning between platforms (Kim et al., 2023).

2.3.3 Cybersecurity Threats

Metaverse platforms face growing vulnerability to data breaches that can compromise highly sensitive user information. These threats are increasingly driven by sophisticated attack vectors that exploit metaverse-specific features with alarming precision. For example, avatar hijacking exploits weak authentication protocols, allowing malicious actors to impersonate educators and launch phishing attacks that evade conventional email filters (Kim et al., 2023). Unencrypted biometric streams from AR/VR devices pose serious risks, as they can enable malicious actors to infer sensitive information about users' physical environments and behaviors (Kourtesis, 2024).

Moreover, AI-powered social engineering tactics, such as deepfake avatars generated by Generative Adversarial Networks (GANs) or diffusion models, can manipulate student interactions and erode trust in virtual environments (Alkaeed et al., 2024). The adoption of 5G and 6G technologies further compounds these risks, as the high data throughput and broad exposure surface of millimeter-wave transmissions can increase the vulnerability of metaverse platforms to interception and other cyber threats (Alja'Afreh et al., 2024), potentially allowing attacks to occur before detection systems can respond. These developments underscore the urgent need for comprehensive, multi-layered cybersecurity strategies tailored to the metaverse context.

2.3.4 Ethical Considerations

Balancing user autonomy with data collection practices presents a significant ethical challenge in the metaverse, where immersive environments give rise to complex dilemmas. A key concern related to user autonomy is behavioral manipulation, where platform algorithms designed to maximize engagement have been shown to significantly increase compulsive usage among adolescent users. This raises serious questions about the validity of informed consent, particularly in educational contexts where users may not fully understand how their behavior is being influenced (Kourtesis, 2024).

In parallel, cultural bias in affective computing highlights the risks associated with data collection practices. Emotion recognition systems trained predominantly on Western facial expressions frequently misinterpret emotional states in East Asian users, leading to inaccurate assessments and inequitable outcomes in learning environments (Kourtesis, 2024). These concerns are further complicated by the privacy paradox, as the vast majority of students express concern about their privacy, yet many still willingly provide sensitive biometric data in exchange for personalized learning features (Alkaeed et al., 2024). Collectively, these ethical tensions underscore the need for transparent, user-centered design practices and culturally responsive systems that respect individual autonomy while ensuring responsible data collection and protection.

2.3.5 Regulatory Compliance

Privacy regulations are continuously evolving, and educational metaverse platforms must stay current to ensure compliance with the standards discussed earlier in this chapter, including GDPR, FERPA, COPPA, and other relevant frameworks. Ensuring compliance with privacy regulations across different jurisdictions is complex, especially for platforms with a global user base.

Jurisdictional conflicts pose major challenges for regulatory compliance in educational metaverse platforms. For example, a European university operating a metaverse-based learning environment may be required to simultaneously comply with the GDPR's "right to be forgotten," California's Consumer Privacy Act (CCPA) disclosure obligations, and China's Personal Information Protection Law (PIPL). Recent studies indicate that the complexity of meeting these overlapping requirements can be financially burdensome, with compliance costs consuming a substantial portion of operational budgets for edtech startups (Kim et al., 2023).

The absence of metaverse-specific legislation compels institutions to rely on existing frameworks such as FERPA, which are often insufficient for addressing novel privacy concerns. For instance, FERPA's directory information exceptions do not account for avatar-generated behavioral data, leaving significant gaps in student data protection (Alja'Afreh et al., 2024).

2.3.6 Data Ownership and Control

Determining data ownership in the metaverse is challenging, as users generate vast amounts of data through their interactions. Implementing mechanisms that allow users to control their data—such as data portability and the right to be forgotten—is essential yet complex. Emerging approaches, like blockchain-based smart contracts, offer promising models for user-centric data ownership in educational environments (Kim et al., 2023).

However, implementation challenges persist despite the promise of blockchain-enabled privacy solutions in the metaverse. One key issue is data portability, as cross-chain interoperability limitations restrict students' ability to transfer NFT-based academic achievements across platforms (Dastagir et al., 2022).

Another major concern is the right to erasure, since the immutable nature of blockchain ledgers conflicts with legal mandates for data deletion. While this tension remains unresolved in many systems, emerging cryptographic approaches—such as zero-knowledge proofs—now offer potential workarounds by enabling selective data obfuscation without compromising the integrity of the blockchain (Alja'Afreh et al., 2024; Dastagir et al., 2022).

2.3.7 *Technical Limitations*

Scaling privacy-preserving measures in tandem with the expansion of the metaverse presents a significant technical challenge. Solutions must be both efficient and scalable to manage the growing volume of user data without degrading performance. A key difficulty lies in implementing robust privacy protections without compromising the real-time, immersive experience that defines the metaverse, as many privacy-preserving techniques inherently introduce latency.

For instance, homomorphic encryption introduces significant computational overhead, which can adversely affect VR interactions. This added latency may exceed the commonly cited 200-millisecond threshold necessary to maintain a seamless sense of immersion (Alja'Afreh et al., 2024). While edge computing can reduce this latency, it demands more GPU resources, placing it beyond the financial reach of many underfunded educational institutions (Kim et al., 2023).

2.3.8 *User Biometric Data Protection*

As previously discussed in this chapter, technologies such as AR and VR inherently capture users' biometric data—including facial gestures, eye movements, and body posture—which can be misused if not adequately protected. Designing these technologies with privacy as a core principle is essential to prevent such misuse and ensure user trust. However, biometric safeguards continue to lag behind data collection capabilities.

For instance, while the GDPR classifies gait analysis as sensitive personal data, only a small portion of educational metaverse platforms currently encrypt motion capture streams (Kourtesis, 2024). Emerging solutions such as multimodal anonymization and federated learning show promise, with pilot programs demonstrating significant reductions in re-identification risks through decentralized model training (Alja'Afreh et al., 2024; Alkaeed et al., 2024).

2.3.9 *Identity and Access Management*

Managing digital identities and ensuring secure access control in the metaverse is complex. Unauthorized access to user accounts can lead to identity theft and other privacy breaches. Implementing strong authentication mechanisms and monitoring access patterns are essential to mitigate these risks.

One promising approach to strengthening identity management in this context is the use of Decentralized Identifiers (DIDs) combined with Zero-Knowledge

Succinct Non-Interactive Arguments of Knowledge (zk-SNARKs). These technologies allow for critical functions such as age verification—essential for COPPA compliance—without exposing sensitive personal data like birthdates (Dastagir et al., 2022; Kim et al., 2023). However, despite their potential, most institutions lack the necessary infrastructure to audit third-party verifiers, introducing significant supply-chain vulnerabilities (Kim et al., 2023).

2.3.10 Legal and Intellectual Property Issues

The metaverse introduces a range of complex legal and intellectual property challenges, particularly regarding the ownership of digital assets and the protection of user-generated content. These issues are especially pronounced in educational contexts, where credentials and learning artifacts often span platforms and jurisdictions. For instance, NFT-based credentialing systems operate within jurisdictional grey areas, raising concerns about legal recognition and dispute resolution. To address this, proposed solutions such as blockchain-based timestamping could empower learners to verify the provenance of their achievements across diverse legal systems (Alja'Afreh et al., 2024).

However, resolving such challenges requires more than technical fixes; it calls for adaptive legal frameworks that balance innovation with user rights and accountability. These complexities further underscore the importance of multidisciplinary approaches to privacy and data protection in educational metaverse environments. Emerging solutions—such as quantum-resistant encryption and AI-driven governance frameworks—offer promising directions, but their successful implementation will depend on close collaboration among educators, technologists, and policymakers (Alja'Afreh et al., 2024; Alkaeed et al., 2024; Kim et al., 2023).

3 Scenario-Based Analyses

The following three scenario-based analyses highlight the critical importance of implementing robust privacy measures to safeguard user data and build trust across diverse metaverse applications. In light of the limited availability of real-world data and comprehensive case studies, these hypothetical scenarios serve as a valuable approach for exploring emerging privacy challenges. By simulating realistic contexts and conceptually testing solutions, the analyses offer practical insights for developers, educators, and policymakers seeking to navigate the complex landscape of privacy in metaverse environments.

3.1 Scenario-Based Analysis 1: VLE

This scenario-based analysis examines the privacy challenges and potential solutions associated with a university's implementation of a VLE, including a virtual lab, designed to enhance remote learning experiences. The platform integrated VR technology to create immersive, real-time classroom settings, enabling students to interact with instructors and peers in dynamic, spatially enriched learning environments. While effective in promoting engagement and presence, this environment also introduces a range of privacy concerns, as outlined below.

Privacy Problems: One major concern was biometric data sensitivity. The VLE collected biometric information such as facial expressions and body movements through VR devices. This raised concerns about the potential misuse or unauthorized access to highly sensitive and personally identifiable data. In addition, there were data security risks associated with the processing of large volumes of academic and behavioral data. The platform faced challenges in securing this data during both transmission and storage, with heightened risks of data breaches or cyberattacks—particularly those targeting academic records and biometric identifiers.

Another important issue involved consent and transparency. Although students and guardians provided consent, ensuring that they fully understood the implications of biometric data collection was difficult. The immersive nature of the VLE meant that data was being collected continuously during interactions, which could blur the boundaries of what qualifies as informed consent in a dynamic digital setting.

Privacy Solutions: To address these challenges, the VLE adopted a data minimization strategy, gathering only essential data needed to deliver personalized learning experiences. Any nonessential or potentially intrusive data was deliberately excluded to reduce privacy exposure. In terms of legal and procedural safeguards, the platform ensured strong regulatory compliance with both FERPA and GDPR requirements. This included implementing data access control mechanisms, managing user rights, and adhering to clear data retention policies.

Informed user consent was another critical element of the solution. Clear and accessible privacy policies were communicated to both students and parents, who were required to provide explicit consent prior to platform use. These consent processes were regularly reviewed and updated to align with best practices. In parallel, the VLE implemented robust data security measures. Encryption protocols were applied to all transmitted and stored data, and RBAC, alongside MFA, was introduced to ensure only authorized personnel could access sensitive information.

Lastly, the system was supported by ongoing privacy audits and feedback loops. These audits helped assess compliance and identify areas for improvement. Furthermore, feedback from students and faculty was actively incorporated into iterative improvements of the platform's privacy and data management practices.

Outcome: The VLE significantly improved student engagement and remote learning outcomes. Through proactive privacy strategies, regulatory alignment, and user-centered consent processes, the platform maintained high data privacy and security standards. The scenario underscores the importance of balancing immersive educational experiences with rigorous privacy-preservation mechanisms.

3.2 Scenario-Based Analysis 2: Virtual Shopping Mall

The second scenario-based analysis focuses on shopping within metaverse environments across different contexts. A retail company launched a virtual shopping mall in the metaverse, allowing users to browse and purchase products within an immersive 3D space. The platform integrated AI algorithms to personalize the shopping experience by analyzing user behavior and preferences, thereby enhancing engagement and tailoring product recommendations in real time. Online shopping already has a long history of using AI to understand buyer behaviors (Biswas & Patra, 2023; Tripathi et al., 2024), and within the metaverse, even more extensive information about potential buyers and users is likely to be exposed. This raises critical questions regarding the key privacy issues in such environments and the corresponding solutions needed to address them.

Privacy Problems: A major concern was user profiling and behavioral tracking. The platform collected extensive behavioral data, such as navigation patterns, gaze tracking, and purchase history, to deliver personalized shopping experiences. Although this enhanced the relevance of content and recommendations, it raised significant privacy concerns, especially related to inferred consumer attributes that users had not explicitly provided. In addition, third-party data sharing risks emerged as a critical issue.

To support personalized advertisements and product recommendations, the platform had to share select user data with external vendors. However, ensuring that these third parties adhered to the same privacy standards introduced a supply chain vulnerability. Another key challenge involved cross-jurisdictional compliance. With users located across different countries, the platform had to navigate varying data protection regulations—including the GDPR in the European Union, the CCPA, and PIPL in China—significantly increasing the complexity of compliance efforts.

Privacy Solutions: To address these issues, the platform adopted a set of robust privacy solutions. First, it established transparent data practices by developing comprehensive and user-friendly privacy policies. These policies clearly outlined the types of data collected, the purposes of use, and the parties with whom the data was shared. Written in plain language, the policies were designed to enhance user understanding and trust. Regarding external data exchange, the company pursued a strategy of limited third-party sharing with data protection agreements. Sharing was restricted to essential vendors and governed by strict contractual terms that included privacy obligations, data minimization clauses, and audit rights to ensure accountability.

The platform also implemented strong user consent and control mechanisms. Users were required to provide informed consent before any data collection began. Additionally, they were given granular control over the types of data shared, including options to opt out of personalization or third-party tracking features. In terms of legal compliance, the platform developed dynamic data handling workflows to adjust practices based on each user's location. This allowed the company to meet region-specific legal requirements such as GDPR's "right to be forgotten" and CCPA's "opt-out of sale" provisions. Lastly, the platform reinforced its security infrastructure by deploying advanced encryption protocols and tokenizing payment information. Regular penetration testing and internal audits were conducted to ensure data integrity and uncover potential vulnerabilities.

Outcome: The virtual shopping mall achieved strong customer engagement by providing personalized, immersive experiences while maintaining trust through transparent privacy practices. Its privacy-by-design approach allowed it to grow a global user base while complying with diverse legal requirements.

3.3 Scenario-Based Analysis 3: Digital Virtual Archive

This scenario-based analysis explores the privacy challenges associated with a national archive's launch of a metaverse platform designed to digitize and showcase historical records. The platform enables users to explore interactive timelines, virtually reconstruct historical sites, and annotate primary documents using VR and AI-enhanced features. While these tools enhance engagement and public access to cultural heritage, they also introduce privacy risks.

Privacy Problems: One of the primary concerns was user interaction tracking. To personalize experiences and improve interface design, the platform monitored behavioral data such as time spent on specific artifacts, gaze patterns, and movement pathways. However, this raised privacy concerns, especially when such interaction data was linked to demographic or login credentials, potentially compromising user anonymity. Another significant challenge involved sensitive historical content and ethical boundaries. Some digitized records depicted culturally sensitive or politically charged material. The immersive representation of such content brought ethical concerns regarding cultural ownership, informed consent, and data stewardship—especially from communities historically marginalized or misrepresented.

Additionally, the platform faced data integrity and preservation risks. The goal of enabling dynamic user interaction—such as annotation, zooming, and voice commentary—needed to be balanced with the archival responsibility of maintaining authenticity. There was a risk that unauthorized modifications or AI-generated overlays could distort or compromise the historical accuracy of the digitized documents.

Privacy Solutions: To address these issues, the platform adopted a data minimization strategy, collecting only anonymized interaction metrics by default. Users could opt into more personalized experiences, but each data point was associated with a clear statement explaining its purpose—for example, navigation improvements or engagement tracking for educational resources. To further secure the system, encryption and access controls were implemented. Strong encryption protected both user data and digital artifacts, while RBACs ensured that only authorized personnel—such as researchers, historians, or curators—could modify or annotate original records.

A third layer of protection involved ethical guidelines and community consultation. Developed in collaboration with historians, Indigenous leaders, and cultural experts, these guidelines helped govern the selection, display, and contextualization of sensitive materials. They ensured cultural respect and prevented the misappropriation of marginalized narratives. In terms of user-generated contributions, the platform introduced consent for interactive contributions. Contributors engaging with storytelling, annotation, or voice uploads were presented with explicit consent forms and retained ownership of their submissions, preserving their agency over personal or cultural narratives.

Finally, to ensure long-term accountability, the platform explored the use of immutable logs for audit and verification. Blockchain-based timestamping was considered to securely log all access and modifications. This created an audit trail that supported both archival transparency and the integrity of preserved materials.

Outcome: The virtual archive successfully enhanced public access to historical records by providing an immersive and interactive platform that engaged users across diverse backgrounds. At the same time, it maintained high privacy standards through robust data protection measures, upheld ethical integrity by addressing culturally sensitive content with care, and ensured cultural sensitivity by involving relevant communities in the digitization and presentation of historical narratives.

4 Discussion

The findings of this chapter reveal a complex duality at the heart of educational metaverse platforms. While these immersive environments offer substantial pedagogical benefits, they also generate unprecedented volumes of sensitive data, demanding comprehensive and forward-thinking privacy-preservation strategies (Alnuaimi & Alawida, 2024). A review of the literature highlights the multifaceted nature of these privacy challenges, ranging from biometric data collection and behavioral tracking to regulatory compliance and ethical concerns (Di Dario et al., 2024; Kourtesis, 2024). As immersive environments increasingly blur the lines between educational content delivery and continuous user surveillance, concerns around informed consent, autonomy, and data misuse become more pressing.

Among the most critical insights emerging from the analysis is the imperative of data minimization. Educational platforms must strike a careful balance between offering personalized learning experiences and restricting data collection to what is strictly necessary (Zhao et al., 2023). Overcollection not only heightens the risk of breaches and misuse but also undermines user trust. The use of data audits, as discussed in this chapter, provides a concrete mechanism to monitor and refine data collection practices, helping platforms align with both privacy objectives and legal standards (Özdemir et al., 2024; Parlar, 2023). By focusing on necessity rather than abundance, platforms can mitigate risk while maintaining educational quality.

Closely tied to data minimization are the principles of user consent and transparency. Ensuring that consent is explicit, informed, and ongoing—rather than static and one-time—is crucial for fostering trust. Transparent privacy policies, presented in accessible language, are not just legal obligations but ethical imperatives. These measures empower users—particularly students and their guardians—to make informed decisions about their data and how it is used (Alkaeed et al., 2024), contributing to a culture of accountability and agency.

Yet transparency alone is insufficient without robust data security. As the technical capabilities of the metaverse expand, so too must the sophistication of its protective measures. Encryption standards such as AES-256 and MFA remain vital for safeguarding data in transit and at rest (Alja'Afreh et al., 2024). However, emerging threats like quantum computing already challenge these standards, necessitating a forward-looking approach that includes the development of quantum-resistant encryption techniques.

To ensure that privacy is not merely a reactive measure but a proactive design principle, the concept of privacy by design must be embedded into platform architecture from the outset (Hamadi, Paik, & Benatallah, 2007). This includes conducting regular privacy audits, training staff on privacy protocols, and building user-facing tools that allow individuals to view, modify, or delete their data (Wang et al., 2022). Such tools reinforce user autonomy and align with global privacy frameworks, while also cultivating institutional practices grounded in ethical responsibility.

Navigating regulatory compliance is another persistent and complex challenge. Educational metaverse platforms often span multiple jurisdictions, making it essential to address the inconsistencies and conflicts across legal frameworks such as GDPR, FERPA, COPPA, and CCPA (Gulati & Pal, 2024). This regulatory fragmentation underscores the need for global harmonization efforts to facilitate consistent privacy standards and reduce compliance burdens.

Technical and ethical risks also intensify as metaverse applications evolve. The vast and varied data collected in immersive environments make anonymization difficult and raise the risk of re-identification. Moreover, the use of AI systems trained on such datasets may unintentionally reinforce cultural biases—particularly in domains like emotion recognition, where non-Western facial expressions may be misinterpreted (Di Dario et al., 2024; Kourtesis, 2024). These issues necessitate more culturally inclusive and technically robust AI training processes.

Additionally, platform interoperability introduces new risks. The absence of standardization in consent and data-sharing protocols increases the likelihood of privacy violations during cross-platform transitions. Blockchain-based decentralized identity systems offer a potential solution (Kim et al., 2023; Saparamadu et al., 2025), though practical implementation remains limited by infrastructure and governance challenges.

Looking forward, the integration of emerging technologies will be critical in mitigating future privacy threats. AI can support anomaly detection and privacy threat mitigation, while federated learning offers a model for decentralizing data processing without sacrificing personalization (Alkaeed et al., 2024). The convergence of 6G networks and blockchain technologies may further facilitate secure and privacy-respecting AR/VR experiences (Bhattacharya et al., 2021). Blockchain, with its ability to support tamper-proof data records and smart contract execution, can strengthen data integrity (Dastagir et al., 2022), while zero-knowledge proofs and selective disclosure can allow users to interact privately across decentralized systems.

Techniques like differential privacy, which introduce statistical noise to obscure individual identities, also show promise for conducting research and analytics without compromising user data (Alkaeed et al., 2024). However, these approaches must be carefully calibrated; over-application can reduce the utility of the data, while under-application may fail to ensure adequate protection. Hence, context-specific risk-benefit analyses are essential.

The development of metaverse-specific privacy frameworks is both necessary and overdue. These should address challenges such as cross-border data transfers, nuanced consent management, and rights to access or erasure—issues inadequately covered by existing legacy regulations. A phased implementation roadmap could support this transition. In the short term, platforms can focus on basic protections such as encryption, consent flows, and user dashboards. Medium-term goals might include pilot testing of decentralized identity systems and scalable architectures. Over the long term, sustained investment in privacy research and the establishment of dedicated ethical review boards for the metaverse will be vital.

Ultimately, safeguarding privacy in the educational metaverse requires a coordinated effort among developers, educators, policymakers, and researchers. Developers must embrace privacy by design and build adaptive consent systems; institutions should promote data literacy and offer privacy education; and policymakers must create regulatory environments that reward best practices and enable compliance at scale. However, it is important to acknowledge the trade-offs: blockchain's immutability can conflict with users' right to erasure, federated learning may strain under-resourced institutions, and excessive anonymization can reduce data utility.

5 Conclusion

While the metaverse continues to offer promising advancements in immersive learning, this chapter has emphasized that privacy is not merely a technical hurdle but a foundational pillar for building ethical and sustainable educational ecosystems. As users increasingly engage in metaverse environments with real-world credentials and biometric identifiers, the nature of data collection shifts from transactional to deeply personal. This reality demands that privacy concerns be treated as integral to system architecture, not as afterthoughts.

By adopting an analytical lens and exploring both literature and scenario-based analyses, this chapter has demonstrated that privacy in the educational metaverse is a multifaceted challenge requiring holistic and forward-looking strategies. Rather than focusing solely on specific threats or technologies, it has illustrated the need for systemic solutions—rooted in transparency, user empowerment, regulatory awareness, and cross-sector collaboration.

Moving forward, the responsible development of educational metaverse platforms will depend on proactive alignment with privacy principles, including dynamic consent, user control, and culturally responsive data practices. Developers, educators, and policymakers must work together to establish privacy as a default setting in virtual learning environments. Only through this collective and anticipatory approach can we ensure that the metaverse evolves into an inclusive, secure, and trustworthy space for education.

References

Ali, M., Naeem, F., Kaddoum, G., & Hossain, E. (2023). Metaverse communications, networking, security, and applications: Research issues, state-of-the-art, and future directions. *IEEE Communications Surveys & Tutorials*, *26*(2), 1238–1278.

Alja'Afreh, M., Hasan, O., Al-Mallah, R., & Karime, A. (2024). Blockchain-driven networking and communications in the metaverse. In *2024 6th International Conference on Blockchain Computing and Applications (BCCA)* (pp. 826–833). IEEE.

Alkaeed, M., Qayyum, A., & Qadir, J. (2024). Privacy preservation in Artificial Intelligence and Extended Reality (AI-XR) metaverses: A survey. *Journal of Network and Computer Applications*, *231*, 103989. https://doi.org/10.1016/j.jnca.2024.103989

Alnuaimi, R., & Alawida, M. (2024). A multi-layered strategy for addressing privacy and data security issues in metaverse. In *2024 2nd International Conference on Intelligent Metaverse Technologies & Applications (iMETA)* (pp. 113–118). IEEE.

Amano, T., Mizumoto, T., Kala, S. M., Yamaguchi, H., Matsui, T., & Yasumoto, K. (2024). *Visual privacy control for metaverse and the beyond.* IEEE Pervasive Computing.

Arya, V., Chui, K. T., Hsu, C. H., Gupta, B. B., & Zhang, J. (2022, December). Novel entropy based technique for Distributed Denial of Services (DDoS) attack detection in metaverse. In *International Conference on Smart Systems and Advanced Computing* (pp. 189–193). Springer Nature Switzerland.

Bhattacharya, P., Saraswat, D., Dave, A., Acharya, M., Tanwar, S., Sharma, G., & Davidson, I. E. (2021). Coalition of 6G and blockchain in AR/VR space: Challenges and future directions. *IEEE Access*, *9*, 168455–168484.

Biswas, K., & Patra, G. (2023). Role of artificial intelligence (AI) in changing consumer buying behaviour. *International Journal of Research Publication and Reviews*, *4*, 943–951.

Chen, C., Wu, T., Ren, C., & Sun, C. (2024, December). A scalable and timely data auditing scheme for metaverse. In *2024 IEEE Smart World Congress (SWC)* (pp. 2245–2252). IEEE.

Darejeh, A. (2023). Empowering education through eerp: A customizable educational vr escape room platform. In *2023 IEEE International Symposium on Mixed and Augmented Reality Adjunct (ISMAR-Adjunct)* (pp. 764–766). IEEE.

Darejeh, A., & Mashayekh, S. (2023). Mind-object control as a new form of interaction in metaverse for users with motor disability. *Journal of Biomedical Physics & Engineering*, *13*(1), 257.

Darejeh, A., Huynh, L., Ji, X., Marcus, N., & Polly, P. (2024a). Assessing the efficacy of virtual reality-based laboratories versus real-life settings for PCR procedures, within medical science laboratories. In *2024 IEEE International Symposium on Mixed and Augmented Reality Adjunct (ISMAR-Adjunct)* (pp. 489–492). IEEE.

Darejeh, A., Mondal, M., Marcus, N., & Vassar, A. (2024b). Exploring the cognitive load effects of diverse virtual reality interaction methods in interactive educational platforms. In *2024 IEEE International Symposium on Mixed and Augmented Reality Adjunct (ISMAR-Adjunct)* (pp. 435–438). IEEE.

Darejeh, A., Moghadam, T. S., Delaramifar, M., & Mashayekh, S. (2024c). A framework for AI-powered decision making in developing adaptive e-learning systems to impact learners' emotional responses. In *2024 11th International and the 17th National Conference on E-Learning and E-Teaching (ICeLeT)* (pp. 1–6). IEEE.

Dastagir, M. B. A., Tariq, O., & Han, D. (2022). A smart card based approach for privacy preservation authentication of non-fungible token using non-interactive zero knowledge proof. In *2022 IEEE Smartworld, Ubiquitous Intelligence & Computing, Scalable Computing & Communications, Digital Twin, Privacy Computing, Metaverse, Autonomous & Trusted Vehicles (SmartWorld/UIC/ScalCom/DigitalTwin/PriComp/Meta)* (pp. 2428–2435). IEEE.

De Felice, F., Petrillo, A., Iovine, G., Salzano, C., & Baffo, I. (2023). How does the metaverse shape education? A systematic literature review. *Applied Sciences*, *13*(9), 5682.

Di Dario, D., Pentangelo, V., Colella, M. I., Palomba, F., & Gravino, C. (2024). Collecting and implementing ethical guidelines for emotion recognition in an educational metaverse. In *Adjunct Proceedings of the 32nd ACM Conference on User Modeling, Adaptation and Personalization* (pp. 549–554).

Falchuk, B., Loeb, S., & Neff, R. (2018). The social metaverse: Battle for privacy. *IEEE Technology and Society Magazine*, *37*(2), 52–61.

Frosio, G., & Obafemi, F. (2025). *Regulating data access in the metaverse*. Queen's University Belfast.

Garon, J. M. (2023). Legal considerations for offering metaverse-based education. *JL & Education*, *52*, 53.

Guhathakurta, A. (2024). Data privacy, data protection and user rights in the world of metaverse: addressing the legal challenges in the EU, the US and India. *Data Protection and User Rights in the World of Metaverse: Addressing the Legal Challenges in the EU, the US and India (October 21, 2024)*.

Gulati, P., & Pal, K. (2024). The notion of privacy rights in the metaverse: Examining legal hemispheres in India's digital era. In *Metaverse Driven Intelligent Information Systems: Emerging Trends and Future Directions* (pp. 159–176). Springer Nature Switzerland.

Hamadi, R., Paik, H. Y., & Benatallah, B. (2007). Conceptual modeling of privacy-aware web service protocols. In *Advanced Information Systems Engineering: 19th International Conference, CAiSE 2007, Trondheim, Norway, June 11–15, 2007. Proceedings 19* (pp. 233–248). Springer Berlin Heidelberg.

Hector, L. I., Mendana-Cuervo, C., & Juan Luis, C. C. (2024). The Metaverse: Privacy and Information Security Risks. Available at SSRN 4803584.

Kim, M., Oh, J., Son, S., Park, Y., Kim, J., & Park, Y. (2023). Secure and privacy-preserving authentication scheme using decentralized identifier in metaverse environment. *Electronics*, *12*(19), 4073.

Kourtesis, P. (2024). A comprehensive review of multimodal XR applications, risks, and ethical challenges in the metaverse. *Multimodal Technologies and Interaction*, *8*(11), 98.

Kovacova, M., Horak, J., & Popescu, G. H. (2022). Haptic and biometric sensor technologies, deep learning-based image classification algorithms, and movement and behavior tracking tools in the metaverse economy. *Analysis and Metaphysics*, *21*, 176–192.

Trigo Kramcsak, P. R., & Papakonstantinou, V. (2024). *Personal data processing within immersive virtual worlds: Privacy challenges in the interconnected data-driven metaverse - MetaverseUA Chair Research Paper #2.* https://papers.ssrn.com/sol3/papers.cfm?abstract_id=5030132

Kumar, P., Kumar, R., Aloqaily, M., & Islam, A. N. (2023). Explainable AI and blockchain for metaverse: A security and privacy perspective. *IEEE Consumer Electronics Magazine*, *13*, 90–97. https://doi.org/10.1109/MCE.2023.3237344

Li, M., & Yu, Z. (2023). A systematic review on the metaverse-based blended English learning. *Frontiers in Psychology*, *13*, 1087508.

Liu, H., Xu, D., Tian, Y., Peng, C., Wu, Z., & Wang, Z. (2023). Wasserstein generative adversarial networks based differential privacy metaverse data sharing. *IEEE Journal of Biomedical and Health Informatics, 28*(11), 6348–6359. https://doi.org/10.1109/JBHI.2023.3287092

Moerel, L. (2024). Metaverse and data protection. In *Research Handbook on the metaverse and law* (pp. 118–157). Edward Elgar Publishing.

Mohammed, Z. K., Mohammed, M. A., Abdulkareem, K. H., Zebari, D. A., Lakhan, A., Marhoon, H. A., Nedoma, J., & Martinek, R. (2024). A metaverse framework for IoT-based remote patient monitoring and virtual consultations using AES-256 encryption. *Applied Soft Computing*, *158*, 111588.

Nkoro, E. C., Nwakanma, C. I., Lee, J.-M., & Kim, D.-S. (2024). Detecting cyberthreats in metaverse learning platforms using an explainable DNN. *Internet of Things*, *25*, 101046. https://doi.org/10.1016/j.iot.2023.101046

Novianti, D. N., Shiddieq, D. F., Roji, F. F., & Susilawati, W. (2024). Komparasi algoritma support vector machine dan naïve Bayes untuk analisis sentimen pada metaverse: Comparison of support vector machine and naïve Bayes algorithms for sentiment analysis of the metaverse. *MALCOM: Indonesian Journal of Machine Learning and Computer Science*, *4*, 231–239.

Özdemir, K., Çelebi, Ö. F., & Nacar, R. (2024). Consumer security and privacy in the metaverse: A bibliometric analysis. *Dokuz Eylül Üniversitesi İşletme Fakültesi Dergisi*, *25*(1), 111–130.

Pałczyńska, J. (2023). *Surveillance of employees in the Metaverse: Data protection and privacy challenges [Master's thesis, Jagiellonian University]. Jagiellonian University Institutional Repository. https://ruj.uj.edu.pl/entities/publication/15eec6c9-6a0d-492d-bd9a-8bd7d3149ab4*

Paneva, V., Strauss, M., Winterhalter, V., Schneegass, S., & Alt, F. (2024). Privacy in the metaverse. *IEEE Pervasive Computing*, *23*(3), 73–78.

Pappachan, P., & Bansal, R. (2024). *Metaverse security paradigms*. IGI Global.

Parlar, T. (2023). Data privacy and security in the metaverse. *Studies in Big Data* (Vol. 133). Springer Science and Business Media Deutschland GmbH. https://doi.org/10.1007/978-981-99-4641-9_8

Raab, S. (2021). Protection of children's personal data and where reforms are needed. *Journal of Data Protection & Privacy*, *4*(4), 346–361.

Saparamadu, P. V. I. N., Sepasgozar, S., Guruge, R. N. D., Jayasena, H. S., Darejeh, A., Ebrahimzadeh, S. M., & Eranga, B. A. I. (2025). Optimising contract interpretations with large language models: A comparative evaluation of a vector database-powered Chatbot vs. ChatGPT. *Buildings*, *15*(7), 1144.

Sargazi Moghadam, T., Darejeh, A., Delaramifar, M., & Mashayekh, S. (2024). Toward an artificial intelligence-based decision framework for developing adaptive e-learning systems to impact learners' emotions. *Interactive Learning Environments*, *32*(7), 3665–3685.

Sciolla, J. (2024). Marketing in the metaverse: Transparency obligations for restoring the basic truth-in advertising principle. In *Proceedings of the International Congress 'Towards a Responsible Development of the Metaverse', Alicante 13–14 June 2024* (pp. 1–27). Universidad de Alicante.

Skowronski, D. S. (2022). Coppa and educational technologies: The need for additional online privacy protections for students. *Georgia State University Law Review*, *38*(4), 12.

Tripathi, S., Tiwari, A., Singh, S., Roy, A., Singla, J., & Chaudhry, R. (2024). Analyzing the consumer buying behavior by adapting artificial intelligence (AI). In *2024 International Conference on Healthcare Innovations, Software and Engineering Technologies (HISET)* (pp. 377–379). IEEE. https://doi.org/10.1109/HISET58917.2024.10356599

Truong, V. T., Le, L., & Niyato, D. (2023). Blockchain meets metaverse and digital asset management: A comprehensive survey. *IEEE Access*, *11*, 26258–26288. https://doi.org/10.1109/ACCESS.2023.3260459

Upadhyay, U., Kumar, A., Sharma, G., Saini, A. K., Arya, V., Gaurav, A., & Chui, K. T. (2024). Mitigating risks in the cloud-based metaverse access control strategies and techniques. *International Journal of Cloud Applications and Computing (IJCAC)*, *14*(1), 1–30.

Wang, Y., Su, Z., Zhang, N., Xing, R., Liu, D., Luan, T. H., & Shen, X. (2022). A survey on metaverse: Fundamentals, security, and privacy. *IEEE Communications Surveys & Tutorials*, *25*(1), 319–352.

Westin, A. F. (1968). Privacy and freedom. *Washington and Lee Law Review*, *25*(1), 166.

Wodi, A. (2023). Data privacy and security: Fair information principles and data management strategies in a changing world. *SSRN*. https://doi.org/10.2139/ssrn.4601171

Xianing, X. (2024). Sustaining international education with metaverse-based learning for Chinese: A systematic review (2019–2023). *Education towards a Global Citizen Generation, 13*(3), 690–704. http://dx.doi.org/10.6007/IJARPED/v13-i3/20906

Yang, K., Zhang, Z., Touliang, Y., & Ma, J. (2023). A secure authentication framework to guarantee the traceability of avatars in metaverse. *IEEE Transactions on Information Forensics and Security, 18*, 3817–3832. https://doi.org/10.1109/TIFS.2023.3294860

Yang, L., Ni, S. T., Wang, Y., Yu, A., Lee, J. A., & Hui, P. (2024). Interoperability of the metaverse: A digital ecosystem perspective review. *arXiv preprint arXiv, 2403*, 05205.

Yinhui, K. N., Md Yunus, M., M. Rafiq, K. R., & Awang, M. M. (2024). Sustaining STEAM education and creativity through Metaverse applications: A systematic review (2019–2023). *International Journal of Academic Research in Business and Social Sciences, 14*(5), 664–682. https://doi.org/10.6007/IJARBSS/v14-i5/21592

Zhang, L., Ou, Z., Hu, C., Kan, H., & Zhang, J. (2024). *Data sharing in the metaverse with key abuse resistance based on decentralized CP-ABE*. IEEE Transactions on Computers.

Zhao, R., Zhang, Y., Zhu, Y., Lan, R., & Hua, Z. (2023). Metaverse: Security and privacy concerns. *Journal of Metaverse, 3*(2), 93–99.

10 Generative Artifical Intelligence for Education and the Metaverse

Intelligent Virtual Instructors, Chatbots, and Recommender Systems

Jihyun Lee, Ali Darejeh, Arifah Anwar Fadzil, Stephen Lukins, Rubin Roy, Nemat Bhullar, Michael Shi, and Samad M.E. Sepasgozar

1 Introduction

Since OpenAI launched ChatGPT in November 2022 as a conversational AI application based on GPT-3.5 (and later GPT-4), generative artificial intelligence (GenAI) has captured the attention of innovators and educators worldwide. This release marked a significant milestone in the public's engagement with AI, triggering widespread interest in its capabilities. Initially, access to these technologies was limited, but over time, tools like ChatGPT became widely available and began reshaping public and professional discourse (Saparamadu et al., 2025).

While GenAI tools are now commonly used for general communication tasks, their effectiveness in addressing more specific or complex educational challenges remains an area of active exploration. This raises critical questions about how AI can be better understood and more effectively integrated into educational practice. Clarifying the perspective from which AI is viewed—whether as a concept, a technology, an algorithm, a system, a tool, or a product—is a necessary step in this process.

Definitions of AI vary widely, each emphasizing different aspects of its functionality. Some researchers describe AI as technology in which software and machines mimic human intelligence (Fetzer, 1990). Others focus on its ability to simulate human cognitive functions such as perception, reasoning, language interaction, and problem-solving (Markauskaite et al., 2022; Sargazi Moghadam et al., 2024). A more formal definition frames AI as "machine-based systems that can, given a set of human-defined objectives, make predictions, recommendations, or decisions that influence real or virtual environments" (Darejeh et al., 2024c; UNICEF, 2021, p. 16). Building on these foundations, GenAI has been more recently defined as a class of algorithms capable of generating diverse content—including audio, code, images, text, 2D/3D simulations, and video—in response to user prompts (Peñalvo & Ingelmo, 2023). Notable examples include ChatGPT and DALL·E. These evolving definitions reflect AI's historical trajectory: from general-purpose technology to systems capable of sophisticated human–machine interaction and content generation.

DOI: 10.1201/9781003557555-10

This expanding capability is largely driven by rapid advancements in natural language processing (NLP) and machine learning. While early conversational agents such as ELIZA (developed in the 1960s) laid the foundation for human–computer interaction, today's GenAI systems train on massive datasets and leverage self-supervised learning to produce outputs that are increasingly coherent and human-like (Perrault & Clark, 2024). For example, Meta's LLaMA 3 models, released in July 2024, were trained on approximately 120 terabytes of text—equivalent to two or three times the content of all published books (Dubey et al., 2024; Taycher, 2010). The progress in GenAI capabilities has been so rapid that many systems now outperform humans in tasks such as image classification, reading comprehension, handwriting recognition, and even domain-specific expertise across fields like medicine, law, and engineering (Kiela et al., 2023). This has prompted researchers to develop more robust benchmarks to avoid "benchmark saturation."

In parallel with its technical growth, GenAI adoption has accelerated at an unprecedented rate—surpassing even the early adoption rates of personal computers and the internet. Unlike previous technologies that faced resistance, GenAI has been met with rapid acceptance. A 2024 U.S. survey found that 39% of adults aged 18–64 had already used GenAI tools (Bick, Blandin, & Deming, 2024). Comparable adoption levels have been observed internationally, though most users still report infrequent use (Fletcher & Nielsen, 2024). Industry sectors such as finance, pharmaceuticals, law, and education are already leveraging GenAI for content creation, technical documentation, and decision support. Its potential economic impact is substantial, with projections estimating contributions of up to $4.4 trillion annually to the global economy (Bick, Blandin, & Deming, 2024; Perrault & Clark, 2024).

Among all sectors, education stands out as both a key beneficiary and a complex testing ground for GenAI. While integration has largely been driven from the ground up—by individual teachers and students rather than institutions—early adoption has yielded both enthusiasm and uncertainty. A 2023 survey reported that 63% of K–12 teachers in the U.S. were using ChatGPT to assist with lesson planning, generate instructional ideas, and expand content knowledge (Bagley, 2024). Both students and teachers reported overwhelmingly positive perceptions of GenAI's educational value. However, institutional frameworks have struggled to keep pace: most schools lack formal AI policies, and 75% of respondents indicated they had received no training in AI use (Bagley, 2024).

In light of these trends, this chapter explores the transformative potential of GenAI in education and the metaverse by examining its emerging applications, implementation challenges, and future directions. It demonstrates how GenAI tools are reshaping educational practices by enabling personalized learning through AI-driven tutors, enhancing the understanding of complex and technical concepts through simulated environments, and supporting course and career decision-making via intelligent recommendation systems. This chapter also underscores the integration of GenAI with immersive technologies—particularly virtual reality (VR)—as a means to create engaging, adaptive, and scalable learning experiences. Through four applied empirical studies—including virtual laboratory instruction, critical thinking development, career guidance, and university admission prediction—it bridges theoretical perspectives with practical implementations, offering a comprehensive analysis of GenAI's role in redefining educational futures and contributing to broader societal advancement.

2 Literature Review

This section explores the evolving role of GenAI in educational contexts, with a focus on three interconnected domains. First, it examines how GenAI enhances learning experiences through personalization, inclusivity, and system-level support. Second, it reviews the integration of GenAI in the metaverse, emphasizing its role in generating dynamic content, realistic avatars, and immersive interactions. Finally, it investigates the transformative potential of combining GenAI and VR in education, highlighting their impact on learner engagement, skill development, and personalized instruction within immersive environments.

2.1 Enhancing Learning Experiences with GenAI

One of the most compelling advantages of GenAI in educational contexts is its capacity to enhance personalization. This includes delivering timely feedback, tailoring support to individual learning needs, and enabling curriculum differentiation and assessment adjustments that encourage deeper student engagement. According to Vincent-Lancrin and Van Der Vlies (2020), five key areas in the field of AI in education demonstrate significant promise for advancing personalized learning.

First, online and blended learning environments are expected to benefit substantially from AI-powered tools such as chatbots, learning analytics, and intelligent tutoring systems. Unlike traditional platforms, which tend to be relatively static, AI-driven systems can dynamically adapt content, pacing, and feedback in response to individual learner profiles (Peiqi, 2024).

Second, AI technologies are making education more inclusive by supporting students with diverse needs. For instance, AI-powered tools assist visually impaired students by reading and summarizing text aloud in real time. Speech recognition software allows students to dictate input, with AI refining grammar, punctuation, and sentence structure for clarity. Additionally, emotion recognition systems can detect signs of anxiety or disengagement, enabling timely mental health interventions. In more immersive contexts, VR applications combined with AI enable students with mobility impairments to engage in simulations that would otherwise be physically inaccessible. Collectively, these tools promote more equitable access to learning by adapting instruction and support to a wide range of learner needs (Acosta-Vargas et al., 2024; AlAli & Wardat, 2024).

Third, foreign language education has been significantly enriched through AI integration. Tools employing speech analysis, pronunciation correction, and grammar feedback make language acquisition more accessible and responsive to learners' real-time input and progress. Fourth, AI supports system-level decision-making within educational institutions. In the United States, for example, AI solutions are widely adopted by school districts and state agencies to address dropout prevention. Predictive analytics within early warning systems identify at-risk students and facilitate timely, targeted interventions (Cong-Lem & Daneshfar, 2024; Pérez-Núñez, 2023).

Fourth, there has been system-level support for schools and institutions assisted by AI technology. For example, AI solutions are increasingly used to support school principals and district leaders in addressing student dropout prevention through real-time data analysis and early intervention strategies. These solutions provide early feedback to reduce dropout rates by identifying at-risk students through early warning systems (Córdova-Esparza et al., 2025; Diaz & Moreira, 2024).

Finally, AI is increasingly employed for behavioral monitoring and classroom evaluation. By utilizing sensor-based data and real-time analytics, these systems can assess student engagement and interaction patterns. This enables educators to receive actionable insights and recommendations, contributing to improved classroom management and more effective teaching strategies (Jaboob, Hazaimeh, & Al-Ansi, 2025).

2.2 Immersive Content Generation with GenAI in the Metaverse

The metaverse is a virtual universe that blends physical reality with digital virtuality, offering users a space to create, explore, interact, and collaborate. Within this immersive environment, individuals can engage in entertainment, communication, education, and a wide range of experiential activities. GenAI plays a pivotal role in both the creation and operation of the metaverse, significantly enhancing content generation, personalization, and real-time interactivity (Fadhel et al., 2024).

One of GenAI's most impactful contributions is the development of realistic and dynamic avatars, environments, and multimedia assets. In VR settings, for instance, GenAI-powered avatars appear less mechanized and more human-like, increasing user trust and facilitating more natural interactions (Ahuja et al., 2023; Darejeh, Marcus, & Sweller, 2022). In terms of visual design, GenAI can autonomously produce art, landscapes, virtual effects, image morphing, fashion designs, and highly detailed environments—such as digital cities, forests, and architectural structures—enhancing immersion while dramatically reducing manual design time (Chamola et al., 2024a). In addition, GenAI facilitates the efficient generation of 3D assets for applications such as virtual sculpture and object reconstruction, further streamlining the metaverse development process (Jauhiainen, 2024).

Beyond visuals, GenAI enhances the auditory dimension of the metaverse. It can generate personalized soundscapes, simulate virtual music concerts, and synthesize emotionally adaptive audio for deeper user engagement. Real-time voice synthesis for non-player characters (NPCs) allows for more immersive interaction, while AI-driven translation and interpretation tools support seamless multilingual communication in shared virtual spaces (Chamola et al., 2024b; Nagarkar, 2024).

Text-based GenAI applications also play a critical role. These include intelligent chatbots for dynamic conversations, interactive storytelling that adapts to user decisions, and real-time language translation to facilitate cross-cultural engagement (Godwin-Jones, 2023). When embedded into NPCs, GenAI enables natural, context-aware responses that mimic human dialogue, further blurring the line between artificial agents and human participants (Jauhiainen, 2024).

Looking forward, the integration of multimodal GenAI models is expected to unlock even more advanced and cohesive experiences. For instance, in virtual retail environments, users could try on digital garments, receive peer feedback in real time, and access AI-generated video or text-based customer support—all within a single, immersive session (Chamola et al., 2024b; Dang Quan et al., 2024). Such innovations highlight GenAI's evolving capacity to enrich interaction, creativity, and accessibility within the metaverse.

2.3 GenAI-Enhanced Learning in Educational Metaverse Environments

The metaverse is increasingly adopted in education due to its potential to enhance learning outcomes, boost engagement, and offer experiential learning opportunities (Onu et al., 2023). When combined with GenAI, immersive virtual environments can provide learners with a more interactive, personalized, and adaptive educational experience than either technology could achieve alone. For example, GenAI can create interactive visual stories to help students grasp new concepts and scientific principles more effectively (Chamola et al., 2024b).

A key application of GenAI in educational metaverse environments is the generation of intelligent NPCs. These NPCs can respond to student input, offer real-time feedback, and assist in the learning process, making virtual spaces more dynamic and responsive. Operating on predefined pedagogical rules or adaptive AI algorithms, these agents allow students to repeatedly practice diverse scenarios, applying their learning in realistic simulations. NPCs in education generally fall into three categories: tutors/advisors, tutees/students, and peers, each serving a distinct instructional function.

- **NPC tutors** offer personalized support, including tailored explanations, study plans, reminders, resources, mock exams, and progress assessments (Chamola et al., 2024b).
- **NPC tutees**, particularly in teacher training, provide novice educators with simulated classroom practice. TeachLive, for example, is a mixed-reality system that enables educators to refine their classroom management skills by interacting with avatar students before teaching in real settings (Larson et al., 2019).
- **NPC peers** simulate collaborative learners progressing alongside the user, enabling interaction, discussion, and teamwork in virtual group settings (Hwang & Chien, 2022).

One of the most prominent technologies underpinning these educational metaverse experiences is VR. Widely used in fields such as industrial simulation, driving, and medical training, VR has shown increasing promise in education (Darejeh,

2023; Hamad & Jia, 2022). It supports practical skill development, knowledge retention, and the cultivation of soft skills such as communication, collaboration, and leadership (Darejeh et al., 2024a; Holuša et al., 2023). VR enables learners to experiment in safe, cost-effective environments, making it particularly beneficial for high-risk domains like science laboratories, where traditional instruction may involve expensive equipment and safety concerns (Darejeh et al., 2024b). Studies have even reported high approval of VR among primary school students, reinforcing its appeal across age groups (Paszkiewicz et al., 2021).

Despite these advantages, important research gaps remain—particularly in understanding the long-term educational impacts of VR and GenAI integration. Poorly designed virtual reality learning environments (VRLEs) can disrupt rather than enhance learning (Oje et al., 2023). Thus, there is a pressing need to establish design principles that optimize instructional value in immersive settings. Drawing on Moore's (1989) framework for distance learning, three types of interaction are critical in VRLEs: learner–content, learner–instructor, and learner–learner. Yet, many studies have focused primarily on learner–content interactions, often overlooking the equally vital learner–instructor dynamic—especially in AI-enhanced environments (Fink et al., 2024).

Current VRLEs often rely on static hinting systems, such as on-screen text or scripted narration, which lack the context sensitivity and responsiveness of human instruction. In this context, GenAI-powered educational avatars offer a promising solution. These AI-driven agents can act as instructors, guiding students through virtual tasks, adapting feedback in real time, and reducing cognitive load during complex interactions (Darejeh, Marcus, & Sweller, 2022; Fink et al., 2024).

The next section presents a series of empirical studies that investigate how GenAI is currently being used in education. The first study specifically addresses the gap identified here—evaluating the role of GenAI-powered tutoring avatars in supporting students within immersive virtual learning environments.

3 Empirical Studies

This section presents four studies illustrating the application of GenAI in educational contexts within the metaverse. The first study explores the use of a speaking avatar instructor in a metaverse-based electrical engineering laboratory, highlighting its potential to enhance instructional support and reduce cognitive load. The second study examines how GenAI tools can foster critical thinking skills among high school students in social science classrooms. The third study introduces a GenAI-driven system that provides students with personalized predictions regarding their university admission prospects. Lastly, the fourth study showcases a GenAI-powered platform designed to assist students in making informed career decisions by analyzing their interests, abilities, and value profiles to generate customized recommendations.

All participants across the studies were recruited in accordance with a human research ethics protocol approved by the University of New South Wales and sanctioned by its institutional review board.

3.1 Empirical Study One: A GenAI-Powered Talking Avatar in a VR-Based Electrical Engineering Laboratory

The first study presents the development and implementation of a GenAI-based talking avatar designed to support undergraduate learning in a VR electrical engineering laboratory, developed as part of the University of New South Wales (UNSW) metaverse platform. Engineering concepts are often abstract and conceptually complex, which can make it challenging for students to construct coherent mental models—a difficulty that frequently hinders deep understanding (Streveler et al., 2008). The integration of GenAI talking avatars into VR learning environments offers promising opportunities to address these challenges by providing personalized, interactive instructional support tailored to students' cognitive needs.

Recent studies have demonstrated the benefits of VR-based learning environments in enhancing students' knowledge, motivation, and cognitive engagement, particularly when learning to operate electronic laboratory hardware (Singh et al., 2020; Stefan et al., 2024). For instance, VR-based training in low-voltage (LV) electrical safety was shown to improve participants' ability to acquire and retain critical safety information (Stefan et al., 2024). Another notable example, the Virtual Electrical Laboratory, combines narrative-based and text-based instruction with interactive puzzle-based tasks to teach theoretical concepts (Albarracin-Acero et al., 2024). Evaluation results were highly encouraging: 100% of expert reviewers and 94% of student participants affirmed the tool's effectiveness in demonstrating key electrical concepts, especially in the context of DC circuits. Students also reported increased confidence, reduced anxiety about damaging physical equipment, and lower cognitive load—key indicators of improved learning experiences in immersive environments (Singh et al., 2020).

3.1.1 Application Development for Study One

Our VR electrical engineering laboratory was developed using Unreal Engine 5 and designed for HTC VIVE Pro 2 headsets. The GenAI-powered talking avatar was built using the OpenAI API and powered by the ChatGPT-4 language model. It was further enhanced with domain-specific knowledge to ensure accurate and context-aware responses during open-ended student interactions. Rather than overwhelming students with dense information, the avatar provides concise, step-by-step instructions, answers technical questions, and actively guides learners to the appropriate laboratory equipment, thereby improving engagement and promoting active participation. Beyond instructional support, the avatar can simulate real-world laboratory dynamics by selecting or generating specific laboratory items in response to student actions, interacting with other virtual instructors or students, and initiating context-based dialogues.

Within the virtual laboratory, students engaged in both hands-on circuit-building activities and theoretical assessments, with the AI avatar instructor providing real-time guidance and feedback. The VR laboratory is structured into five sequential levels, each designed to represent a distinct phase of the learning process. These levels are aimed at helping students transfer the skills acquired in the virtual environment to real-world laboratory settings, thereby building both confidence and competence for physical laboratory operations.

In Level One (see Figure 10.1), students are introduced to essential safety protocols modeled on those used in actual electrical engineering laboratories. They also learn to identify potential hazards, helping to ensure that the safety knowledge gained in the VR setting is directly applicable in real-world contexts. Additionally, this level familiarizes students with the VR interface and hand controllers, easing the transition into the immersive environment.

Figure 10.1 A user explores the environment and listens to Avatera's safety guidance during the induction session in the virtual laboratory.

From Levels Two to Four, students engage in a series of progressively challenging activities designed to build foundational knowledge in circuit design and interpretation. These stages focus on the identification and use of key electronic components commonly found in electrical engineering laboratories. Students construct circuits based on schematic diagrams, applying their understanding of components such as breadboards, light-emitting diodes (LEDs), and batteries. As shown in Figure 10.2, learners explore correct methods for connecting components and practice identifying the anode and cathode terminals of LEDs—essential skills for assembling functional circuits. To reinforce conceptual understanding, the system incorporates interactive visual feedback. For example, Figure 10.3 illustrates how incorrect connections result in simulated outcomes such as the failure or explosion of an integrated circuit (IC), accompanied by visual effects like smoke.

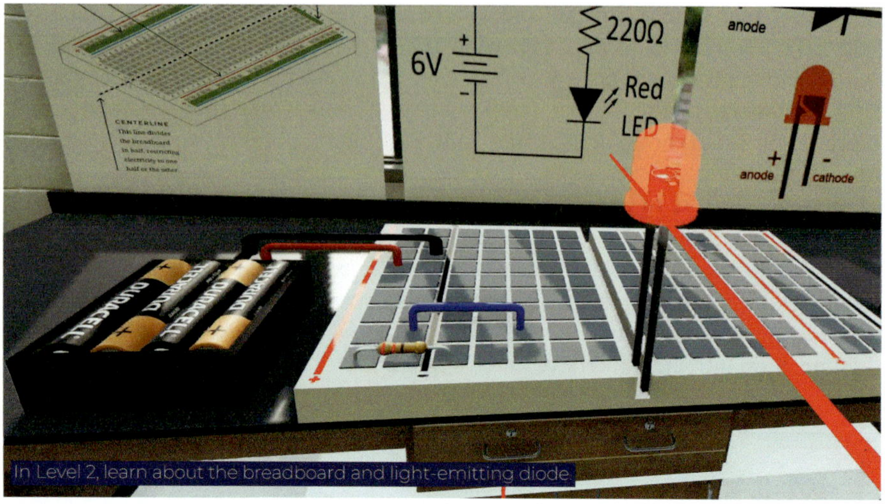

Figure 10.2 Identifying and connecting basic electronic components in a VR laboratory.

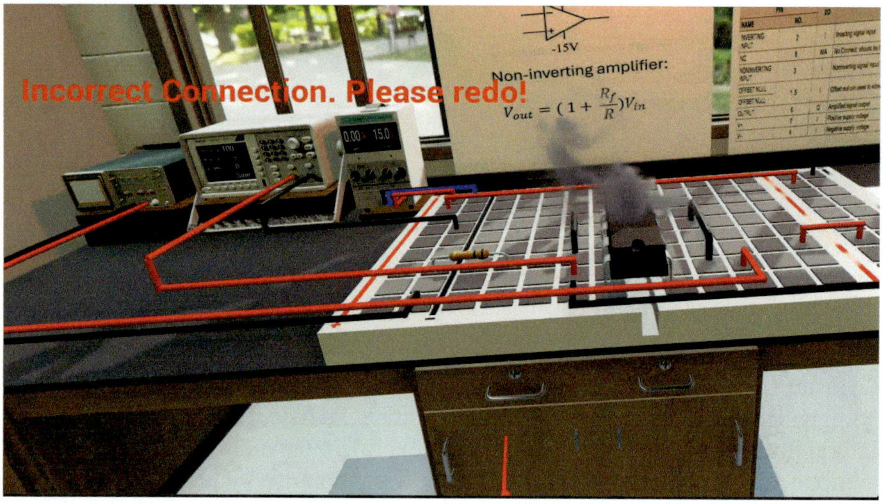

Figure 10.3 Simulated feedback on incorrect circuit connections with visual effects.

In Level 5 (see Figures 10.4 and 10.5), students complete a structured quiz that incorporates embedded questions to evaluate their understanding of key electronics concepts. The GenAI avatar provides personalized, context-aware feedback based on individual responses, helping students troubleshoot errors and accurately revise their circuit designs. Adapted from common assessment formats used in electronics education—such as pre-lab, post-lab, and final exams—this quiz is designed to assess core theoretical knowledge that may not be effectively measured through circuit-building tasks alone.

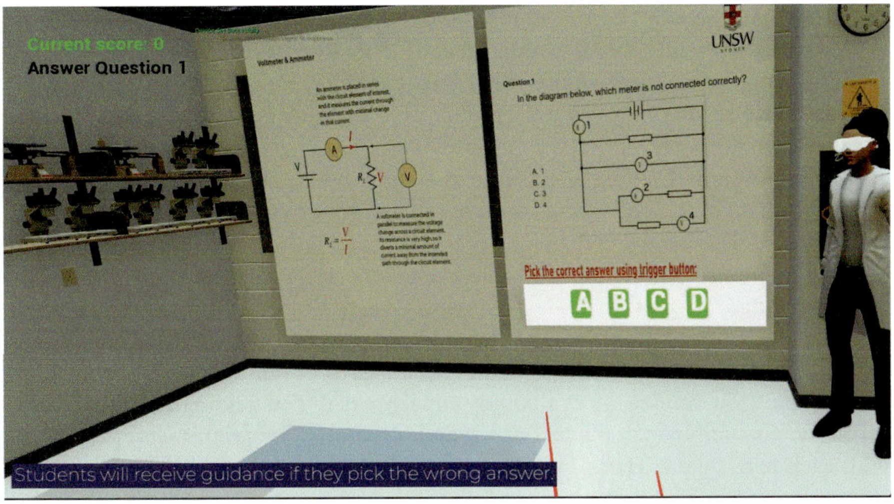

Figure 10.4 GenAI avatar offering real-time feedback and assessing circuit knowledge through quiz questions.

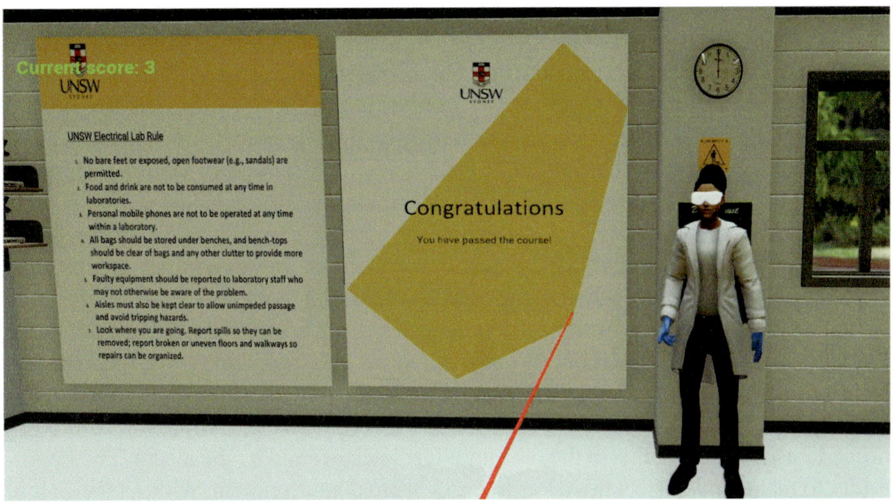

Figure 10.5 The avatar provides feedback after the quiz.

3.1.2 Methodology for Conducting Study One

Ten undergraduate students enrolled in an introductory electrical engineering course participated in the study. The objective was to evaluate the impact of the GenAI-powered speaking avatar, embedded within our VR laboratory environment, on students' conceptual understanding and procedural knowledge. After signing consent forms, students completed a pre-test designed to assess their baseline knowledge in three key areas: identification of circuit components, interpretation of schematic diagrams, and understanding of laboratory safety protocols.

Following the pre-test, students participated in a 25-minute session using the VR electrical engineering laboratory via HTC VIVE headsets, where they learned to build circuits in an immersive environment. During this session, they received guided instruction from a GenAI-powered talking avatar, which responded to spoken student questions related to laboratory procedures, circuit assembly, and safety protocols. The avatar provided real-time, interactive support to help students complete laboratory tasks, reinforce safety practices, and deepen their conceptual understanding.

After the session, students completed a post-test of comparable structure and difficulty to the pre-test. Both assessments consisted of multiple-choice and short-answer questions aimed at evaluating conceptual understanding and procedural accuracy. To complement the test data, semi-structured interviews were conducted with each participant. These interviews explored user perceptions of the system's responsiveness, clarity, realism, and overall usefulness in supporting learning through interactive questioning.

3.1.3 Results from Study One

A paired-sample t-test was conducted to compare students' scores on the pre-test and post-test, each scored out of 100. The score improved from M = 42.2, SD = 10.6 on the pre-test to M = 58.4, SD = 9.3 on the post-test, reflecting a learning gain of 18.2 points. This difference was statistically significant: $t(9) = 5.71$, $p < .001$, demonstrating the effectiveness of the AI-driven VR laboratory environment in enhancing students' understanding of electrical engineering concepts.

Interview responses provided further insight into these outcomes. Sixty percent of students reported that the GenAI avatar's explanations improved their comprehension of tasks, particularly when working with unfamiliar components or interpreting schematic diagrams. The avatar's ability to break down instructions into step-by-step guidance, respond to spoken questions in real time, and highlight relevant equipment locations helped make the learning experience feel personalized and manageable.

Fifty percent of students also noted increased confidence in completing laboratory procedures, attributing this to the real-time feedback offered by the avatar. The opportunity to make and correct mistakes in a safe virtual environment, without the risk of damaging equipment, was frequently cited as a key benefit. Several students highlighted the usefulness of visual and auditory cues, including simulations of incorrect actions such as wire overheating or component failure. These elements reinforced key safety concepts and made abstract principles more tangible and memorable.

Beyond cognitive support, the avatar served an affective role by reducing anxiety and supporting self-paced learning. Participants appreciated being able to revisit instructions without judgment, and four students from non-English-speaking backgrounds found the avatar's pronunciation, pacing, and multilingual capabilities particularly helpful for understanding technical terminology. This contributed to a more inclusive and accessible learning experience.

3.1.4 Insights from Study One

This pilot study highlights the potential of integrating a GenAI avatar within a VR laboratory to enhance student learning in electrical engineering. The significant improvement in test scores indicates that even brief, immersive instruction can lead to substantial gains in conceptual and procedural understanding.

The VR environment offered a safe, controlled space for experimentation. Students were able to make and correct mistakes without fear of damaging equipment, promoting deeper engagement. The intelligent avatar served as a responsive tutor, guiding learners through tasks with verbal explanations, visual cues, and real-time corrective feedback. This multimodal support was particularly valuable for students with limited prior laboratory experience.

Participants reported increased confidence and reduced anxiety, attributing this to the avatar's clear instructions and the opportunity to learn at their own pace. The simulation of realistic consequences—such as wire overheating or incorrect connections—made abstract principles tangible and memorable.

The system also supported inclusivity. Students from non-English-speaking backgrounds found the avatar's pronunciation and pacing helpful and appreciated the ability to revisit instructions without judgment. These affordances contributed to a more accessible and equitable learning experience.

Pedagogically, the AI avatar demonstrated its value as a scalable digital instructor, offering timely guidance in place of human tutors. While further development is needed—such as incorporating adaptive learning paths and emotional responsiveness—the results suggest strong potential for VR- and AI-enhanced instruction in technical education.

3.2 Empirical Study Two: GenAI for Critical Thinking in Social Science High-School Classroom

The second study we present investigates the use of GenAI to enhance high school students' critical thinking skills, with a particular focus on strengthening their reasoning and argumentation abilities. Critical thinking is widely recognized as a vital skill in both academic and real-world contexts. As modern societies become increasingly complex and change more rapidly, relying on learned, intuitive responses to novel problems can lead to suboptimal outcomes (Stanovich, 2016). Therefore, analyzing arguments, evaluating evidence, and making reasoned decisions are essential for navigating the flood of information, including misinformation and pseudoscience.

Despite its recognized importance, studies show that educational institutions frequently struggle to effectively develop students' critical thinking skills, leaving many graduates ill-prepared for the complexities of the modern world (Roksa & Arum, 2015). Teaching critical thinking is inherently challenging, as it requires both the acquisition of specific cognitive tools—referred to as *mindware* (Stanovich, 2018)—and the cultivation of thinking dispositions, such as actively open-minded thinking (Baron, 2019). However, traditional instructional methods often fall short in supporting this dual development, with research showing that many students exhibit only minimal gains in critical thinking over the course of their education (Belkin, 2022).

AI-powered pedagogical conversational agents (PCAs), enabled by recent advancements in GenAI, provide an engaging platform for structured argumentation and dialogue. These systems have demonstrated strong potential for delivering personalized, scalable instruction (Schlimbach et al., 2022) and for fostering the development of critical thinking through guided, interactive exchanges (Altay et al., 2022; Papathomas & Kuhn, 2017).

3.2.1 Application Development for Study Two

To address the gap in effective critical thinking instruction, we developed a GenAI-powered PCA using the OpenAI API, powered by the ChatGPT-4 language model. The agent was further enhanced with domain-specific knowledge to ensure accurate and context-aware responses during open-ended student interactions. The system was implemented as a full-stack TypeScript/JavaScript application using React, Node.js, and PostgreSQL. The interface of the application is shown in Figure 10.6. The PCA was designed to engage students in naturalistic dialogue, adopting the persona of a peer with whom students could interact. The intervention focused on two critical components of critical thinking: argument analysis and argument evaluation. These skills were chosen due to their foundational role in reasoning and decision-making and their well-documented importance in the literature (Stanovich & West, 1997).

The PCA presented students with arguments and claims, prompting them to identify key components such as the main claim, supporting reasons, and unstated assumptions. Students were also tasked with evaluating various claims' epistemic strength, ranging from scientifically valid to pseudoscientific. The learning experience was gamified, with students receiving points for correct responses to increase engagement and motivation (Bai et al., 2022).

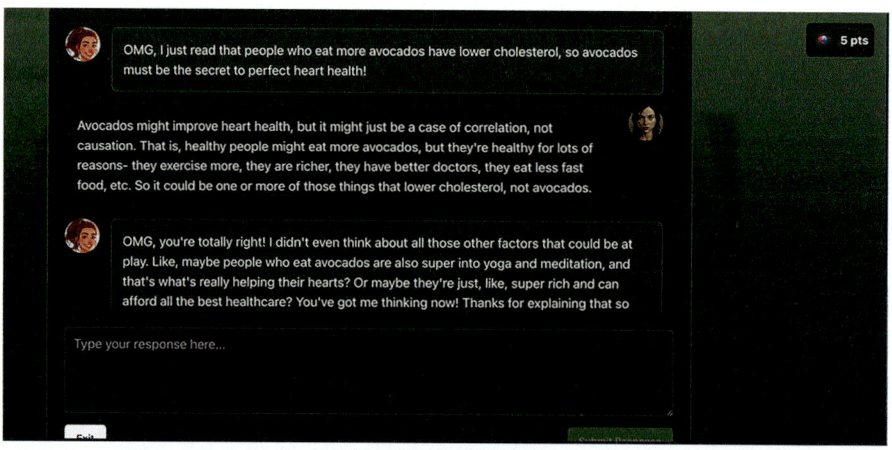

Figure 10.6 The critical thinking chat application.

3.2.2 Methodology for Conducting Study Two

The pilot study ran for two weeks and involved 15 high school students (aged 14–15). The students received five hours of instruction and practice, divided into two to three one-hour sessions per week. Instruction was delivered through direct teaching and interaction with the PCA. The time allocation was about 50/50 between direct teaching and PCA interaction. The PCA was programmed to engage students in two primary activities:

- Argument Component Identification: The PCA presented arguments on various topics, and students were prompted to identify components such as the main claim, supporting reasons, and objections.
- Argument Evaluation: The PCA, adopting the persona of a health-obsessed friend, presented students with various health-related arguments, some valid and others pseudoscientific. Students were tasked with evaluating the strength of each claim and identifying errors in the reasoning connecting claims.

Three assessments were administered to assess students' progress: a pretest before the intervention, a test after one week of instruction (Test 1), and a final test after two weeks (Test 2). These assessments measured students' ability to analyze and evaluate arguments, focusing on their accuracy in identifying argument components and evaluating claims.

3.2.3 Results from Study Two

Results showed a clear improvement in performance from the pretest (M = 50.15%, SD = 18.09, n = 13) to Test 1 (M = 64.53%, SD = 32.07, n = 15), and further improvement from Test 1 to Test 2 (M = 86.15%, SD = 13.76, n = 13). Paired t-tests were conducted to examine whether the differences in performance between the pretest and Test 1, and between Test 1 and Test 2, were statistically significant. Given that two comparisons were made, a Bonferroni correction was applied to adjust the significance level, with the adjusted alpha (α) set to .025 for each test.

The paired-sample t-test to compare the percentage of correct responses in the pretest and Test 1 showed a statistically significant increase in scores from the pretest to Test 1, $t(11) = 2.64$, $p < .025$. The mean difference between the pretest and test 1 was 14.38% (pooled SD = 28.51), indicating a substantial improvement in students' critical thinking skills after just one week of instruction. A second paired-sample t-test was conducted to compare the percentage of correct responses in Test 1 and Test 2. Again, scores were statistically significantly increased from Test 1 to Test 2, $t(14) = 4.07$, $p < .005$. The mean difference between Test 1 and Test 2 was 21.62% (pooled SD = 25.82), suggesting that students improved significantly during the second week of instruction. The Bonferroni correction was applied to account for the two comparisons made. The original significance level (α =.05\alpha =.05α = .05) was adjusted to .025 for both tests. The pretest vs. Test 1 comparison remained statistically significant at the .025 level. However, it was not significant at the .005 level (corrected for α=.01\alpha =.01α =.01). The Test 1 vs. Test 2 comparison remained statistically significant at both the 0.025 and the more stringent .005 level, indicating strong evidence of improvement over the second week of instruction.

3.2.4 Insights from Study Two

The findings from this pilot study provide preliminary evidence that AI-powered PCAs can significantly improve students' critical thinking skills, at least in argument analysis and evaluation. The statistically significant gains from pretest to test 2 suggest that direct instruction and AI-facilitated dialogue effectively promote these skills. Informal communication with the participants suggests that gamification may have further enhanced engagement, which likely contributed to the positive results.

These findings provide some reason to be optimistic about the tractability of teaching important critical thinking skills, given both the fundamentality of argument evaluation to critical thinking (Stanovich & West, 1997) and the potential of AI to support personalized learning (Tapalova & Zhiyenbayeva, 2022). The results demonstrate the feasibility of using AI-powered PCAs as a scalable, low-cost solution for teaching these skills in educational settings. Future research should involve larger, more diverse samples and, ideally, a randomized control design to validate these findings and explore the long-term impact of AI-facilitated critical thinking instruction. Additionally, further investigation could investigate the optimal duration and intensity of such interventions and their effectiveness in different educational contexts.

3.3 Empirical Study Three: AI-Chatbot Development for Career Advising for High-School Students

In the third study, we explore the potential of GenAI to support students in making informed and personalized career choices. While adaptive learning platforms have demonstrated that personalized pathways enhance student engagement and performance (Rivera et al., 2018), few existing systems integrate GenAI and conversational agents specifically for career guidance. This gap is significant, as many students enter tertiary education without clear goals, leading to frequent changes of major and high dropout rates. For example, in Australia, 25.4% of students who began university in 2017 had dropped out by 2023 due to dissatisfaction or lack of direction (Australian Department of Education, 2024; Gordon & Steele, 2015).

Career decision-making is particularly challenging for high school and university students, who often lack self-awareness and exposure to potential career paths aligned with their interests, abilities, and values. Traditional advising methods frequently fall short by offering rigid, one-size-fits-all recommendations that fail to engage students in reflective, meaningful dialogue (Aragon-Hahner et al., 2023). Access to personalized in-person counselling is also limited—many students receive only generic guidance through group sessions or broad personality tests, which may not address their specific needs. For instance, assessments like the Holland Code test classify users into general archetypes without offering the adaptive support needed for nuanced decision-making (Wei, 2024).

Although more personalized systems such as the Personalized Career-Path Recommender System (PCRS) have been developed (Qamhieh et al., 2020), these tools primarily target adult or university-level users, leaving younger students underserved. This lack of tailored support is problematic, as students often require both exploratory and structured guidance during a critical developmental period (Aragon-Hahner et al., 2023).

GenAI-powered conversational agents offer a compelling alternative. These systems can simulate the responsiveness of a human advisor and scale efficiently to support large numbers of students. Through dynamic, context-aware dialogue, they deliver both cognitive and emotional support, making them well-suited to facilitate meaningful career exploration during high school and university years (Adamopoulou & Moussiades, 2020; Nguyen et al., 2022).

3.3.1 Application Development for Study Three

The chatbot we have developed, FindMyFuture, is a GenAI-driven career decision support tool designed to assist high school and undergraduate students in exploring and identifying potential career paths. The system integrates self-discovery activities—including assessments of interests, value preferences, and aptitudes—and generates personalized career recommendations based on these inputs. Powered by OpenAI's GPT-4o-mini, which has been customized for this application, the chatbot processes user data to create individual career profiles and tailored guidance. The software architecture of the application is illustrated in Figure 10.7.

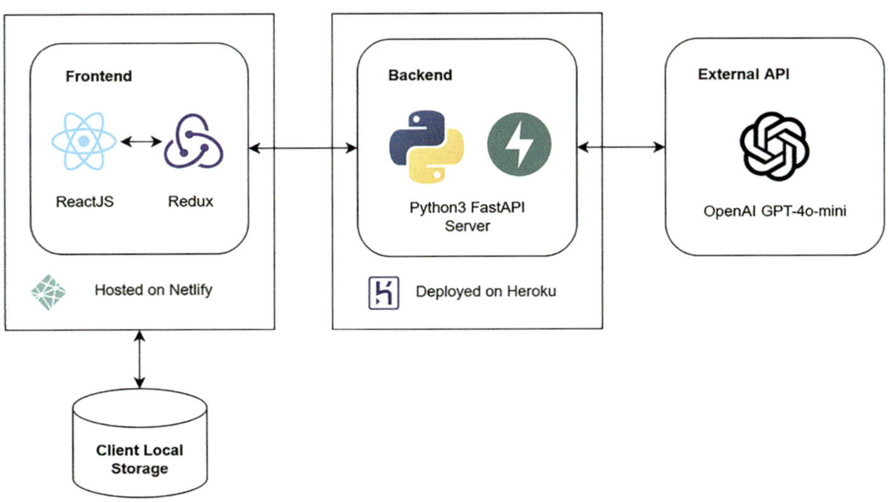

Figure 10.7 Software architecture of our career advisor Chatbot.

The system is designed to allow users to effortlessly input their areas of interest and existing skill sets through an intuitive interface. This user-centric design lowers the cognitive barrier for students, many of whom may be unfamiliar with digital career advisory platforms. Upon launch, the application's home screen (Figure 10.8) presents users with a clean layout, guiding them to initiate the process of career exploration. The system prompts students to reflect on and input their hobbies, preferred work values (e.g., creativity, stability, financial reward), and personal aspirations. These inputs serve as the foundation for constructing a dynamic user profile.

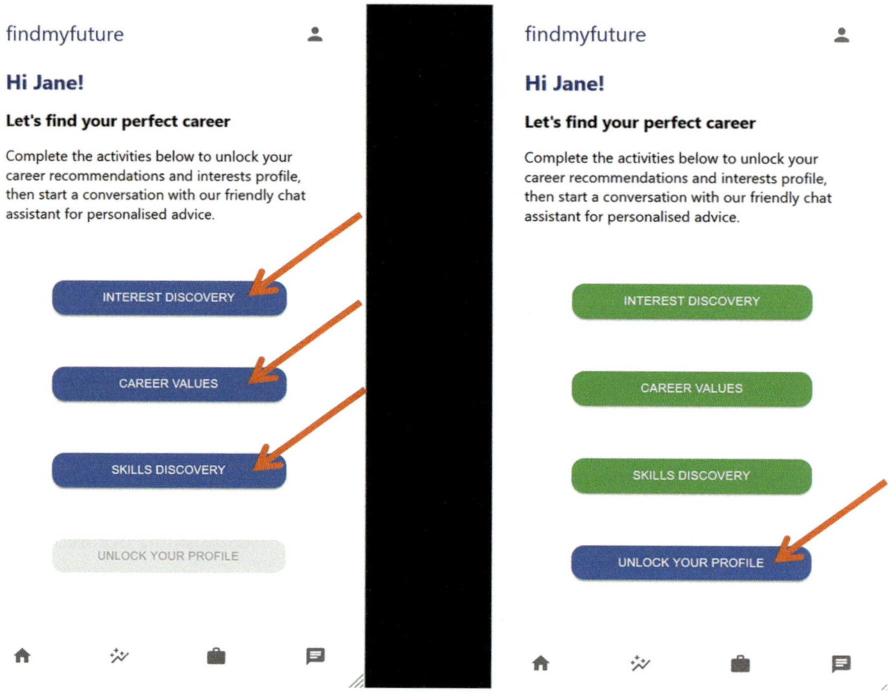

Figure 10.8 The career advisor home screen.

As illustrated in Figure 10.9, the platform allows students to input both qualitative and quantitative information—ranging from personality traits and leisure activities to career-related values. This step goes beyond the rigid categorizations used in traditional career quizzes by enabling users to describe themselves in more nuanced and context-sensitive terms. Through this design, the system fosters greater self-awareness and encourages students to think critically about what truly motivates and defines them.

Once the necessary information is entered, the system processes the data and generates a personalized profile of the user. As shown in Figure 10.10, this profile is visualized using a range of interactive and accessible charts. These graphical summaries help students identify patterns in their interests, strengths, and value orientations, facilitating a clearer understanding of how these traits align with

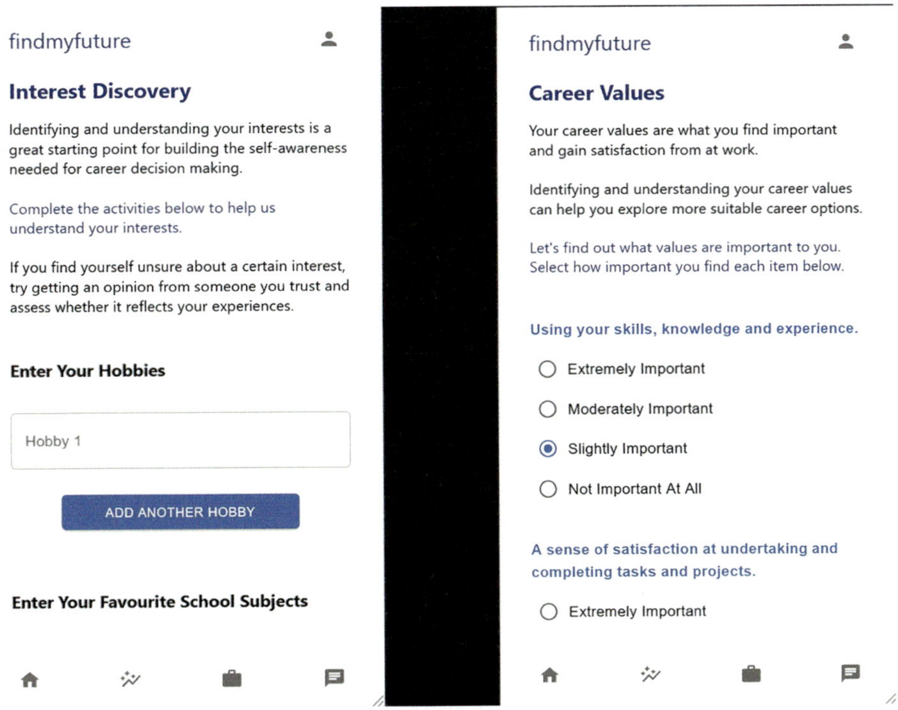

Figure 10.9 Entering hobbies and career values.

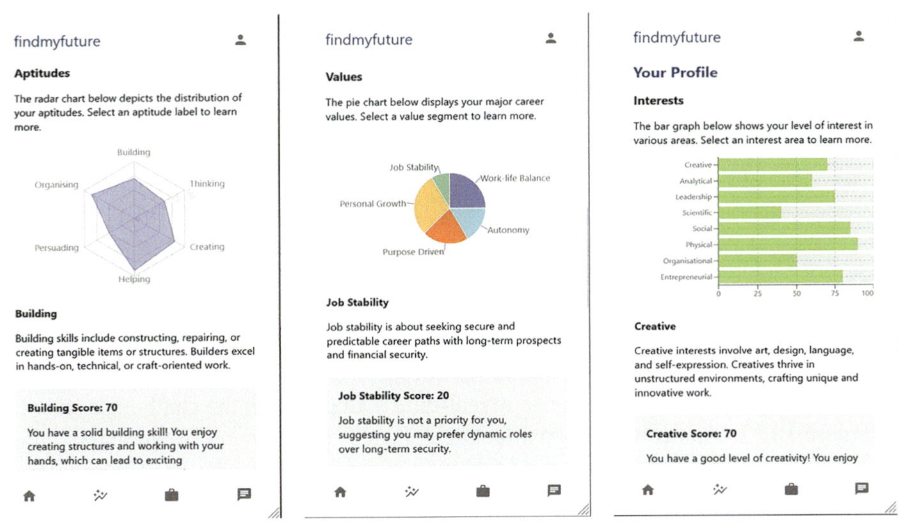

Figure 10.10 Analysis of users' characteristics.

potential careers. Visualizing this information supports metacognitive reflection, empowering students to make more informed and confident decisions about their futures.

Following this analysis, the system utilizes its GenAI capabilities to generate a tailored list of career recommendations. These suggestions are not only aligned with the user's profile but are also accompanied by detailed job descriptions, as illustrated in Figure 10.11. Each recommendation includes key information such as required qualifications, typical work environments, projected job growth, and skill development pathways. This holistic presentation ensures that students are not only exposed to familiar career paths but are also encouraged to consider lesser-known but potentially rewarding options aligned with their profile.

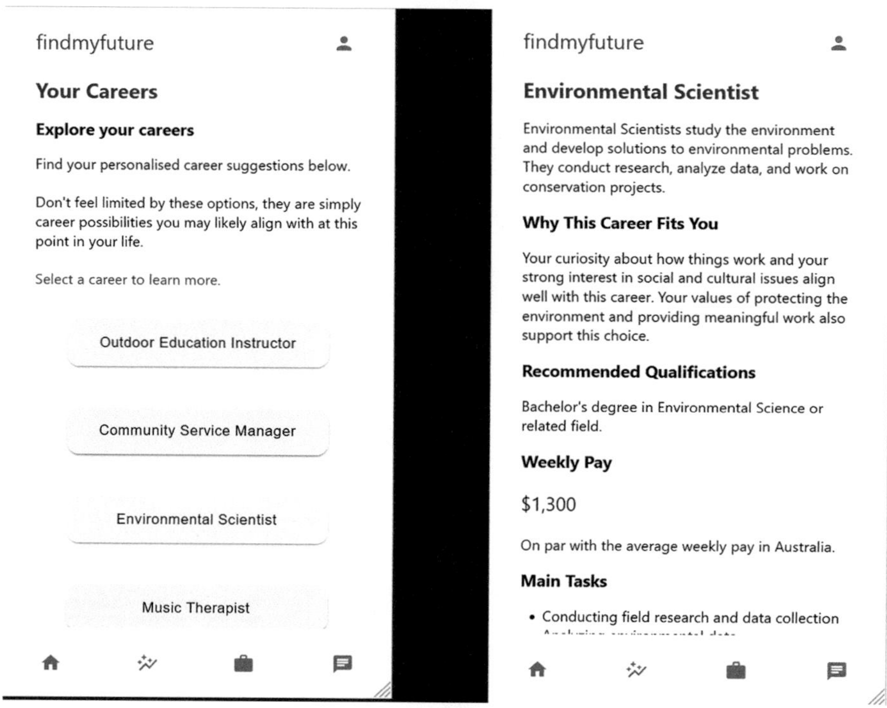

Figure 10.11 Job recommendation screen.

To further support exploration and clarify uncertainties, the platform features a conversational chatbot. If users have additional questions or wish to delve deeper into specific career options, they can engage in real-time dialogue with the chatbot, as shown in Figure 10.12. This conversational feature enhances engagement by mimicking the responsive and empathetic nature of a human career advisor. The chatbot provides personalized guidance, answers complex questions, and even suggests next steps such as courses, certifications, or internship opportunities related to a chosen career path.

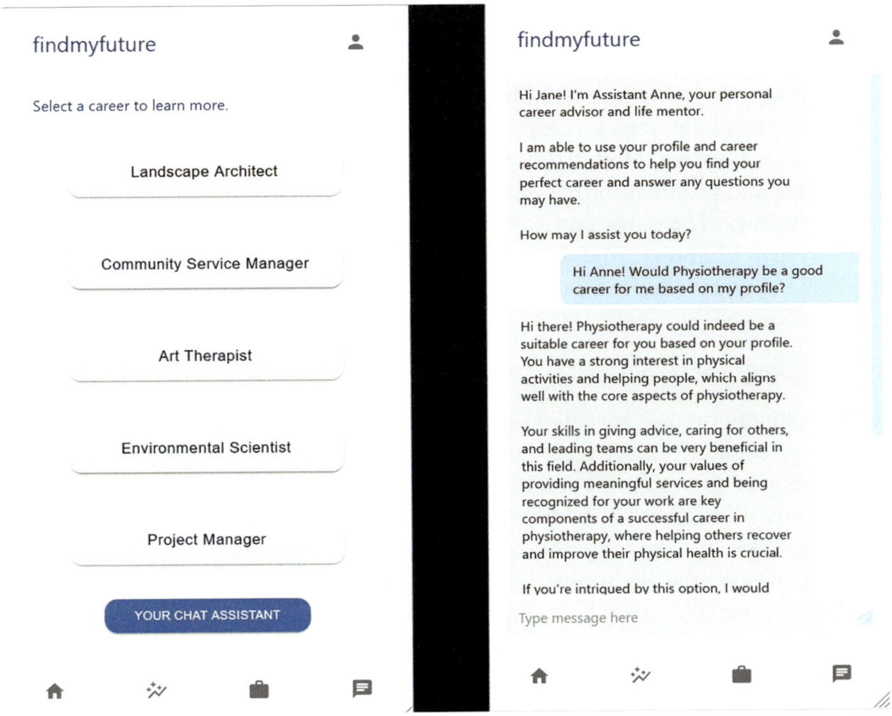

Figure 10.12 System Chatbot.

3.3.2 Methodology for Conducting Study Three

Our career recommendation application was tested with 22 individuals (N = 22), primarily students along with a few early-career professionals, all aged 18 or older. Prior to participation, each individual was briefed on the study's purpose, provided with an overview of the application, informed about the research ethics clearance, and introduced to the experimental procedure. The experiment consisted of four components: participants completed a pre-questionnaire, undertook a traditional job quiz and personality test, interacted with the career recommendation application, and then completed a post-questionnaire. Each session lasted approximately one hour per participant.

- In the pre-questionnaire, the participants provided information about their academic and demographic backgrounds, whether they knew what career to pursue after high school, and their level of self-awareness in their interests, values, and skills. We also asked if they had previously used career decision support tools, taken part in a career development program, or received consultations from career advisors.

- In the post-questionnaire, the participants indicated:
  - User Satisfaction: The perceived benefit gained after using the system.
  - Accuracy: The extent to which career recommendations align with the participant's intentions and aspirations.
  - Personalization: How relevant the participant finds the career recommendations and profile information.
  - Diversity: The range and novelty of career recommendations provided.
  - System Usability: The degree to which the application was easy to navigate and operate independently.

3.3.3 Results from Study Three

Post-questionnaire data revealed that 76% of participants found FindMyFuture to offer more accurate career recommendations than traditional job quizzes, a result attributable to the system's ability to generate context-aware and personalized suggestions using GenAI. Additionally, 80% of users reported that the platform provided more eye-opening and diverse career insights, expanding their understanding of potential pathways. Furthermore, 71% of participants indicated they would recommend FindMyFuture to a friend over conventional tools like standard quizzes or personality tests. These findings underscore the system's effectiveness and perceived value. Qualitative feedback further reinforced these outcomes, with participants noting:

> *The introduction of AI was brilliant, and having a conversation about my skills and passions is impressive*. [Participant 20]
>
> *The response was more organic and understandable compared to [the] general personality test*. [Participant 7]
>
> *The recommended jobs suggested both occupations I had previously been employed in or had researched extensively, as well as other jobs that, while in my periphery, I had not fully considered before*. [Participant 5]

3.3.4 Insights from Study Three

The pilot study revealed that FindMyFuture effectively enhances career exploration by offering more personalized and diverse recommendations than traditional tools. A majority of participants reported that the GenAI-generated suggestions more accurately reflected their interests and aspirations, while many noted that the system introduced them to career paths they had not previously considered. This highlights the chatbot's potential to expand users' awareness beyond conventional job options.

Participants also found the conversational interface to be a major strength. Unlike static assessments, the chatbot enabled users to ask questions and receive tailored responses in real time, fostering deeper engagement. Seventy-one percent of users indicated they would recommend FindMyFuture to others, reflecting strong user satisfaction. Qualitative feedback further emphasized the system's role in promoting self-reflection. By requiring students to articulate their interests, values, and skills, and then presenting a personalized profile and recommendations, the system supported greater clarity in career decision-making. Users appreciated the transparent rationale behind suggestions, which increased trust and confidence.

Overall, the findings suggest that GenAI-powered tools like FindMyFuture can supplement traditional career advising by offering accessible, adaptive, and student-centered guidance—especially for users with limited access to individualized support.

3.4 Empirical Study Four: AI-Chatbot Development for University Admission Prediction

In this study, we developed and evaluated an AI-powered system designed to predict high school students' likelihood of success in university entrance applications. University admissions processes are inherently complex and continually evolving to reflect shifting societal, economic, and political priorities. Certain disciplines, such as medicine, remain especially competitive, requiring applicants to demonstrate academic excellence, complete prerequisite coursework, and undertake additional assessments such as the University Clinical Aptitude Test (UCAT). Preparing these application packages can be financially burdensome—due to costs associated with test registration and study materials—and psychologically taxing, as applicants often face significant uncertainty while trying to align their applications with personal goals and aspirations (Al-Surmi et al., 2022).

In Australia, while support is available through high school advisors and university services, access is often limited. High student-to-staff ratios and appointment backlogs frequently delay or restrict applicants' opportunities for meaningful consultation (Thottoli et al., 2023). In recent years, AI models have been introduced to mitigate some of these challenges (e.g., Dierckx de Casterlé et al., 2021). However, most existing models focus narrowly on quantitative indicators such as exam scores, standardized test results, and the number of awards, essays, or recommendation letters. This exclusive reliance on numerical data may overlook essential contextual elements that reflect a student's broader potential (Schussler & Weatherton, 2021).

Qualitative dimensions—such as the depth of extracurricular involvement, the narrative richness of personal essays, and the insight offered by recommendation letters—are often crucial in articulating a student's unique voice, strengths, and readiness for higher education. AI systems that disregard these aspects risk excluding applicants who may excel in university environments but fall outside conventional academic profiles.

To address these limitations, we developed an AI system offering personalized predictions of a student's admission prospects to the Medical School at UNSW. We focused on Medical School admissions due to their high competitiveness and the substantial psychological pressure they place on applicants. By delivering accurate, data-driven predictions, the system seeks to reduce student stress, recalibrate unrealistic expectations, enable early intervention, and help guide students toward educational and career pathways that are better aligned with their individual strengths and profiles.

3.4.1 Application Development for Study Four

The university admission prediction system was developed using the OpenAI API, which enabled advanced NLP and intelligent decision-making capabilities. These capabilities allowed the system to interpret both structured and unstructured data, accommodating a variety of input formats and sources. Users could upload a broad range of application materials, including (a) age, (b) gender, (c) ethnicity, (d) high school academic records, (e) ATAR score, (f) curriculum vitae (CV), (g) high school report, (h) personal statement, and (i) any additional documents the applicant deemed relevant. This flexibility ensured that the system could process diverse inputs beyond the limitations of traditional form-based systems.

Unlike static rule-based models, this system employed neural networks and other machine learning models capable of learning complex patterns from data. The training pipeline followed a standard supervised learning process, as outlined below and illustrated in Figure 10.13 (cf., Murel & Kavlakoglu, 2024).

1 *Raw data and target*: The training data set is gathered at this stage, and the model's output type is configured.
2 *Feature engineering* involves selecting unique characteristics and transforming the raw data into relevant data for the NN model.
3 *Training, validation, and test set*: All sets resulting from step 2 (Feature Engineering) are then used to train, validate, and test the NN model. Neural networks produce high accuracy and can process different data types. The test set seeks to assess the accuracy of the neural network model.
4 *Test the resultant model* to evaluate the prediction outcomes for new input data.

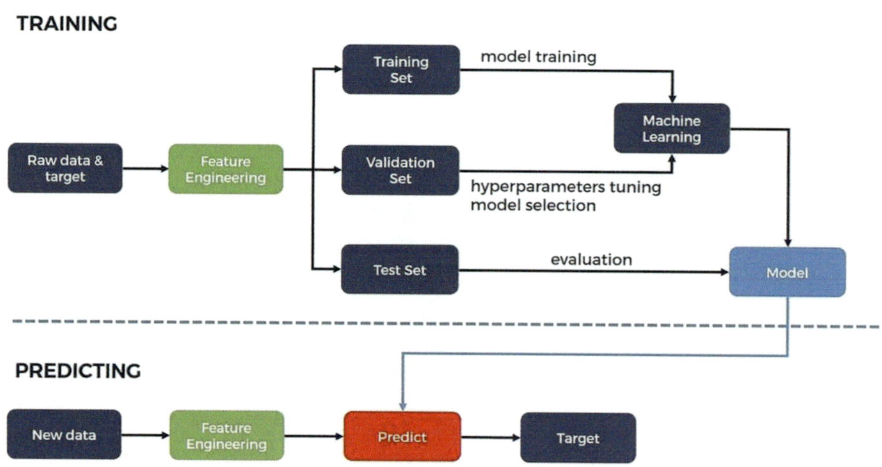

Figure 10.13 The process of training a neural network model.

The application was deployed on Google Cloud using the Vertex AI platform, which offered scalable compute resources and streamlined model deployment pipelines. This cloud infrastructure enabled real-time inference and integration into a user-friendly web interface. The architecture of the deployed system is depicted in Figure 10.14, which outlines data flow from user input to model inference and output visualization.

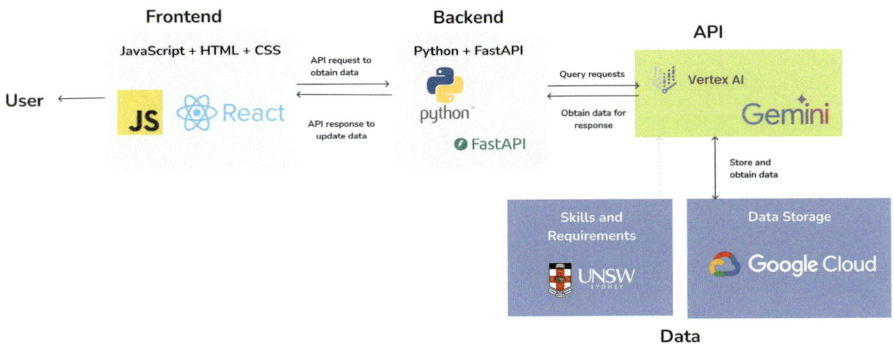

Figure 10.14 The software architecture framework of our university admission prediction application.

Due to the absence of real application data from UNSW, a synthetic dataset was generated to simulate various applicant profiles. This dataset reflected the characteristics of both successful and unsuccessful applicants, including academic achievement, extracurricular involvement, and personal narrative strength. The use of synthetic data allowed for controlled experimentation and early-stage validation of the system's predictive framework. In practice, the system was designed to evaluate a student's admission prospects holistically, moving beyond numeric scores to consider qualitative indicators of motivation, leadership, and alignment with program goals.

The process of interacting with the university admission prediction system is designed to be intuitive, informative, and user-friendly, allowing applicants to receive real-time insights into their admission prospects. Figures 10.15–10.18 illustrate the key steps involved in using the system. As shown in Figure 10.15, users begin by entering their personal information, including demographic details such as age, gender, and ethnicity. This step ensures that the system accounts for background information that may influence application outcomes, especially in programs that prioritize diversity, outreach, or specific quotas.

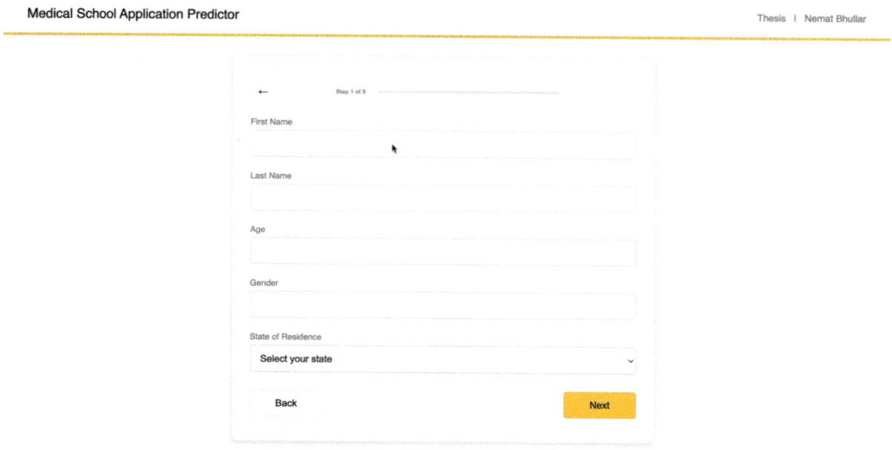

Figure 10.15 Entering personal information.

In Figure 10.16, users proceed to input their academic background, including high school grades, subjects taken, and ATAR score.

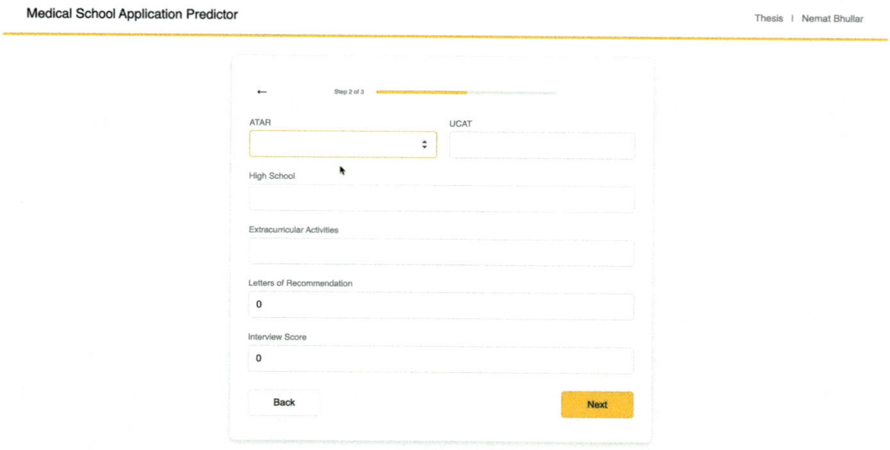

Figure 10.16 Entering high school information.

Next, as shown in Figure 10.17, applicants are prompted to upload supporting documents, such as their curriculum vitae (CV), high school reports, and personal statements. These qualitative documents are processed using NLP models capable of extracting relevant themes, sentiments, and indicators of motivation, leadership, and alignment with the program's values.

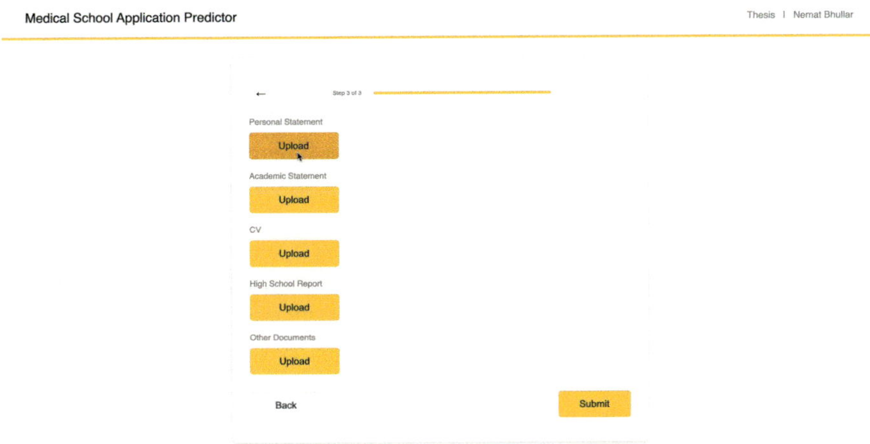

Figure 10.17 Uploading documents.

Once all data have been submitted, the system analyzes the applicant's full profile using the trained neural network model. In Figure 10.18, users receive a personalized prediction of their likelihood of admission. This includes a probability score along with interpretive feedback explaining how different elements of the application influenced the result. Where appropriate, the system may also suggest areas for improvement—such as strengthening the personal statement or gaining relevant experience—making the tool not only predictive but also developmental.

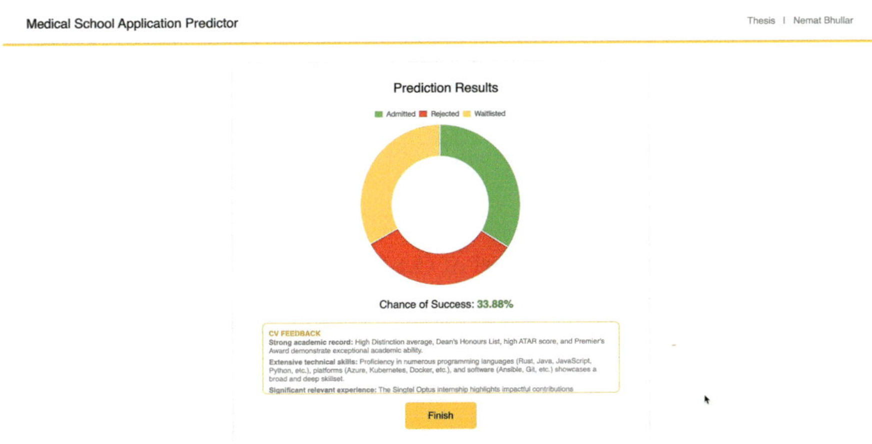

Figure 10.18 Receiving the prediction of the acceptance rate to the university.

3.4.2 Methodology for Conducting Study Four

A pilot study was conducted with seven high school students interested in applying to medical school to evaluate our GenAI-powered university admission prediction system. Each participant took part in a 30-minute session that included a brief orientation, system interaction, and a post-questionnaire. During the session, participants entered personal and academic details, uploaded relevant documents (such as CVs and personal statements), and received feedback and admission probability estimates from the system. After using the system, participants completed a questionnaire assessing its clarity, usefulness, accuracy, ease of use, and overall satisfaction. Also, our AI model performance was evaluated using F1-score, ROC AUC, and Precision-Recall AUC metrics.

3.4.3 Results from Study Four

Qualitative findings from the post-questionnaire revealed positive user perceptions. Students reported that the system helped them better understand how various components of their application—such as their personal statement, ATAR, and CV—were evaluated. Many noted that receiving predictions and feedback reduced their anxiety and increased their confidence in planning university applications. The form-based interface was particularly well received, with participants highlighting its intuitive design. Participants felt that the application provided a more streamlined, focused, and purposeful experience compared to traditional guidance sessions or unstructured chatbot interactions due to its clear structure, intuitive interface, and tailored feedback mechanisms.

In addition to gathering user feedback, the model's performance was evaluated using standard machine learning metrics to assess predictive reliability. The model achieved an F1-score (macro and micro) of 0.5, a Precision-Recall AUC of 33.1%, and a ROC AUC of 49.6%, indicating performance slightly above random classification. The confusion matrix showed a 44% true positive rate for applicants who were actually admitted. While the model captured some relevant patterns, the results suggest moderate predictive capability and highlight the need for further refinement—particularly in improving precision and recall for high-stakes predictions such as medical school admission.

3.4.4 Insights from Study Four

This study demonstrates the potential of GenAI to augment university admission prediction systems by incorporating both quantitative and qualitative dimensions of an applicant's profile. Unlike existing tools that primarily rely on test scores and grades, this system processes personal statements, CVs, and other narrative-based documents to provide more holistic recommendations.

Participants reported high levels of satisfaction, with several expressing appreciation for receiving tailored feedback that helped them better understand their competitive standing. The real-time feedback feature also reduced their anxiety regarding uncertain admissions decisions, which aligns with prior literature on AI's role in reducing cognitive load in complex decision-making (Schussler & Weatherton, 2021).

One of the most significant insights from this pilot is the value of combining narrative content with structured academic records. By doing so, the model was able to identify students who may not be top performers in standardized tests but show exceptional potential in other areas, such as leadership, community engagement, or scientific curiosity—attributes highly valued by medical schools.

However, this system also highlights ethical considerations. For instance, how data is sourced, interpreted, and protected will influence both fairness and trust. Moreover, transparency in model decisions is vital if applicants are to rely on these predictions when planning their futures. Future iterations of the system could incorporate explainable AI components to offer justifications for recommendations and support counsellors in interpreting outcomes (Hashemi, Darejeh, & Cruz, 2024, 2025). Additional testing with more diverse student groups, including applicants from rural or disadvantaged backgrounds, would ensure greater inclusivity and generalizability of results.

4 Discussion

This chapter has demonstrated the transformative potential of GenAI in education. Across a diverse set of applications presented in the four empirical studies—including laboratory instruction, critical thinking development, career exploration, and university admission prediction—GenAI systems consistently provided personalized, data-informed experiences that meaningfully supported learners. The VR-based electrical engineering laboratory, for instance, leveraged a GenAI-powered avatar to deliver real-time, context-sensitive feedback, enabling students to complete complex tasks with greater confidence and reduced anxiety. These outcomes align with prior research emphasizing the importance of personalized, responsive guidance in reducing cognitive load and enhancing engagement in immersive learning environments (Darejeh et al., 2022; Fink et al., 2024).

Similarly, the AI-powered PCA for critical thinking instruction addressed a long-standing pedagogical gap by enabling structured dialogue and guided reasoning practice. Students demonstrated significant gains in their ability to analyze and evaluate arguments—skills that are difficult to cultivate through conventional instruction. The PCA's capacity to scaffold reasoning processes supports Stanovich's (2016) emphasis on "mindware" development, while the gamified elements reinforced motivation and engagement, in line with findings on the benefits of educational gamification (Bai et al., 2022).

In the domain of career guidance, the FindMyFuture chatbot illustrated how GenAI can offer nuanced and adaptive support that goes beyond traditional tools. Unlike static questionnaires or typology-based career quizzes, the system dynamically integrated both qualitative and quantitative inputs to generate tailored recommendations. Participants reported that the system not only offered more accurate and diverse career suggestions but also helped them reflect more deeply on their values and aspirations. These findings are consistent with the literature advocating for context-aware and dialogue-based support in educational decision-making (Nguyen et al., 2022; Qamhieh et al., 2020).

The university admission prediction system extended these capabilities by modelling holistic applicant profiles that incorporate academic performance, extracurricular achievements, and narrative-based components such as personal statements. This approach acknowledges that admissions decisions are multifaceted and that qualitative indicators often capture dimensions of student potential that quantitative metrics alone cannot. While the model's predictive performance was moderate, users still found the system helpful in clarifying expectations and reducing application-related anxiety—an important finding in light of research linking GenAI to cognitive load reduction and confidence-building in high-stakes settings (Schussler & Weatherton, 2021).

A unifying strength across all four systems is their combination of GenAI's NLP capabilities with human-centered design principles, resulting in tools that are intuitive, adaptive, and inclusive. These systems created safe spaces for exploration—environments where students could ask questions, reflect on feedback, and learn from mistakes without fear of judgment. This is particularly significant for students from culturally and linguistically diverse backgrounds, who may face additional barriers in traditional educational settings (Vincent-Lancrin & Van Der Vlies, 2020). Features such as multilingual avatars and culturally responsive recommendation engines highlight the capacity of GenAI to support equitable learning environments.

Another core strength lies in the integration of qualitative data, such as narrative reflections and open-ended responses, enabling the systems to deliver highly context-aware feedback. This shift away from purely quantitative evaluation represents an important evolution in educational technology—one that recognizes the learner as a complex, multifaceted individual whose educational needs cannot be reduced to test scores alone.

However, these advancements come with challenges. Ethical considerations around data privacy, algorithmic transparency, and potential bias must be addressed. Systems like those for career recommendation and admission prediction require access to sensitive personal data, underscoring the need for robust data governance frameworks (Al-Surmi et al., 2022). Moreover, algorithmic decisions must be interpretable and auditable to ensure fairness and build user trust—especially in systems that can influence high-stakes decisions.

There are also pedagogical risks. While GenAI-powered avatars and chatbots can provide scalable instructional support, they must complement—not replace—human educators. The relational, emotional, and contextual insights offered by teachers remain irreplaceable. Striking the right balance between technological efficiency and human connection is essential to preserve the depth and social richness of the educational experience (Fink et al., 2024).

Looking ahead, the findings of this chapter highlight the need for continued innovation and research in this space. Future developments should focus on refining adaptive learning systems that respond to individual cognitive styles and emotional states. For example, advanced NLP models can be trained to recognize when a learner is struggling with a concept and offer differentiated explanations tailored to their background knowledge (Dubey et al., 2024).

Moreover, GenAI-based virtual tutors can be embedded within metaverse environments, where groups of students interact through their own avatars in collaborative tasks. These intelligent tutors can observe group dynamics, offer timely guidance, and intervene with personalized suggestions to improve teamwork and learning outcomes. For instance, during a simulated engineering project in the metaverse, a GenAI tutor could highlight discrepancies in group problem-solving strategies, prompt quieter participants to contribute, or recommend resources relevant to a group's emerging challenges. This integration not only supports knowledge acquisition but also cultivates interpersonal skills such as collaboration, negotiation, and leadership—critical competencies in both academic and professional contexts.

In summary, the four studies affirm that GenAI—especially when integrated with immersive and adaptive interfaces—can deliver scalable, personalized, and contextually rich educational experiences. Yet, realizing its full potential will depend on responsible design, continuous evaluation, and thoughtful alignment with pedagogical principles that center on learner agency, inclusion, and long-term development.

5 Conclusion

This chapter has explored the transformative role of GenAI in education, focusing on five key trends: enhancing personalization, empowering students through AR/VR technology, facilitating foreign language learning and exchange, and advancing student monitoring and evaluation. These trends were illustrated through four practical empirical studies, each showcasing the integration of GenAI in different educational contexts—from VR-based laboratory instruction and critical thinking development to personalized career guidance and university admission prediction.

A consistent theme across the studies is GenAI's capacity to prioritize personalization by dynamically tailoring outputs to individual user inputs. While VR technology was explicitly employed in the electrical engineering laboratory in study 1, its pedagogical value extends beyond this application. VR-enabled avatars could facilitate real-time interactions between students and educators, offering personalized guidance and formative assessment. Similarly, in career advising, VR could be used to simulate immersive job shadowing or internship experiences, allowing students to explore potential career paths in realistic, hands-on environments. These applications underscore the inclusive potential of GenAI and VR systems, particularly in accommodating diverse linguistic, cultural, and learning needs, thereby promoting global accessibility in education.

The studies also highlight how GenAI systems can enhance system-level educational infrastructure. Tools for career counselling and university admission prediction can be integrated into school and government platforms, offering scalable solutions to meet growing demands for personalized student support (Majjate et al., 2024). The inclusion of interactive features—such as self-paced quizzes, real-time feedback, and adaptive conversational agents—reduces the reliance on constant tutor supervision while preserving instructional quality. These developments open new possibilities for reaching underserved populations, especially in regions with limited access to human advisors or high-quality educational resources.

Looking ahead, future innovations could build on these foundations by integrating immersive remote internships, collaborative metaverse classrooms, and emotionally responsive AI tutors. Such enhancements will not only broaden access to authentic learning experiences but also contribute to more equitable and efficient education systems. As educational technologies evolve, it is crucial to ensure that GenAI remains aligned with pedagogical values, ethical standards, and learner-centered design principles—ultimately fostering inclusion, engagement, and lifelong learning.

References

Acosta-Vargas, P., Acosta-Vargas, G., Salvador-Acosta, B., & Jadán-Guerrero, J. (2024, June). Addressing web accessibility challenges with generative artificial intelligence tools for inclusive education. In *2024 Tenth International Conference on eDemocracy & eGovernment (ICEDEG)* (pp. 1–7). IEEE.

Adamopoulou, E., & Moussiades, L. (2020). Chatbots: History, technology, and applications. *Machine Learning with Applications*, *2*, 100006.

Ahuja, A. S., Polascik, B. W., Doddapaneni, D., Byrnes, E. S., & Sridhar, J. (2023). The digital metaverse: applications in artificial intelligence, medical education, and integrative health. *Integrative Medicine Research*, *12*(1), Article 100917. https://doi.org/10.1016/jimr.2022.100917

AlAli, R., & Wardat, Y. (2024). Opportunities and challenges of integrating generative artificial intelligence in education. *International Journal of Religion*, *5*(7), 784–793.

Albarracin-Acero, D. A., Romero-Toledo, F. A., Saavedra-Bautista, C. E., & Ariza-Echeverri, E. A. (2024). Virtual reality in the classroom: Transforming the teaching of electrical circuits in the digital age. *Future Internet*, *16*(8), 279–279. https://doi.org/10.3390/fi16080279

Al-Surmi, A., Bashiri, M., & Koliousis, I. (2022). AI based decision making: Combining strategies to improve operational performance. *International Journal of Production Research*, *60*(14), 4464–4486.

Altay, S., Schwartz, M., Hacquin, A. S., Allard, A., Blancke, S., & Mercier, H. (2022). Scaling up interactive argumentation by providing counterarguments with a chatbot. *Nature Human Behaviour*, *6*(4), 579–592.

Aragon-Hahner, S., Körber, J., Schneller, M., & Strickroth, S. (2023). FindMyself: A mobile self-reflection app to support students in career decision-making. *Proceedings of the 22nd International Conference on Mobile and Ubiquitous Multimedia (MUM '23)*, 463–476. https://doi.org/10.1145/3626705.3627774

Australian Department of Education. (2024). *Higher education statistics*. Retrieved November 15, 2024, from https://www.education.gov.au/higher-education-statistics,

Bagley, N. (2024). *AI Chatbots in schools: Findings from a Poll of K-12 teachers, students, parents, and college undergraduates*. Insight Research. Retrieved from https://youthtoday.org/2024/06/ai-chatbots-in-schools-findings-from-a-poll/

Bai, S., Hew, K. F., Gonda, D. E., Huang, B., & Liang, X. (2022). Incorporating fantasy into gamification promotes student learning and quality of online interaction. *International Journal of Educational Technology in Higher Education*, *19*(1), 29.

Baron, J. (2019). Actively open-minded thinking in politics. *Cognition*, *188*, 8–18.

Belkin, I. (2022). Organization and analysis of a pedagogical experiment in the conditions of student training in institutions of higher education. In *The analysis of the characteristic concepts of the establishment and development of Ukrainian society in the context of the historical, scientific and philosophical approaches* (pp. 71–90). Primedia eLaunch.

Bick, A., Blandin, A., & Deming, D. J. (2024). *The rapid adoption of generative ai* (No. w32966). National Bureau of Economic Research.

Chamola, V., Bansal, G., Das, T. K., Hassija, V., Sai, S., Wang, J., Zeadally, S., Hussain, A., Yu, F., Guizani, R. M., & Niyato, D. (2024a). Beyond reality: The pivotal role of generative AI in the metaverse. *IEEE Internet of Things Magazine, 7*(4), 126–135. https://doi.org/10.1109/iotm.001.2300174

Chamola, V., Sai, S., Bhargava, A., Sahu, A., Jiang, W., Xiong, Z., Dusit Niyato, & Hussain, A. (2024b). A comprehensive survey on generative AI for metaverse: Enabling immersive experience. *Cognitive Computation, 16*, 3286–3315. https://doi.org/10.1007/s12559-024-10342-9

Cong-Lem, N., & Daneshfar, S. (2024). Generative AI and second/foreign language education from Vygotsky's cultural-historical perspective. In *Innovations in Technologies for Language Teaching and Learning* (pp. 175–188). Springer Nature Switzerland.

Córdova-Esparza, D. M., Terven, J., Romero-González, J. A., Córdova-Esparza, K. E., López-Martínez, R. E., García-Ramírez, T., & Chaparro-Sánchez, R. (2025). Predicting and preventing school dropout with business intelligence: Insights from a systematic review. *Information, 16*(4), 326.

Dang Quan, T., Wei-Han Tan, G., Aw, E. C. X., Cham, T. H., Basu, S., & Ooi, K. B. (2024). Can you resist the virtual temptations? Unveiling impulsive buying in metaverse retail. *Asia Pacific Journal of Marketing and Logistics, 36*(10), 2259–2280.

Darejeh, A. (2023). Empowering education through eerp: A customizable educational vr escape room platform. In *2023 IEEE International Symposium on Mixed and Augmented Reality Adjunct (ISMAR-Adjunct)* (pp. 764–766). IEEE.

Darejeh, A., Huynh, L., Ji, X., Marcus, N., & Polly, P. (2024a). Assessing the efficacy of virtual reality-based laboratories versus real-life Settings for PCR procedures, within medical science laboratories. In *2024 IEEE International Symposium on Mixed and Augmented Reality Adjunct (ISMAR-Adjunct)* (pp. 489–492). IEEE.

Darejeh, A., Marcus, N., & Sweller, J. (2022). Increasing learner interactions with E-learning systems can either decrease or increase cognitive load depending on the nature of the interaction. *L'Année psychologique, 122*(3), 405–437.

Darejeh, A., Moghadam, T. S., Delaramifar, M., & Mashayekh, S. (2024c). A framework for AI-powered decision making in developing adaptive e-learning systems to impact learners' emotional responses. In *2024 11th International and the 17th National Conference on E-Learning and E-Teaching (ICeLeT)* (pp. 1–6). IEEE.

Darejeh, A., Mondal, M., Marcus, N., & Vassar, A. (2024b). Exploring the cognitive load effects of diverse virtual reality interaction methods in interactive educational platforms. In *2024 IEEE International Symposium on Mixed and Augmented Reality Adjunct (ISMAR-Adjunct)* (pp. 435–438). IEEE.

Diaz, J., & Moreira, F. (2024). Toward educational sustainability: An AI system for identifying and preventing student dropout. *IEEE Revista Iberoamericana de Tecnologias del Aprendizaje, 19*, 100–110.

Dierckx de Casterlé, B., De Vliegher, K., Gastmans, C., & Mertens, E. (2021). Complex qualitative data analysis: Lessons learned from the experiences with the qualitative analysis guide of Leuven. *Qualitative Health Research, 31*(6), 1083–1093.

Dubey, A., Jauhri, A., Pandey, A., Kadian, A., Al-Dahle, A., Letman, A., Mathur, A., Schelten, A., Yang, A., & Fan, A. (2024). The llama 3 herd of models. *arXiv preprint arXiv, 2407,* 21783.

Fadhel, M. A., Duhaim, A. M., Albahri, A. S., Al-Qaysi, Z. T., Aktham, M. A., Chyad, M. A., Wael Abd-Alaziz, Albahri, O. S., A.H. Alamoodi, Laith Alzubaidi, Gupta, A., & Gu, Y.

(2024). Navigating the metaverse: Unraveling the impact of artificial intelligence—A comprehensive review and gap analysis. *Artificial Intelligence Review*, *57*, Article 264. https://doi.org/10.1007/s10462-024-10881-5

Fetzer, J. H. (1990). What is artificial intelligence? In J. H. Fetzer (Ed.), *Artificial intelligence: Its scope and limits* (pp. 3–27). Springer.

Fink, M. C., Robinson, S. A., & Ertl, B. (2024). AI-based avatars are changing the way we learn and teach: Benefits and challenges. *Frontiers in Education, 9*, Article 1416307. https://doi.org/10.3389/feduc.2024.1416307

Fletcher, R., & Nielsen, R. (2024). What does the public in six countries think of generative AI in news? *Reuters Institute for the Study of Journalism.*

Godwin-Jones, R. (2023). Emerging spaces for language learning: AI bots, ambient intelligence, and the metaverse. *Language Learning & Technology*, *27*(2), 6–27.

Gordon, V. N., & Steele, G. E. (2015). *The undecided college student: An academic and career advising challenge.* Charles C Thomas Publisher.

Hamad, A., & Jia, B. (2022). How virtual reality technology has changed our lives: An overview of the current and potential applications and limitations. *International Journal of Environmental Research and Public Health*, *19*(18), Article 11278. https://doi.org/10.3390/ijerph191811278

Hashemi, M., Darejeh, A., & Cruz, F. (2024, October). A user-centric exploration of axiomatic explainable AI in participatory budgeting. In *Companion of the 2024 on ACM International Joint Conference on Pervasive and Ubiquitous Computing* (pp. 126–130).

Hashemi, M., Darejeh, A., & Cruz, F. (2025). Understanding user preferences in explainable artificial intelligence: A mapping function proposal. *ACM Transactions on Intelligent Systems and Technology*, *16*(4), 1–37. https://doi.org/10.1145/3733837

Holuša, V., Vaněk, M., Beneš, F., Švub, J., & Staša, P. (2023). Virtual reality as a tool for sustainable training and education of employees in industrial enterprises. *Sustainability*, *15*(17), 12886. https://doi.org/10.3390/su151712886

Hwang, G.-J., & Chien, S.-Y. (2022). Definition, roles, and potential research issues of the metaverse in education: An artificial intelligence perspective. *Computers and Education: Artificial Intelligence, 3,* Article 100082. https://doi.org/10.1016/j.caeai.2022.100082

Jaboob, M., Hazaimeh, M., & Al-Ansi, A. M. (2025). Integration of generative AI techniques and applications in student behavior and cognitive achievement in Arab higher education. *International Journal of Human–Computer Interaction*, *41*(1), 353–366.

Jauhiainen, J. S. (2024). The Metaverse: Innovations and generative AI. *International Journal of Innovation Studies, 8*(3), 262–272. https://doi.org/10.1016/j.ijis.2024.04.004

Kiela, D., Thrush, T., Ethayarajh, K., & Singh, A. (2023). Plotting progress in AI. *Contextual AI Blog*. https://contextual.ai/blog/plotting-progress

Larson, K. E., Hirsch, S. E., McGraw, J. P., & Bradshaw, C. P. (2019). Preparing preservice teachers to manage behavior problems in the classroom: The feasibility and acceptability of using a mixed-reality simulator. *Journal of Special Education Technology*, *35*(2), 63–75. https://doi.org/10.1177/0162643419836415

Majjate, H., Bellarhmouch, Y., Jeghal, A., Yahyaouy, A., Tairi, H., & Zidani, K. A. (2024). AI-powered academic guidance and counselling system based on student profile and interests. *Applied System Innovation*, *7*(1), 6. https://doi.org/10.3390/asi7010006

Markauskaite, L., Marrone, R., Poquet, O., Knight, S., Martinez-Maldonado, R., Howard, S., ... & Siemens, G. (2022). Rethinking the entwinement between artificial intelligence and human learning: What capabilities do learners need for a world with AI? *Computers and Education: Artificial Intelligence*, *3*, 100056.

Moore, M. G. (1989). Three types of interaction. *American Journal of Distance Education, 3*(2), 1–7. https://doi.org/10.1080/08923648909526659

Murel, J., & Kavlakoglu, E. (2024). What is a feature engineering? Retrieved March 15, 2024, from IBM, www.ibm.com

Nagarkar, S. B. (2024). *Improving realism and interactivity in games: A study of AI-integrated non player characters (NPCs)* (Master's thesis, Universitat Politècnica de Catalunya).

Nguyen, D. C., Dinh, N. H. D., Pham-Nguyen, C., Le Dinh, T., & Nguyen Hoai Nam, L. (2022, November). Itcareerbot: A personalized career counselling chatbot. In *Asian Conference on Intelligent Information and Database Systems* (pp. 423–436). Springer Nature Singapore.

Oje, A. V., Hunsu, N. J., & May, D. (2023). Virtual reality assisted engineering education: A multimedia learning perspective. *Computers & Education: X Reality*, *3*, Article 100033. https://doi.org/10.1016/j.cexr.2023.100033

Onu, P., Pradhan, A., & Mbohwa, C. (2023). Potential to use metaverse for future teaching and learning. *Education and Information Technologies, 29*, 8893–8924. https://doi.org/10.1007/s10639-023-12167-9

Papathomas, L., & Kuhn, D. (2017). Learning to argue via apprenticeship. *Journal of Experimental Child Psychology*, *159*, 129–139.

Paszkiewicz, A., Salach, M., Strzałka, D., Budzik, G., Nikodem, A., Wójcik, H., & Witek, M. (2021). VR education support system—A case study of digital circuits design. *Energies*, *15*(1), Article 277. https://doi.org/10.3390/en15010277

Peiqi, J. (2024). Applications of generative artificial intelligence in online learning and ethical governance framework. *Online Learning*, *7*(5), 206–216.

Peñalvo, F. J. G., & Ingelmo, A. V. (2023). What do we mean by GenAI? A systematic mapping of the evolution, trends, and techniques involved in Generative AI. *IJIMAI*, *8*(4), 7–16.

Pérez-Núñez, A. (2023). Exploring the potential of generative AI (ChatGPT) for foreign language instruction: Applications and challenges. *Hispania*, *106*(3), 355–362.

Perrault, R., & Clark, J. (2024). *Artificial Intelligence Index Report 2024*. AI Index. Retrieved March 15, 2024, from https://aiindex.stanford.edu/report/

Qamhieh, M., Sammaneh, H., & Demaidi, M. N. (2020). PCRS: Personalized career-path recommender system for Engineering students. *IEEE Access*, *8*, 214039–214049. https://doi.org/10.1109/access.2020.3040338

Rivera, A. C., Tapia-Leon, M., & Lujan-Mora, S. (2018). Recommendation systems in education: A systematic mapping study. In Á. Rocha & T. Guarda (Eds.), *Proceedings of the International Conference on Information Technology & Systems (ICITS 2018)* (pp. 937–947). Springer International Publishing. https://doi.org/10.1007/978-3-319-73450-7_89

Roksa, J., & Arum, R. (2015). Inequality in skill development on college campuses. *Research in Social Stratification and Mobility*, *39*, 18–31.

Saparamadu, P. V. I. N., Sepasgozar, S., Guruge, R. N. D., Jayasena, H. S., Darejeh, A., Ebrahimzadeh, S. M., & Eranga, B. A. I. (2025). Optimising contract interpretations with large language models: A comparative evaluation of a vector database-powered Chatbot vs. ChatGPT. *Buildings*, *15*(7), 1144.

Sargazi Moghadam, T., Darejeh, A., Delaramifar, M., & Mashayekh, S. (2024). Toward an artificial intelligence-based decision framework for developing adaptive e-learning systems to impact learners' emotions. *Interactive Learning Environments*, *32*(7), 3665–3685.

Schlimbach, R., Rinn, H., Markgraf, D., & Robra-Bissantz, S. (2022). A literature review on pedagogical conversational agent adaptation. *In Pacific Asia Conference on Information Systems* (Vol. 1).

Schussler, E. E., & Weatherton, M. R. (2021). Success for all? A call to re-examine how student success is defined in Higher Education. *CBE—Life Sciences Education, 20*(1), ar15.

Singh, G., Mantri, A., Sharma, O., & Kaur, R. (2020). Virtual reality learning environment for enhancing electronics engineering laboratory experience. *Computer Applications in Engineering Education, 29*(1), 229–243. https://doi.org/10.1002/cae.22333

Stanovich, K. E. (2016). The comprehensive assessment of rational thinking. *Educational Psychologist, 51*(1), 23–34.

Stanovich, K. E. (2018). Miserliness in human cognition: The interaction of detection, override and mindware. *Thinking & Reasoning, 24*(4), 423–444.

Stanovich, K. E., & West, R. F. (1997). Reasoning independently of prior belief and individual differences in actively open-minded thinking. *Journal of Educational Psychology, 89*(2), 342–357.

Stefan, H., Mortimer, M., Horan, B., & McMillan, S. (2024). How effective is virtual reality for electrical safety training? Evaluating trainees' reactions, learning, and training duration. *Journal of Safety Research, 90*, 48–61. https://doi.org/10.1016/j.jsr.2024.06.002

Streveler, R. A., Litzinger, T. A., Miller, R. L., & Steif, P. S. (2008). Learning conceptual knowledge in the engineering sciences: Overview and future research directions. *Journal of Engineering Education, 97*(3), 279–294. https://doi.org/10.1002/j.2168-9830.2008.tb00979.x

Tapalova, O., & Zhiyenbayeva, N. (2022). Artificial intelligence in education: AIEd for personalised learning pathways. *Electronic Journal of e-Learning, 20*(5), 639–653.

Taycher, L. (2010, August 5). *Books of the world, stand up and be counted! All 129,864,880 of you.* Inside Google Books. Retrieved November 15, 2024, from https://booksearch.blogspot.com/2010/08/books-of-world-stand-up-and-be-counted.html

Thottoli, M. M., Alruqaishi, B. H., & Soosaimanickam, A. (2023). Robo academic advisor: Can chatbots and artificial intelligence replace human interaction? *Contemporary Educational Technology, 16*(1), ep485.

UNICEF. (2021). *Policy guidance on AI for children.* United Nations Children's Fund. Retrieved March 15, 2024, from https://www.unicef.org/innocenti/reports/policy-guidance-ai-children

Vincent-Lancrin, S., & Van der Vlies, R. (2020). Trustworthy artificial intelligence (AI) in education: Promises and challenges. *OECD Education Working Papers*, (218), 0_1–17.

Wei, R. (2024). Examining the influence of the RIASEC theory within the holland code on students' academic performance in their chosen disciplines among the context of higher education. *Cogent Education, 11*(1), 2391274.

11 Generative Artificial Intelligence for Image Synthesis

Applications in Construction, Architecture, and Metaverse Design

Ali Darejeh, Sara Mashayekh, and Samad M. E. Sepasgozar

1 Introduction

The integration of generative artificial intelligence (GenAI) into design and visualization processes is reshaping how professionals conceptualize, prototype, and communicate ideas. By leveraging machine learning and advanced neural architectures, GenAI systems can produce different outputs from minimal input—such as text prompts, sketches, or reference images (Saparamadu et al., 2025; Sardar et al., 2019).

Among the most impactful developments in GenAI systems are Generative AI Image Synthesis (GAI-IS) tools, which support the generation of both 2D and 3D content for use across physical and digital environments. These tools offer immense potential not only for enhancing design efficiency but also for democratizing participation—allowing non-experts to contribute meaningfully to visual ideation (Jones, 2023; Kazi et al., 2017; Witteveen & Andrews, 2022).

These tools are proving especially valuable in fields like architecture, construction, and metaverse development, where efficient, high-fidelity content generation is crucial to innovation (Franchi & Ntagiou, 2021; Ghannad & Lee, 2022; Nalbant, Aydın, & Uyanık, 2023).

Despite growing adoption, research on GAI-IS tools remains fragmented. While some studies have examined their use in early-stage architectural design (Beyan & Rossy, 2023; Paananen, Oppenlaender, & Visuri, 2024), few have conducted comprehensive comparisons across real-world and virtual applications—especially within the context of the metaverse. Moreover, key challenges persist, such as limited contextual awareness in generated outputs, ethical considerations, and integration with established design workflows.

This chapter addresses these gaps by examining the foundational technologies, practical applications, and current limitations of GAI-IS tools. It opens with a technical overview of key GenAI model architectures—specifically generative adversarial networks (GANs), diffusion models, and transformer-based architectures—that underpin modern GAI-IS tools, before exploring their applications in architectural, urban, and metaverse design contexts (Baduge et al., 2022; OpenAI, 2023;

DOI: 10.1201/9781003557555-11

Walmsley, 2022; Wang, Xu, & Sun, 2022; Wu, Stouffs, & Biljecki, 2022). The chapter concludes by identifying key challenges and proposing future directions for the integration of GAI-IS tools into both real-world design workflows and virtual environments.

2 Core Technologies Behind GenAI for Image Synthesis

GenAI is a transformative technology that leverages machine learning, data-driven algorithms, and advanced computational techniques to optimize workflows, automate complex tasks, and generate diverse forms of content—including text, images (such as architectural designs, characters, and furniture), and 3D immersive virtual environments. GenAI models human thought processes through statistical approaches, enabling the design of neural networks that mimic human decision-making (Darejeh et al., 2024; Sargazi Moghadam et al., 2024; Saparamadu et al., 2025; Sardar et al., 2019).

Building on this foundation, GAI-IS represents a specialized area within GenAI focused on the automated generation of visual content, including images and 3D representations. GAI-IS relies on foundational model architectures that guide how images are created from various forms of input, such as noise, sketches, or text. Among the most prominent and widely adopted approaches are GAN-based architectures, diffusion models, and, more recently, transformer-based architectures. Each offers distinct approaches, advantages, and limitations in generating photorealistic or stylistically rich visual content (Franchi & Ntagiou, 2021; Nalbant, Aydın, & Uyanık, 2023; Ng, 2022).

GAN-based architectures: Introduced by Ian Goodfellow and his collaborators in 2014, GANs are a class of deep learning architectures designed to generate synthetic data that closely resembles real-world datasets (Goodfellow et al., 2014). Unlike traditional generative models, GANs employ an adversarial training paradigm involving two neural networks: the generator and the discriminator. The generator takes random noise as input and transforms it into data samples—such as images, text, or audio—while the discriminator attempts to classify inputs as real or generated (Antipov, Baccouche, & Dugelay, 2017). This dynamic, zero-sum interaction drives both networks to improve over time: the generator learns to produce increasingly realistic outputs, and the discriminator becomes more adept at identifying fakes. This adversarial learning process is governed by a minimax optimization framework, where the generator minimizes the likelihood of its outputs being detected, and the discriminator maximizes its classification accuracy. The result is a balance in which the generator produces data that is nearly indistinguishable from reality (Goodfellow et al., 2014).

GANs have demonstrated remarkable potential in various applications, including IS, video generation, style transfer, and even scientific discovery, such as drug development and climate modeling (Karras et al., 2019). Beyond these domains, GANs play a critical role in designing metaverse environments, where they are used to generate virtual landscapes, 3D assets, and immersive spaces that mimic or surpass real-world settings (Franchi & Ntagiou, 2021). For example, GANs can create expansive virtual cities, dynamic avatars, and interactive spaces that form the foundation of the metaverse (Franchi & Ntagiou, 2021; Nalbant, Aydın, & Uyanık, 2023). Similarly, GANs are invaluable in the construction industry for simulating buildings, landscapes, and urban areas, offering realistic visualizations that aid in early-stage prototyping and decision-making (Chen et al., 2023).

Diffusion models: In addition to GANs, diffusion models have emerged as a powerful alternative for generative tasks, particularly in the creation of high-quality images and 3D assets. Diffusion models operate by iteratively adding and removing noise to and from data, gradually transforming random noise into coherent and detailed outputs. Unlike GANs, which rely on adversarial training, diffusion models use probabilistic processes to approximate the distribution of training data. This approach has demonstrated superior stability and consistency, making diffusion models well-suited for applications that require fine-grained control over the generation process (Hatamizadeh et al., 2025; Manovich, 2024). Diffusion models have been successfully applied not only to architecture and urban design but also to metaverse development, enabling the creation of intricate virtual environments, realistic textures, and dynamic animations (Gupta et al., 2025; Li et al., 2024).

Transformer-based architectures: Transformers are deep learning architectures originally developed for natural language processing tasks and have since become foundational in many GenAI applications. These models utilize self-attention mechanisms to process sequential data, enabling them to capture contextual relationships and long-range dependencies between elements. In IS, transformer-based architectures leverage this capacity to align textual input with corresponding visual representations.

Unlike GANs, which employ adversarial training, or diffusion models, which rely on iterative denoising, transformers directly learn to map language to images using autoregressive or encoder-decoder structures. These models are trained on vast datasets of text-image pairs, allowing them to produce semantically accurate and visually coherent outputs, even from complex or abstract prompts (OpenAI, 2023; Wang, Xu, & Sun, 2022).

Transformer-based models are known for their scalability, making them suitable across various domains such as architecture, digital art, and industrial design. They also support fine-tuned transfer learning, enabling customization for domain-specific tasks. As the field progresses, hybrid models are emerging that combine transformer architectures with elements from diffusion models to further enhance output quality, fidelity, and controllability (Analytics Vidhya, 2024; Cornia, Baraldi, & Cucchiara, 2022; Malakan, Hassan, & Mian, 2022).

3 Diverse Applications of GAI-IS Tools Across Metaverse and Construction Contexts

GAI-IS tools, powered by GAN-based architectures, diffusion models, and transformer-based architectures, represent a revolutionary advancement in generating high-quality visual content with minimal human input. These tools can generate images based on textual descriptions, sketches, or reference inputs, making them ideal for designing both virtual metaverse spaces and real-world architectural prototypes (Kazi et al., 2017). For instance, in metaverse environments, designers can describe a futuristic city and use diffusion models or transformer-based architectures to generate detailed visuals of landscapes, buildings, and urban features. Similarly, in real-world construction, GAI-IS tools can visualize building designs or modify existing layouts—adjusting colors, textures, or shapes as needed (Fernandez, 2022; Liu et al., 2025; Maganga, 2023; Nalbant, Aydın, & Uyanık, 2023; Yang, Shen, & Zhou, 2021). These systems rely on training datasets to identify patterns, textures, and spatial elements, enabling the generation of content that is both realistic and innovative.

In 2022, GAI-IS models—particularly those utilizing text-to-image or image-to-image technologies based on GAN-based architectures, diffusion models, and transformer-based architectures—emerged as a major trend, enabling the rapid generation of highly detailed and visually compelling images from simple textual prompts (Maganga, 2023; Witteveen & Andrews, 2022). Tools such as MidJourney (diffusion), DALL-E 3 (transformer), and StyleGAN-T (GAN) gained significant popularity due to their ability to produce complex outputs within seconds, greatly benefiting industries like architecture, urban planning, and virtual world-building (Sauer et al., 2023). By automating complex creative tasks, these models not only reduce time and costs associated with prototyping but also expand creative possibilities, allowing designers to iterate quickly and explore diverse options without requiring advanced technical expertise (Beyan & Rossy, 2023; Li et al., 2025; Meeran, 2021).

Beyond text-to-image and image-to-image generation, a wide range of generative techniques—including sketch-to-image—have further expanded the scope of GAI-IS applications (Koley et al., 2023). These innovations allow professionals and non-specialists alike to rapidly visualize ideas, bridging the gap between conceptualization and tangible design representation. Compared to traditional, time-intensive design processes, these AI-driven approaches significantly enhance speed, accessibility, and workflow efficiency.

While GAI-IS technologies are used across various industries—including graphic design, fashion, and gaming—this chapter examines their dual function in shaping both physical environments (architecture and construction) and virtual spaces (metaverse design). In these domains, GAI-IS models support the creation of high-fidelity prototypes of buildings, interiors, and landscapes—helping stakeholders visualize designs more effectively, identify potential flaws early, and make informed modifications prior to construction (Hegazy & Mohamed Saleh, 2023; Jones, 2023; Yan et al., 2022).

Importantly, GAI-IS tools also enhance communication and collaboration across multidisciplinary teams. Many stakeholders—especially those without design expertise—struggle to interpret technical drawings or visualize spatial concepts, often leading to miscommunication. GAI-IS addresses this gap by transforming simple textual or conceptual inputs into photorealistic visuals, making complex designs accessible to a broader audience. This capability is particularly valuable across domains such as interior and building design, landscape architecture, urban planning, accessibility design, and historical restoration (Kahila & Kyttä, 2009; Verheijden & Funk, 2023; Yildirim, 2022a, 2022b).

Within metaverse environments, transformer-based architectures and diffusion models play a pivotal role in generating 3D models of cities, buildings, interactive spaces, and avatars—effectively bridging the digital and physical worlds (Darejeh, Marcus, & Sweller, 2022; Lv, 2023; Ratican, Hutson, & Wright, 2023). While 2D GAI-IS tools primarily support early-stage conceptualization, more advanced 3D GAI-IS systems like NVIDIA GET3D, Meshy, and Autodesk Dream Sketch go beyond ideation by enabling the direct creation of detailed, ready-to-deploy 3D assets. These advancements significantly reduce the time and effort required for manual asset modeling, streamlining workflows from initial concept to full implementation.

The following sections provide a detailed exploration of how GAI-IS tools are applied within construction and metaverse design, highlighting their transformative impact on architectural visualization, urban planning, interior design, and virtual environment development.

3.1 Building Design

GAI-IS tools play a pivotal role in expanding creative possibilities in building design by enabling architects and metaverse designers to explore a wide range of innovative concepts. These tools facilitate design ideation by rapidly generating visual representations of alternative solutions, allowing users to test and refine their ideas more effectively than through traditional methods. One key application is shape and layout exploration, where GAI-IS tools assist in producing diverse architectural forms and spatial configurations, encouraging experimentation and the development of unique building concepts.

Another critical use is material selection, as these tools allow designers to virtually simulate and evaluate a variety of interior and exterior material options. This helps ensure that selected materials meet both aesthetic preferences and functional requirements. Furthermore, GAI-IS tools support the integration of sustainable design features such as green roofs, energy-efficient windows, and renewable energy systems. By incorporating these elements early in the design process, architects can create buildings that are both environmentally responsible and energy-efficient (Abdallah & Estévez, 2023; Paananen, Oppenlaender, & Visuri, 2024). Figure 11.1 illustrates four eco-friendly buildings generated using MidJourney. Unless otherwise cited, all images in this chapter were created by the authors.

Figure 11.1 An eco-friendly building (generated by Midjourney).

3.2 Interior Design

GAI-IS tools offer valuable capabilities for creating realistic visualizations of building interiors, enabling designers to experiment with various layouts, materials, and furnishings. This facilitates the development of interior environments that are both functional and visually appealing, while also aligning with the specific needs of occupants. GAI-IS tools enhance the interior design process through several key features. First, in terms of space planning, these tools allow designers to generate and compare different configurations of furniture and décor, helping to optimize space usage and improve overall functionality. Second, for color scheme selection, GAI-IS tools equipped with color-matching algorithms enable rapid testing of various color combinations to identify those most suitable for a particular room. Third, when it comes to furniture selection, designers can use GAI-IS tools to generate and assess different pieces of furniture, finishes, and decorative elements—allowing them to refine their choices before making final decisions (Abdallah & Estévez, 2023; Hegazy & Mohamed Saleh, 2023).

Figure 11.2 displays four variations of an interior design generated from the prompt "Interior design for a living room with a black and white sofa for ten people, a large chandelier, and a TV screen on the wall". Figure 11.3 displays four interior designs for a bedroom with a superhero theme. Midjourney generated all the designs.

Figure 11.2 Interior design for a living room with a black and white sofa for ten people, a large chandelier, and a TV screen on the wall (generated by Midjourney).

The combination of GAI-IS tools and Mixed Reality (MR) offers significant value to interior designers. These advanced technologies enable the creation of realistic virtual representations that can be seamlessly integrated with real-world environments, allowing designers to visualize, refine, and present interior concepts with enhanced spatial awareness and interactivity (Zhu, Fukuda, & Yabuki, 2024).

Figure 11.3 Interior design for a boy's bedroom with a superheroes theme (generated by Midjourney).

By utilizing these technologies, designers can effectively address complex renovation challenges such as wall removal, structural modifications, and layout changes—enabling clients to visualize proposed transformations in an immersive and interactive manner. This not only enhances the decision-making process but also fosters stronger collaboration between designers, architects, and clients, ultimately leading to more efficient and well-informed design outcomes.

Additionally, the combination of augmented reality (AR) and GAI-IS tools allows virtual furniture, such as chairs and tables, to be placed within the actual environment of living rooms and bedrooms. This enables clients to better visualize the final results and make more confident decisions regarding spatial arrangement and design preferences.

3.3 Landscape Design

GAI-IS tools offer significant value in landscape design by supporting the creation of outdoor environments that are both functional and aesthetically appealing. These tools enable designers to experiment with a diverse range of plant species tailored to specific site conditions, facilitating the selection of sustainable, climate-responsive, and visually cohesive planting schemes. This capability not only enhances environmental performance but also aligns designs with ecological goals such as biodiversity and low-maintenance landscaping.

In addition to vegetation planning, GAI-IS tools assist in modeling and visualizing various hardscape elements—such as pathways, retaining walls, water features, and architectural structures—allowing designers to explore alternative spatial configurations that improve usability, accessibility, and spatial harmony (Belesky, 2023; Wen & Li, 2025; Zhou & Xiang, 2024). This integrated design approach enhances decision-making and promotes more innovative and effective landscape solutions. Figure 11.4 presents four variations of a landscape that is designed based on this phrase, "Landscape design with a walking path, tall palm trees, red flowers, and a lighthouse", all generated using MidJourney.

Figure 11.4 Landscape design with a walking path, tall palm trees, and red flowers around and a lighthouse far away (generated by Midjourney).

3.4 Urban Design

GAI-IS tools can play a transformative role in urban design, helping designers create attractive, livable cities that meet the needs of both residents and visitors. By generating prototypes of urban areas, designers can experiment with various layout options, planting schemes, and hardscape elements to enhance urban functionality and visual appeal. One powerful application is in shaping city structures—GAI-IS tools allow designers to generate visually striking and culturally resonant urban forms based on text-based inputs. This enables the creation of cityscapes that are not only aesthetically compelling but also contextually appropriate to the culture and identity of a given location.

Additionally, GAI-IS tools can support the development of optimized lighting designs tailored to the specific needs of public spaces, pedestrian areas, and commercial zones. By factoring in variables such as spatial use, time of day, and natural light availability, these tools generate lighting configurations that are both functional and visually harmonious. This capability is particularly valuable in metaverse environments, where lighting design in game engines often presents a significant challenge—GAI-IS tools can streamline this process and offer valuable insights for virtual cityscapes.

Furthermore, similar to the landscape design, GAI-IS tools can assist in designing and visualizing green spaces within urban settings. By modeling different planting layouts and vegetation types, designers can create functional and inviting parks, gardens, and natural areas that contribute to the environmental and social well-being of urban populations (Kamrowska-Załuska, 2021; Wu, Stouffs, & Biljecki, 2022; Ye, Du, & Ye, 2022). Figure 11.5 illustrates four variations of an urban landscape generated using MidJourney, based on the prompt "urban design with a garden and curved futuristic buildings around".

Figure 11.5 Urban design with a garden and curved futuristic buildings around (generated by Mid-journey).

3.5 Accessibility Design

GAI-IS tools can be leveraged to generate building and urban space models that are optimized for accessibility, promoting more inclusive design practices. By incorporating data related to accessibility requirements—such as wheelchair access, tactile navigation, and accommodations for visual impairments—these tools can help architects and planners design environments that serve a broader range of users (Hosseini et al., 2022). This contributes to the creation of equitable urban spaces and buildings that support social inclusion and universal usability.

While accessibility features are often associated with real-world infrastructure, they are equally relevant in virtual environments. Although they may not seem essential at first glance, accessibility considerations are crucial when designing metaverse spaces intended to mirror real-world experiences. Many users who engage with the metaverse may have disabilities, and providing them with inclusive virtual spaces ensures that their digital experiences are as equitable as their physical ones. Moreover, if users choose to create avatars that reflect their real-world physical conditions, metaverse environments should be designed to accommodate these realities. Ensuring accessibility within virtual spaces not only enhances user engagement but also reinforces the importance of inclusive design in both physical and digital realms.

Figure 11.6 displays four variations generated by MidJourney based on the prompt "a wheelchair pathway and elevator designed for people with mobility issues in a train station."

Figure 11.6 Wheelchair pathway and elevator in a train station (generated by Midjourney).

3.6 Historical Restoration

In the field of historical restoration, GAI-IS tools provide powerful capabilities for visualizing and reconstructing historical sites with greater accuracy and authenticity. These tools enable restorers to translate their interpretations of a site's original appearance into detailed digital models, which can guide restoration efforts and enhance historical fidelity. One major application of GAI-IS in this context is reconstruction, where detailed models generated by AI help identify missing or damaged elements of a structure or landmark, offering restorers a more precise blueprint for rebuilding efforts (Wu, Stouffs, & Biljecki, 2022).

Additionally, GAI-IS tools assist with material selection, analyzing characteristics of original construction materials and proposing visually and historically appropriate substitutes. This ensures that restoration work maintains aesthetic coherence with the original structure while preserving its historical integrity (Abdallah & Estévez, 2023; Sun, Zhou, & Han, 2022). Figures 11.7 and 11.8 illustrate reconstructed models of Persepolis in Iran and the Pyramids in Egypt, respectively, both generated using the GAI-IS tool MidJourney.

Figure 11.7 Persepolis Iran—Reconstructed using Midjourney.

Figure 11.8 Egypt Pyramid—Reconstructed using Midjourney.

4 A Comparison Between Different GAI-IS Tools

GAI-IS tools have rapidly evolved over the years, with numerous products in the market offering unique features and capabilities. However, each tool has its own set of strengths and limitations, making it crucial to understand its underlying technologies before selecting the most suitable one for a specific use case. Some of the most powerful GAI-IS tools available today include Janus-Pro-7B, Adobe Firefly, Google's Imagen 3, Stable Diffusion, DALL-E, StyleGAN, MidJourney, and DeepDream, among others (Alcaide-Marzal & Diego-Mas, 2025; Guinness, 2023).

Although numerous GAI-IS tools are available, this review focuses on four representative models, each reflecting a distinct generative architecture and methodological approach. Specifically, DALL·E 3 is examined as an end-to-end transformer-based model that directly maps text to images. MidJourney, while not officially documented, is widely believed to use a latent diffusion model; StyleGAN represents a GAN-based architecture, and DeepDream is based on a Convolutional Neural Network (CNN)—though its latest version, which includes text-to-image capabilities, likely incorporates more advanced models such as diffusion or transformer-based architectures, as CNNs alone cannot generate images from scratch. By analyzing these tools, the review aims to highlight their core technologies, practical applications, and comparative strengths in areas such as image generation quality, customization capabilities, and the degree of creative control they offer to users.

4.1 DALL·E

DALL·E, developed by OpenAI, is a generative model that creates images from natural language prompts. DALL·E 2 is a diffusion-based model, meaning it generates images by iteratively denoising random noise, guided by textual input. It incorporates CLIP (Contrastive Language–Image Pretraining) embeddings in two key ways: (1) to encode text prompts into a latent representation that conditions the diffusion process, and (2) to rank or select the most semantically aligned outputs. This combination enables DALL·E 2 to produce images that are coherent with both simple and complex textual descriptions (Ding et al., 2021; Reddy et al., 2021; Temsah et al., 2024).

In contrast, DALL·E 3 represents a major architectural shift. It is an end-to-end transformer-based model that does not rely on a diffusion process or CLIP embeddings during inference. Instead, it directly models the relationship between text and image using a unified transformer architecture trained on large-scale text–image pairs. This design allows DALL·E 3 to better handle detailed, nuanced prompts and generate more semantically accurate and visually consistent images. While CLIP-like models may still play a role during training (e.g., for supervision or evaluation), they are excluded from the actual image generation pipeline at runtime (Analytics Vidhya, 2024; OpenAI, 2023).

OpenAI has optimized DALL·E 3's architecture to produce images that are not only photorealistic and stylistically diverse but also tightly aligned with textual descriptions. To access DALL·E, users simply need to request image generation through a prompt in ChatGPT (see Figure 11.9).

Figure 11.9 Accessing DALL·E via ChatGPT.

4.2 MidJourney

MidJourney is a diffusion-based model designed for text-to-image generation. Developed by MidJourney, Inc., it is particularly well-suited for producing realistic and stylistically rich interior design concepts. The model combines machine learning techniques with human-curated feedback to generate visually compelling images. While MidJourney has not publicly disclosed its architecture, it is widely believed to use a latent diffusion approach similar to models like Stable Diffusion, which progressively transforms random noise into coherent images based on the semantics of a given text prompt (Buzzaccarini et al., 2024; Borji, 2022; Oppenlaender, 2022; Park, 2024). Unlike GAN-based tools such as StyleGAN, which rely on adversarial training between a generator and a discriminator, MidJourney employs iterative denoising processes to enhance image quality and textual alignment.

MidJourney is trained on a large dataset of image–text pairs, enabling it to interpret prompts and generate highly detailed, visually rich outputs. The model is particularly praised for its speed, efficiency, and ability to produce artistic and stylized visuals. However, its performance can vary depending on prompt clarity and may struggle with more complex or technical design specifications (Borji, 2022; Oppenlaender, 2022; Roose, 2022). Although specific implementation details remain proprietary, MidJourney is widely believed to use a latent diffusion model for image generation while leveraging transformer-based models—such as CLIP—for interpreting textual prompts. In addition, it incorporates custom fine-tuning techniques that prioritize creativity and stylistic variation, making it especially popular for concept art, digital illustration, and interior design visualization.

Figure 11.10 displays the MidJourney interface, where users can input text prompts to generate multiple variants of a given image. These variants reflect differences in design rather than external factors such as weather or lighting. One notable feature of MidJourney is its ability to refine specific outputs: users can select a preferred image—such as variant #4—and generate a new set of variations based on that selection, allowing for more targeted and iterative design exploration.

Figure 11.10 Mid-Journey user interface.

4.3 DeepDream

DeepDream, developed by Google, utilizes a CNN model to generate images by enhancing and incorporating specific patterns and features. Originally designed for AI interpretability research, DeepDream is based on an early version of Google's Inception model, which was initially developed for image classification tasks. Instead of classifying images, DeepDream applies feature visualization techniques to enhance and amplify patterns detected by the neural network. The process begins by passing an input image through different layers of the CNN, where low-level layers detect basic features like edges and textures, while deeper layers identify complex structures such as faces and objects. Rather than suppressing or ignoring these patterns, DeepDream modifies the backpropagation algorithm to recursively reinforce them through gradient ascent, iteratively exaggerating their presence and creating highly surreal, dreamlike visuals (Martinez et al., 2024).

DeepDream is implemented using TensorFlow and Python, allowing for extensive customization through parameters such as layer selection, iteration depth, and amplification intensity. Its artistic and hallucinatory effects have made it a popular tool for AI-generated art and creative design, although its strong focus on

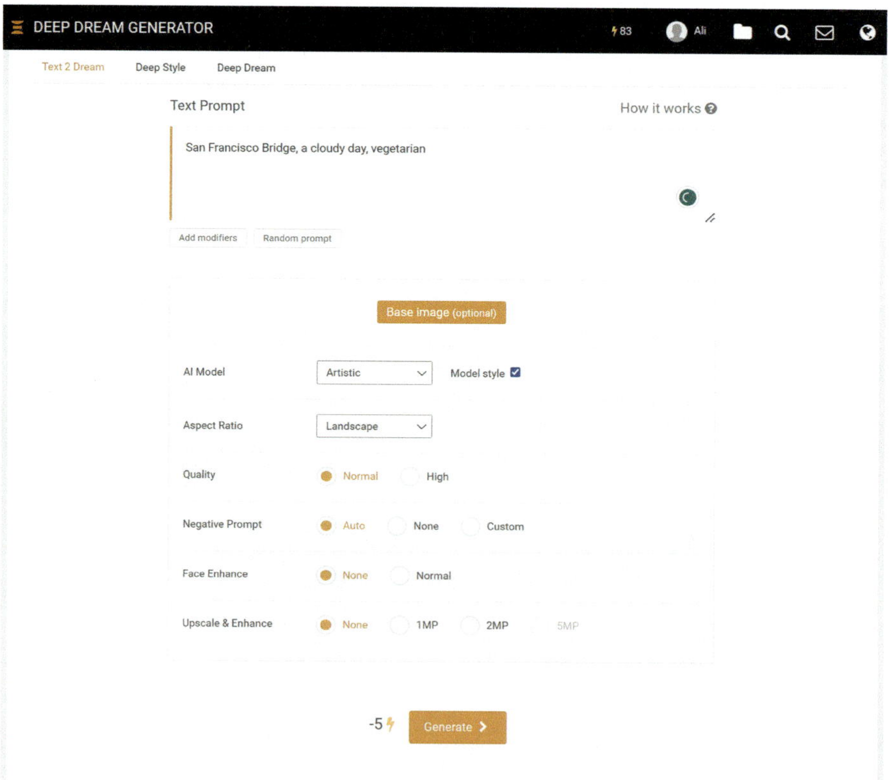

Figure 11.11 Deep Dream user interface.

enhancing specific patterns may limit its applicability in some domains (Giri et al., 2023; Toğaçar, Cömert, & Ergen, 2021).

As previously explained, the newer version of DeepDream includes text-to-image functionality, making it unlikely to rely solely on a CNN-based architecture. While the original CNN architecture may still be used for modifying existing images, the text-to-image functionality likely incorporates more advanced models—such as diffusion models or transformer-based architectures—because CNNs alone are not capable of generating images from scratch. Figure 11.11 depicts the interface of DeepDream, where users can input descriptive text to guide the image generation process. In addition to providing text prompts, users have the option to upload a base image, allowing DeepDream to combine the textual description with visual reference data to generate a new, stylized image.

4.4 StyleGAN

StyleGAN is a GAN architecture developed by NVIDIA that can generate highly realistic and diverse images from latent vectors. Unlike diffusion-based models, StyleGAN does not natively generate images from textual descriptions; rather, it synthesizes images by sampling and manipulating points within a latent space. While extensions such as StyleCLIP enable limited text-guided editing, the original StyleGAN models—including StyleGAN2 and StyleGAN3—are primarily designed for high-fidelity, unsupervised image generation (Sauer et al., 2023).

StyleGAN can also generate variations of an existing image when combined with specialized encoders that map real images into its latent space. However, it is not inherently designed for direct image-to-image translation, which is more effectively handled by conditional GANs such as Pix2Pix or CycleGAN, or by modern diffusion-based tools that can better map source images to target outputs based on specific semantic or stylistic goals. The model is computationally intensive, requiring significant GPU resources and time to produce high-quality outputs, which can be a barrier for users who need rapid or large-scale image generation. To use StyleGAN, users typically start with a base latent vector and adjust parameters such as style layers, interpolation settings, and stochastic noise to generate multiple image variants (Fox et al., 2021).

In recent years, the popularity of StyleGAN has declined due to the rise of more advanced diffusion models and transformer-based architectures, which offer greater flexibility, finer control, and more natural integration of text prompts into IS. While StyleGAN was once the leading architecture for high-quality synthetic imagery, it now struggles with issues such as mode collapse—resulting in repetitive or less diverse outputs—and lacks direct text-to-image generation capabilities. In contrast, diffusion models and transformer-based architectures excel at producing highly customized and semantically rich visuals simply by processing descriptive language inputs. As a result, StyleGAN is now primarily used in specialized domains such as face generation, medical imaging, and scientific research, whereas diffusion- and transformer-based models have become the dominant approaches for mainstream AI-generated imagery (Riemer & Peter, 2024).

4.5 Choosing the Right GAI-IS Tool for the Task

When selecting a GAI-IS tool for construction, landscape, and urban design applications, several factors should be considered. These include the quality of the generated images, the model's generation speed and efficiency, and the flexibility and customizability of the outputs. Based on these criteria, all four GAI-IS tools discussed—StyleGAN, MidJourney, DeepDream, and DALL·E—are strong contenders for both real-world and metaverse-based prototyping in architectural and urban planning contexts.

StyleGAN is particularly well-suited for generating images of buildings and environments under different conditions, as it can produce highly realistic and diverse outputs (Patashnik et al., 2021). For example, designers can use StyleGAN to generate cityscapes and architectural forms from various angles, lighting scenarios, and weather conditions, enabling them to explore and communicate multiple design alternatives.

Another advantage of StyleGAN is its latent space flexibility and capacity for fine-tuned customization (Karras et al., 2020). Although StyleGAN does not support direct input of semantic parameters like "material" or "color scheme" in natural language, users can control visual features by manipulating latent vectors, style layers, and stochastic noise. This capability enables the creation of personalized and context-specific visuals, which is particularly valuable in urban design, where factors such as environmental sustainability, cultural sensitivity, and social inclusivity must be addressed.

Figure 11.12 presents a series of images generated using a StyleGAN-based pipeline, depicting the San Francisco Bridge under cloudy skies with varying vegetation patterns. This model demonstrates strong capability in generating diverse yet coherent visual outputs, offering nine distinct environmental variants with subtle differences in cloud cover, lighting, and landscape foliage. In contrast, Figures 11.13–11.15—produced by DALL·E, MidJourney, and DeepDream, respectively—provide stylistically rich but differently controlled representations of the same scene. DALL·E (Figure 11.13) delivers photorealistic outputs that are semantically aligned with the prompt but show limited variation across samples. MidJourney (Figure 11.14), known for its artistic depth, generates moody and dramatic compositions, emphasizing contrast and stylization over strict environmental realism.

DeepDream (Figure 11.15) generates visually striking outputs with high compositional consistency, closely resembling the stylistic quality of images produced by DALL·E. Both models yield images with natural lighting, coherent vegetation, and accurate structural rendering, highlighting a surprising convergence in visual realism despite their different architectural foundations. While transformer-based and diffusion-based models like DALL·E and MidJourney excel in semantic and stylistic fidelity, and DeepDream offers unexpectedly realistic interpretations, StyleGAN remains superior in its ability to simulate a broader range of vegetation conditions and environmental variability, making it particularly valuable for tasks that require fine-grained environmental control.

Figure 11.12 StyleGAN—San Francisco bridge on a cloudy day and vegetation (Yang, Shen & Zhou, 2021).

Figure 11.13 DALL-E—San Francisco bridge on a cloudy day and vegetation.

Figure 11.14 Mid-Journey—San Francisco bridge on a cloudy day and vegetation.

Figure 11.15 DeepDream—San Francisco bridge on a cloudy day and vegetation.

MidJourney, DALL·E, and DeepDream are valuable tools for rapidly generating visual content from textual or visual cues, particularly in collaborative design scenarios involving non-technical stakeholders.

MidJourney excels at producing stylized, high-resolution imagery, making it particularly effective for interior design concepts such as furniture arrangements, lighting aesthetics, and mood visualization. Its creative outputs are widely used during the early ideation phase for both residential and commercial spaces. DALL·E, on the other hand, is known for its strong semantic alignment with text prompts, making it useful for generating conceptual urban environments, street-level perspectives, and public infrastructure scenarios with relatively high prompt accuracy (Ploennigs & Berger, 2022; Yildirim, 2022a, 2022b).

DeepDream, traditionally recognized for its surreal, hallucinatory image transformations, has more recently shown promise in architectural and urban design applications. DeepDream can be leveraged to generate realistic architectural structures and environmentally integrated urban buildings, especially when adapted or fine-tuned for design-oriented applications. This illustrates that DeepDream is not limited to abstract artistic enhancement; it can be applied to visualize design concepts such as green towers, waterfront developments, and sustainable urban prototypes. Its utility, therefore, extends beyond inspiration and into visual prototyping when the system is configured to emphasize structural coherence and architectural detail (Giri et al., 2023; Toğaçar, Cömert, & Ergen, 2021).

While MidJourney and DALL·E are known for their speed and ease of use—allowing real-time exploration of design prompts—DeepDream also offers a unique advantage in its ability to reinterpret existing images, which can be useful for reimagining or enhancing architectural concepts. Each tool, however, comes with limitations. MidJourney may lack precise architectural logic, DALL·E sometimes struggles with detailed spatial arrangements, and DeepDream, in its standard form, may require customization to ensure semantic and structural accuracy.

To illustrate these comparisons, Figures 11.16–11.24 present images generated from similar design prompts using four GAI-IS tools: DALL·E, MidJourney, DeepDream, and StyleGAN.

For building design scenarios, the prompt "A tall building made of glass and a steel frame with a green roof, located on a waterfront at sunrise, with water in the foreground and building reflections" was used to evaluate how different GAI-IS tools interpret detailed architectural elements (Figures 11.16–11.18). Figure 11.16, generated by DALL·E, provides a semantically consistent but somewhat abstract depiction of the scene. While the outputs capture the general form and waterfront reflections, they lack precision in architectural features and variation across images. Figure 11.17, created using MidJourney, offers visually dramatic and stylistically rich results, with compelling lighting effects and imaginative building forms; however, its interpretations tend to prioritize artistic flair over prompt-specific accuracy. In contrast, Figure 11.18, generated with DeepDream, delivers the most accurate and contextually grounded visualizations. It clearly incorporates the specified green rooftops, glass façades, and steel framing, while maintaining realistic proportions and environmental integration. Among the three, DeepDream outperforms by providing the most faithful and detailed representation of the design prompt, making it particularly suitable for applications that require visual reliability and architectural clarity.

Figure 11.16 DALL-E—A tall building made of glass and steel frame with green roof, located on Waterfront, sunrise, water in the foreground, reflection of buildings on water.

Figure 11.17 Mid-Journey—A tall building made of glass and steel frame with green roof, located on Waterfront, sunrise, water in the foreground, reflection of buildings on water.

Figure 11.18 Deep Dream—A tall building made of glass and steel frame with green roof, located on Waterfront, sunrise, water in the foreground, reflection of buildings on water.

For urban design scenarios, the prompt "A modern city with a curved building made of blue and green glass and a river in between: realistic" was used to evaluate how different GAI-IS tools interpret large-scale architectural environments (Figures 11.19–11.21). Figure 11.19, generated by DALL·E, demonstrates strong adherence to the semantic elements of the prompt. It accurately depicts curved buildings with blue-green glass façades and a river seamlessly integrated into the cityscape. The composition reflects a realistic urban setting with consistent lighting, material textures, and structural clarity. While the buildings exhibit slight stylization typical of generative outputs, the overall image retains architectural plausibility, making it suitable for early-stage design ideation.

Figure 11.20, created using MidJourney, offers the most visually compelling and imaginative output. It features intricately detailed curved towers, atmospheric lighting effects, and rich urban textures, including transportation networks and dynamic reflections on the river. This output balances semantic precision with artistic creativity, producing a conceptually rich and immersive urban scene. Such visual quality and spatial coherence make Midjourney particularly well-suited for exploring futuristic urban planning concepts and immersive metaverse environments.

Figure 11.21, produced by DeepDream, presents a realistic urban composition with consistent spatial structure and recognizable architectural elements. The buildings maintain material fidelity and spatial logic, though the designs tend to be more conventional and less varied than those generated by MidJourney. While DeepDream ensures architectural clarity and realism, it appears more constrained in creative exploration.

Overall, all three tools successfully interpret the core elements of the prompt, but with distinct strengths: DALL·E excels in semantic alignment and architectural coherence, MidJourney leads in visual richness and conceptual innovation, and DeepDream offers realism with a focus on structural consistency.

Finally, in the context of interior design, the prompt "A bedroom with a wooden queen bed, timber floor, and a large window during daytime" was used to evaluate

Figure 11.19 DALL-E—A modern city with a curved building made of blue and green glass and a river between: realistic.

Figure 11.20 Mid-Journey—A modern city with a curved building made of blue and green glass and a river between: Realistic.

Figure 11.21 Deep Dream—A modern city with a curved building made of blue and green glass and a river between: Realistic.

how different GAI-IS tools interpret residential scenarios (Figures 11.22–11.24). Figure 11.22, generated using DALL·E, provides a semantically accurate and proportionally coherent visualization of the bedroom, demonstrating strong alignment with the textual prompt. Figure 11.23, created with MidJourney, delivers a more expressive and stylistically enriched depiction, featuring ambient lighting and refined textures that enhance the spatial atmosphere. In contrast, Figure 11.24, produced by DeepDream, presents a creatively interpreted yet structurally coherent visualization. While it accurately reflects key spatial elements, it lacks the design variety and stylistic depth exhibited by Midjourney.

Figure 11.22 DALL-E—A bedroom with a wooden queen bed and timber floor and a large window daytime.

Figure 11.23 Mid-Journey—A bedroom with a wooden queen bed and timber floor and a large window daytime.

Figure 11.24 Deep Dream—A bedroom with a wooden queen bed and timber floor and a large window daytime.

Ultimately, the choice between StyleGAN, MidJourney, DeepDream, and DALL·E depends on the specific needs and objectives of the user. For tasks requiring stylistic diversity and visually compelling outputs—particularly in interior and urban design—MidJourney consistently outperforms with its atmospheric depth, rich textures, and expressive compositions, making it ideal for early-stage ideation and conceptual exploration. DeepDream and DALL·E demonstrate strong reliability and detail accuracy in architectural visualization, particularly in scenarios such as building design where they effectively capture prompt-specific features like green rooftops and structural realism. Their ability to balance creativity with spatial coherence makes them strong contenders for detailed design representation.

StyleGAN, although we did not have access to test different prompt-based variations in this study, is known for producing high-fidelity synthetic images and allows for fine-grained control through latent space manipulation. It remains a valuable option for use cases that demand controlled visual outputs based on pre-encoded parameters rather than natural language prompts.

5 3D GAI-IS Tools in the Metaverse and Construction Industry

While earlier tools like StyleGAN, MidJourney, DeepDream, and DALL·E are useful for conceptual visualization in real-world construction and the metaverse, they lack the ability to create fully interactive and structurally accurate 3D assets, limiting their practical applications in architectural design, construction prototyping, and immersive virtual environments. To overcome these challenges, new GAI-IS tools have emerged, capable of generating high-fidelity, fully textured 3D models that are directly applicable to virtual reality (VR), metaverse development, and digital twin simulations. Among the most notable of these tools are NVIDIA GET3D, Meshy, Masterpiece X, Autodesk Project Bernini, and Autodesk Dream Sketch, each offering distinct functionalities while sharing core AI-driven modeling principles.

5.1 NVIDIA GET3D

A prime example of a GAI-IS tool is NVIDIA GET3D, which specializes in high-resolution 3D model generation, making it particularly effective for creating individual assets such as buildings, vehicles, characters, and urban elements. Unlike traditional procedural modeling techniques, GET3D is designed to generate detailed, fully textured 3D models from 2D image datasets. This tool does not require 3D object scans for training, making it a cost-effective and scalable solution for VR content creation and metaverse asset development. At its core, GET3D employs a dual-branch neural network architecture: one branch focuses on geometry generation using a signed distance field (SDF) representation, while the other generates high-quality textures through a tri-plane neural network.

To convert the continuous SDF-based geometry into a usable 3D mesh, GET3D incorporates the Deep Marching Tetrahedra (DMTet) algorithm—a mesh extraction technique that reconstructs explicit surface geometry from the learned SDF.

This ensures that the final output models capture fine structural detail and maintain physical accuracy. A differentiable renderer is then used to refine the generated meshes by producing multi-view 2D projections, which are fed into the training process to optimize texture realism through adversarial learning (Gao et al., 2022). GET3D's ability to produce detailed, game-ready assets from minimal input makes it a powerful tool for metaverse development, where large numbers of high-quality and diverse 3D objects are needed for immersive virtual environments.

5.2 Masterpiece X

Another tool that extends the capabilities of AI-driven 3D modeling is Masterpiece X, which eliminates the need for manual sculpting by allowing users to generate 3D assets from either 2D images or text-based prompts. Unlike GET3D, which relies solely on pre-existing 2D image datasets, Masterpiece X employs natural language processing models to interpret textual descriptions and translate them into structured 3D representations. The AI system generates a polygon mesh for geometry and applies AI-powered texture synthesis, ensuring high-quality visual output. Masterpiece X is optimized for rapid 3D content creation, enabling developers to quickly prototype objects such as buildings, characters, and immersive VR environments for applications in gaming, digital prototyping, and architectural visualization (Masterpiece Studio, 2025). Figure 11.25 shows a 3D model of a building with a lion-shaped design created in Masterpiece X.

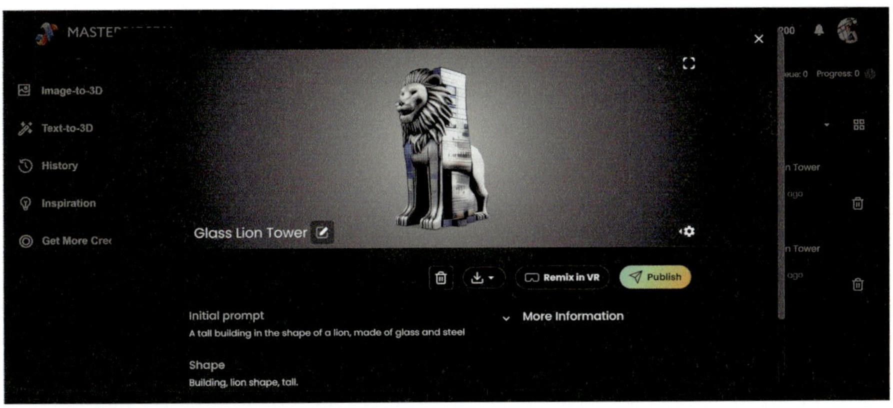

Figure 11.25 A 3D model of a building with a lion-shaped design created in Masterpiece X.

5.3 Meshy

While GET3D and Masterpiece X primarily focus on individual 3D assets, Meshy offers a broader, more flexible approach to AI-powered modeling by incorporating deep generative models to automate 3D asset creation for AR, VR, and digital twin applications. Similar to Masterpiece X, Meshy uses a text-to-3D modeling approach; however, it allows users to define specific design parameters, such as object scale, level of detail, and material properties, providing designers with greater control over asset customization. Additionally, Meshy extends beyond static asset generation, offering an animation component, which allows users to create dynamic 3D models, such as human or animal characters with pre-generated animations (Jie, Shan, & Chung, 2023). This capability makes Meshy an ideal tool for industries requiring interactive content, including gaming, the metaverse, industrial training, and medical simulations. Figure 11.26 shows a 3D model of a building with a lion-shaped design created in Meshy.

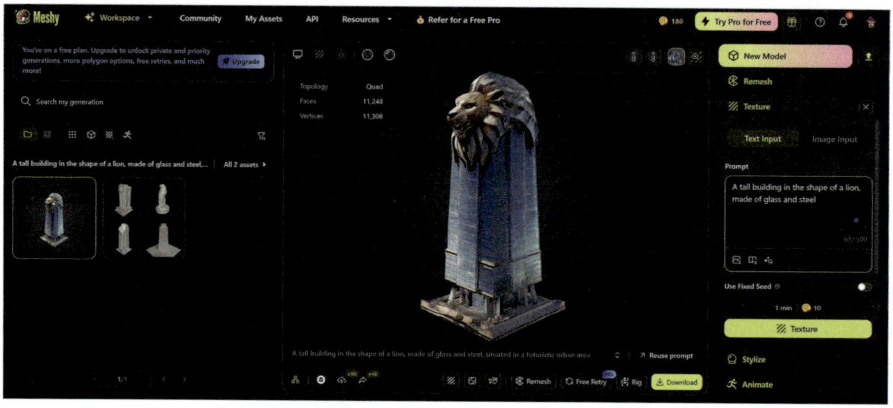

Figure 11.26 A 3D model of a building with a lion-shaped design created in Meshy.

5.4 Autodesk 3D GAI-IS Tools

Another groundbreaking AI-powered tool in 3D modeling is Autodesk Project Bernini, developed by Autodesk Research to advance AI-driven shape generation for architectural design, product development, and entertainment. Unlike text-to-3D or dataset-driven tools, Bernini separates shape and texture generation, ensuring structurally viable and geometrically accurate 3D outputs. The model, trained on 10 million 3D objects, ranging from CAD models to organic forms, can generate functional 3D shapes from 2D images, point clouds, voxels, and textual descriptions. Project Bernini enhances early-stage architectural workflows, allowing designers to experiment with different design alternatives efficiently (Arnold, 2024).

Another innovative solution from Autodesk is Dream Sketch, which provides a unique approach to GenAI-driven 3D modeling by allowing designers to sketch their concepts instead of using text-based prompts (Kazi et al., 2017). This tool blends the expressive nature of freehand sketching with the computational power of AI-based generative algorithms, creating an intuitive and interactive 3D design interface. By simply sketching a rough outline, users can quickly generate multiple AI-generated design alternatives, which are then refined into fully realized 3D models.

Dream Sketch not only enhances conceptual design workflows but also accelerates the transition from idea to prototype, making it an ideal tool for architects, engineers, and metaverse developers who need high-speed iterative modeling (Keshavarzi et al., 2020, 2021). Users of Dream Sketch can navigate the generated models in a virtual 3D environment, experimenting with different angles, perspectives, and modifications (Acosta-Zazueta et al., 2021). This iterative workflow provides greater flexibility and creativity, allowing designers to refine their ideas without being constrained by the rigid nature of traditional 3D modeling software. The ability to seamlessly generate multiple design variations makes Dream Sketch particularly useful for complex building structures such as bridges, high-rise buildings, and urban infrastructure (Baduge et al., 2022). For example, as depicted in Figure 11.27, a designer can sketch a car model, and the GAI-IS tool generates various 3D car models based on that input. Please note that this image is illustrative only and was not created using the Dream Sketch system.

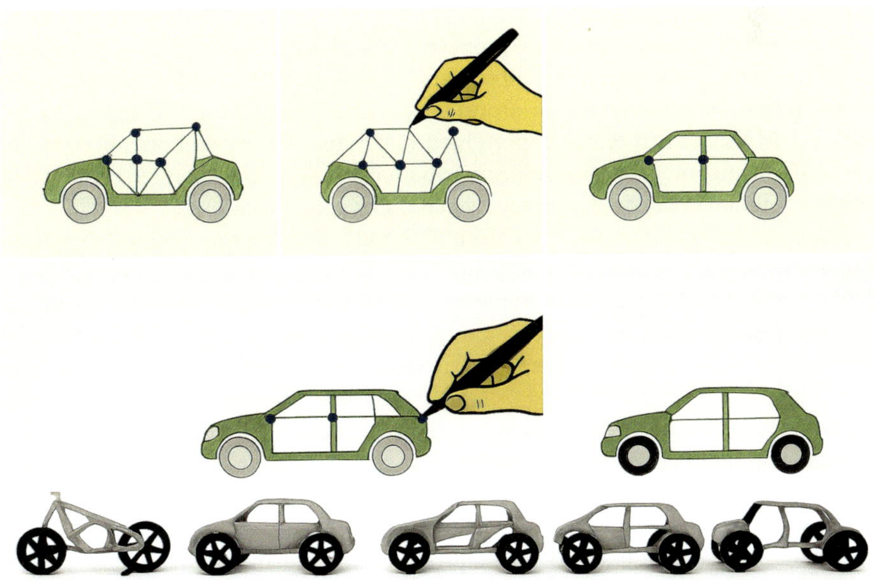

Figure 11.27 Example of 3D car models generated from a hand-drawn sketch.

Despite their distinct functionalities and underlying technologies, tools such as NVIDIA GET3D, Masterpiece X, Meshy, Autodesk Project Bernini, and Autodesk Dream Sketch share a common objective: to automate and streamline 3D content creation for use in construction, virtual reality, and metaverse worldbuilding. NVIDIA GET3D is focused on photorealistic asset generation, making it the preferred choice for high-detail object creation in digital prototyping and VR experiences. Masterpiece X eliminates the technical barrier of 3D modeling expertise, providing a text-to-3D workflow that simplifies content creation for game developers and digital artists. Meshy enhances customization and animation capabilities, making it a versatile tool for both static and dynamic content in virtual spaces. Autodesk Project Bernini pushes AI-driven 3D shape generation forward, catering to architectural, product design, and entertainment industries, while Dream Sketch offers a freeform, intuitive approach to ideation through sketch-based generative modeling.

6 Ethical Considerations and Challenges of GAI-IS Tools

As GAI-IS tools become increasingly integrated into construction and metaverse design industries, it is crucial to address their challenges, limitations, and ethical considerations. While these tools offer unparalleled efficiency, enhance design processes, and facilitate collaboration between designers and stakeholders, they also introduce technological and ethical concerns that must be resolved to ensure responsible and equitable AI adoption.

One of the most significant challenges and ethical concerns surrounding GAI-IS tools is ownership and intellectual property (IP). Many of these tools rely on large datasets containing architectural blueprints, images, and technical drawings, some of which may be copyrighted or proprietary (Amer, 2023; Israel & Amer, 2022). If a GAI-IS tool incorporates copyrighted materials without authorization, it can lead to IP infringement, raising concerns about who owns AI-generated content and whether original creators should receive credit or compensation.

To address ownership concerns, designers and architects must ensure they have legal rights to use any data input into their GAI-IS tools. They should also establish clear licensing and distribution frameworks, ensuring compliance with copyright laws and fair-use policies. Some AI companies have introduced opt-in or opt-out policies for artists and designers contributing data to AI models, while others have implemented watermarking techniques to track AI-generated content and its sources. These strategies help promote ethical AI usage while protecting the rights of original content creators.

Another key challenge of GAI-IS tools is their limited creativity. These models are trained to generate designs based on predefined rules, datasets, and probabilistic patterns, making them less capable of producing truly innovative or unconventional concepts (Oppenlaender et al., 2023; Paananen, Oppenlaender, & Visuri, 2024). Unlike human designers, who apply critical thinking and contextual awareness, AI-generated designs often replicate existing styles and fail to introduce novel architectural solutions. To enhance GAI-IS capabilities, researchers can incorporate advanced machine learning techniques, such as reinforcement learning and GANs, which allow models to learn from real-time data and generate more diverse, creative designs.

Beyond IP and creativity concerns, other challenges are more applicable to real-world construction than to the metaverse, including bias in design, lack of transparency, and environmental sustainability concerns. Since GAI-IS tools are trained on historical and real-world datasets, they may reinforce pre-existing biases in urban planning, architectural styles, and accessibility considerations (Kudless,

2023). If a model is trained on a limited dataset that represents only high-income neighborhoods or modern urban designs, it may struggle to generate inclusive designs for diverse socioeconomic, cultural, and geographic settings. AI-generated layouts may favor high-rise commercial buildings over affordable housing solutions or neglect historically significant architectural elements. Moreover, AI tools may fail to integrate accessibility features, leading to exclusionary environments that are not suitable for people with disabilities.

To address bias-related concerns, it is crucial to train AI models on diverse datasets that reflect various cultural, economic, and geographic contexts. Researchers should implement bias detection and mitigation techniques, such as fairness-aware training algorithms and human-in-the-loop interventions, to ensure GAI-IS-generated designs are equitable and inclusive (Oppenlaender et al., 2023).

A further challenge is contextual awareness. GAI-IS tools can only generate images based on input text, images, and the datasets they have been trained on, meaning they cannot fully account for the surrounding environment of a construction or urban design project (Kudless, 2023). This often results in unrealistic, impractical, or contextually inaccurate designs. Expanding the scope of GAI-IS tools by incorporating a broader range of design parameters, environmental factors, and real-time contextual data could help overcome this limitation.

A promising solution to contextual awareness challenges in AI-driven architectural design is Autodesk Forma, an AI-powered generative design tool that prioritizes environmental efficiency in construction and urban planning. Unlike traditional GAI-IS tools, which generate designs based solely on aesthetic preferences or functional criteria, Autodesk Forma integrates environmental simulations to help architects and city planners create energy-efficient, eco-friendly structures. The tool analyzes factors such as solar radiation, wind flow, and thermal comfort to optimize building placement and orientation, reducing energy consumption and heat retention in urban spaces.

Additionally, Autodesk Forma supports sustainable material selection by evaluating carbon footprints and lifecycle impacts, ensuring that generated designs align with green building standards. By embedding sustainability-aware AI into early-stage planning, Autodesk Forma allows designers to create environmentally responsible urban layouts while balancing functionality, aesthetics, and ecological impact. As AI-driven construction practices evolve, integrating tools like Autodesk Forma will be essential for ensuring long-term sustainability in both real-world architecture and digital metaverse environments (Kurniawan et al., 2024; Nasır et al., 2024).

Transparency is another pressing ethical issue in GAI-IS applications. Many of these tools operate as "black-box" systems, meaning that their decision-making processes are difficult to interpret (Chen, Fu, & Lyu, 2023). Since AI-generated architectural and urban designs can impact large-scale infrastructure projects, a lack of explainability makes it difficult to assess biases, validate structural integrity, or hold designers accountable for errors or unintended consequences (Amer, 2023; Contractor et al., 2022).

To mitigate transparency concerns, designers, architects, and AI developers must prioritize algorithmic transparency by thoroughly documenting data sources, model parameters, and decision-making processes. AI-assisted design workflows should be traceable, allowing experts to audit and validate AI-generated outputs before implementation. Additionally, GAI-IS tools should incorporate explainable AI (XAI) methodologies to clarify the rationale behind their decision-making processes, thereby enhancing interpretability and promoting fairness in AI-driven design outcomes (Hashemi, Darejeh, & Cruz, 2024, 2025).

By addressing these technological and ethical challenges, including creativity limitations, bias, transparency, IP concerns, and context awareness, designers and architects can harness AI's full potential while ensuring responsible and equitable AI adoption.

7 Discussion

As demonstrated throughout this chapter, the emergence of GAI-IS tools has introduced a paradigm shift in how both real-world and virtual environments are visualized, conceptualized, and prototyped. While their impact has been most immediate in early-stage design ideation, their full integration into professional practice remains complex. To realize their long-term potential, it is critical to evaluate these tools not only through their technical strengths but also through their limitations in contextual awareness, interoperability, and compliance with industry norms (Kudless, 2023; Wu, Stouffs, & Biljecki, 2022).

A central challenge lies in the tension between visual expressiveness and architectural executability. Transformer-based and diffusion-based models—such as DALL·E and MidJourney, respectively—have proven highly effective in generating stylized, high-resolution images from text prompts, enabling designers to rapidly iterate through diverse conceptual options (Buzzaccarini et al., 2024; OpenAI, 2023; Park, 2024; Wang, Xu, & Sun, 2022). However, their outputs often lack contextual embedding, failing to incorporate real-world constraints such as building codes, material behavior, or zoning regulations (Kudless, 2023). As such, they serve more as visual ideation tools than instruments for producing constructible architectural documentation.

More advanced 3D GAI-IS tools—such as NVIDIA GET3D and Masterpiece X—offer increased structural fidelity by generating detailed, textured 3D meshes directly from sketches or textual input (Gao et al., 2022; Kazi et al., 2017; Masterpiece Studio, 2025). Yet, their integration with Building Information Modeling (BIM) systems remains limited, creating barriers for real-world implementation. Although recent studies advocate for embedding BIM into GAI-IS pipelines to enhance semantic richness and regulatory compliance (Ghannad & Lee, 2021; Ying, Sacks, & Degani, 2023), most tools still lack the ability to generate outputs that meet industry standards without significant manual intervention.

In contrast, GAI-IS tools have found a more natural fit within metaverse development, where design is not bound by physical laws or local regulations. In this domain, generative models support the rapid creation of immersive environments and assets, enabling non-technical users to contribute meaningfully to digital world-building (Gupta et al., 2025; Nalbant, Aydın, & Uyanık, 2023). However, new technical bottlenecks emerge: high-polygon models generated by tools like Meshy or GET3D often require complex optimization workflows to be functional in real-time rendering engines such as Unity or Unreal (Jie, Shan, & Chung, 2023). Moreover, AI-generated textures frequently exhibit inconsistency or lack material realism, requiring post-processing to achieve interactive coherence (Yan et al., 2022).

To move toward broader adoption, future GAI-IS research must prioritize multi-faceted integration—combining sustainability metrics, legal constraints, and environmental context into generative outputs (Kurniawan et al., 2024; Nasır et al., 2024). Systems must also become more interactive and modular, allowing expert users to refine outputs in real-time while ensuring that non-expert users are guided by safeguards aligned with best practices. Ultimately, the goal is not to replace human designers but to augment their capabilities, enabling co-creative workflows that combine machine-driven synthesis with domain expertise. This vision reframes GAI-IS tools as components of a larger design intelligence ecosystem—adaptive, explainable, and grounded in both computational innovation and contextual understanding.

8 Conclusion

As GAI-IS technologies mature, their influence on design disciplines is shifting from experimental novelty to strategic necessity. The potential of these tools lies not only in accelerating visual ideation but also in reshaping how we conceive, communicate, and collaborate across physical and virtual domains. Their successful deployment, however, demands more than technical advancement—it requires the reconfiguration of design workflows, ethical frameworks, and educational practices to accommodate machine-assisted creativity. Moving forward, interdisciplinary collaboration between AI developers, architects, urban planners, and policymakers will be essential to ensure that these technologies are aligned with societal needs, regulatory frameworks, and cultural values. Rather than replacing human expertise, GAI-IS tools should be cultivated as creative partners—augmenting human insight while maintaining accountability, inclusivity, and purpose in both real and imagined worlds.

References

Abdallah, Y. K., & Estévez, A. T. (2023). Biomaterials research-driven design visualized by AI text-prompt-generated images. *Designs*, *7*(2), 48.

Acosta-Zazueta, G., Alcaide-Marzal, J., & Diego-Más, J. A. (2021). Generative design software and different approaches. *Proceedings of DARCH 2021—1st International Conference on Architecture & Design* (pp. 101–108). OCERINT. https://doi.org/10.46529/darch.202130

Alcaide-Marzal, J., & Diego-Mas, J. A. (2025). Computers as co-creative assistants. A comparative study on the use of text-to-image AI models for computer aided conceptual design. *Computers in Industry*, *164*, 104168.

Amer, S. K. (2023). AI imagery and the overton window. *arXiv preprint arXiv, 2306,* 00080.

Analytics Vidhya. (2024, July). *DALL·E 3: Revolutionizing AI image generation*. Analytics Vidhya. Retrieved from https://www.analyticsvidhya.com/blog/2024/07/dall-e3

Antipov, G., Baccouche, M., & Dugelay, J. L. (2017). Face aging with conditional generative adversarial networks. In *2017 IEEE International Conference on Image Processing (ICIP)* (pp. 2089–2093). IEEE.

Arnold, E. (2024). *Autodesk research unveils project bernini for generative AI 3D shape creation.* Autodesk. Retrieved [2, 1, 2025], from https://www.research.autodesk.com/blog/autodesk-research-unveils-project-bernini-for-generative-ai-3d-shape-creation/

Baduge, S. K., Thilakarathna, S., Perera, J. S., Arashpour, M., Sharafi, P., Teodosio, B., … & Mendis, P. (2022). Artificial intelligence and smart vision for building and construction 4.0: Machine and deep learning methods and applications. *Automation in Construction*, *141*, 104440.

Belesky, P. (2023). So predictable?: AI and landscape architecture. *Landscape Architecture Australia*, https://landscapeaustralia.com/articles/so-predictable-ai-and-landscape-architecture-1/

Beyan, E. V. P., & Rossy, A. G. C. (2023). A review of AI image generator: Influences, challenges, and future prospects for the architectural field. *Journal of Artificial Intelligence in Architecture, 2*(1), 53–65.

Borji, A. (2022). Generated faces in the wild: Quantitative comparison of stable diffusion, midjourney and dall-e 2. *arXiv preprint arXiv, 2210,* 00586.

Buzzaccarini, G., Degliuomini, R. S., Borin, M., Fidanza, A., Salmeri, N., Schiraldi, L., …& Pagliardini, L. (2024). The promise and pitfalls of AI-generated anatomical images: evaluating midjourney for aesthetic surgery applications. *Aesthetic Plastic Surgery*, *48*(9), 1874–1883.

Chen, C., Fu, J., & Lyu, L. (2023). A pathway towards responsible ai generated content. *arXiv preprint arXiv, 2303,* 01325.

Contractor, D., McDuff, D., Haines, J. K., Lee, J., Hines, C., Hecht, B., …& Li, H. (2022, June). Behavioral use licensing for responsible AI. In *Proceedings of the 2022 ACM Conference on Fairness, Accountability, and Transparency* (pp. 778–788). ACM.

Cornia, M., Baraldi, L., & Cucchiara, R. (2022). Explaining transformer-based image captioning models: An empirical analysis. *AI Communications*, *35*(2), 111–129.

Darejeh, A., Marcus, N., & Sweller, J. (2022). Increasing learner interactions with E-learning systems can either decrease or increase cognitive load depending on the nature of the interaction. *L`Année psychologique*, *122*(3), 405–437.

Darejeh, A., Moghadam, T. S., Delaramifar, M., & Mashayekh, S. (2024). A framework for AI-powered decision making in developing adaptive e-learning systems to impact learners' emotional responses. In 2024 *11th International and the 17th National Conference on E-Learning and E-Teaching (ICeLeT)* (pp. 1–6). IEEE.

Ding, M., Yang, Z., Hong, W., Zheng, W., Zhou, C., Yin, D., …& Tang, J. (2021). Cogview: Mastering text-to-image generation via transformers. *Advances in Neural Information Processing Systems*, *34*, 19822–19835.

Fernandez, P. (2022). Technology behind text to image generators. *Library Hi Tech News*, 39(10), 1–4. https://doi.org/10.1108/LHTN-10-2022-0116

Fox, G., Tewari, A., Elgharib, M., & Theobalt, C. (2021). Stylevideogan: A temporal generative model using a pretrained stylegan. *arXiv preprint arXiv, 2107,* 07224.

Franchi, V., & Ntagiou, E. (2021, November). Augmentation of a virtual reality environment using generative adversarial networks. In *2021 IEEE International Conference on Artificial Intelligence and Virtual Reality (AIVR)* (pp. 219–223). IEEE.

Gao, J., Shen, T., Wang, Z., Chen, W., Yin, K., Li, D., …& Fidler, S. (2022). Get3d: A generative model of high quality 3d textured shapes learned from images. *Advances In Neural Information Processing Systems*, *35*, 31841–31854.

Ghannad, P., & Lee, Y. C. (2021). Developing an advanced automated modular housing design system using deep learning and Building Information Modeling (BIM). In *International Conference on Computing in Civil Engineering 2021* (pp. 587–595). ASCE.

Ghannad, P., & Lee, Y. C. (2022). Automated modular housing design using a module configuration algorithm and a coupled generative adversarial network (CoGAN). *Automation in Construction*, *139*, 104234.

Giri, D. P., Kannan, D., Manoj, A., Krishna, A., & Ghosh, A. (2023). A survey on illustration generation using deep dream algorithm. *Indian Journal of Natural Sciences*, *13*(76).

Goodfellow, I., Pouget-Abadie, J., Mirza, M., Xu, B., Warde-Farley, D., Ozair, S., Courville, A., & Bengio, Y. (2014). Generative adversarial nets. *Advances in Neural Information Processing Systems*, 27, 2672–2680.

Guinness, H. (2023). The top AI art generators in 2023. https://zapier.com/blog/ai-art-generator/ (accessed: 2023/5/23)

Gupta, A., Yu, L., Sohn, K., Gu, X., Hahn, M., Li, F. F., Essa, I., Jiang, L., & Lezama, J. (2025). Photorealistic video generation with diffusion models. In *European Conference on Computer Vision* (pp. 393–411). Springer.

Hashemi, M., Darejeh, A., & Cruz, F. (2024, October). A user-centric exploration of axiomatic explainable AI in participatory budgeting. In *Companion of the 2024 on ACM International Joint Conference on Pervasive and Ubiquitous Computing* (pp. 126–130). ACM.

Hashemi, M., Darejeh, A., & Cruz, F. (2025). Understanding user preferences in explainable artificial intelligence: A mapping function proposal. *ACM Transactions on Intelligent Systems and Technology*. *16*(4), 1–37. https://doi.org/10.1145/3733837

Hatamizadeh, A., Song, J., Liu, G., Kautz, J., & Vahdat, A. (2025). Diffit: Diffusion vision transformers for image generation. In *European Conference on Computer Vision* (pp. 37–55). Springer.

Hegazy, M., & Mohamed Saleh, A. (2023). *Evolution of AI role in architectural design: between parametric exploration and machine hallucination. MSA Engineering Journal*. *2*(2). https://doi.org/10.21608/msaeng.2023.291873

Hosseini, M., Saugstad, M., Miranda, F., Sevtsuk, A., Silva, C. T., & Froehlich, J. E. (2022). Towards global-scale crowd+ AI techniques to map and assess sidewalks for people with disabilities. *arXiv preprint arXiv, 2206,* 13677.

Israel, M. J., & Amer, A. (2022). Rethinking data infrastructure and its ethical implications in the face of automated digital content generation. *AI and Ethics*, *3*(2), 427–439.

Jie, P., Shan, X., & Chung, J. (2023). A comparative analysis between< Leonardo. Ai> and< Meshy> as AI texture generation tools. *International Journal of Advanced Culture Technology*, *11*(4), 333–339.

Jones, K. (2023). *Will art generated by artificial intelligence replace the role of a graphic designer?* (Doctoral dissertation, The University of Memphis).

Kahila, M., & Kyttä, M. (2009). SoftGIS as a bridge-builder in collaborative urban planning. In S. Geertman, & J. Stillwell (Eds.), *Planning Support Systems Best Practice and New Methods* (Vol. 95, pp. 389–411). Springer.

Kamrowska-Załuska, D. (2021). Impact of ai-based tools and urban big data analytics on the design and planning of cities. *Land*, *10*(11), 1209.

Karras, T., Laine, S., & Aila, T. (2019). A style-based generator architecture for generative adversarial networks. *Proceedings of the IEEE/CVF Conference on Computer Vision and Pattern Recognition*, 4401–4410.

Karras, T., Laine, S., Aittala, M., Hellsten, J., Lehtinen, J., & Aila, T. (2020). Analyzing and improving the image quality of stylegan. In *Proceedings of the IEEE/CVF Conference on Computer Vision and Pattern Recognition* (pp. 8110–8119).

Kazi, R. H., Grossman, T., Cheong, H., Hashemi, A., & Fitzmaurice, G. W. (2017). DreamSketch: Early stage 3D design explorations with sketching and generative design. In *Proceedings of the 30th Annual ACM Symposium on User Interface Software and Technology (UIST '17)* (pp. 401–414). Association for Computing Machinery, New York, NY, USA. https://doi.org/10.1145/3126594.3126662

Keshavarzi, M., Bidgoli, A., & Kellner, H. (2020). V-dream: Immersive exploration of generative design solution space. In *HCI International 2020-Late Breaking Papers: Multimodality and Intelligence: 22nd HCI International Conference, HCII 2020,* Copenhagen, Denmark, *July 19–24, 2020, Proceedings 22* (pp. 477–494). Springer International Publishing.

Keshavarzi, M., Hotson, C., Cheng, C. Y., Nourbakhsh, M., Bergin, M., & Rahmani Asl, M. (2021, May). Sketchopt: Sketch-based parametric model retrieval for generative design. In *Extended Abstracts of the 2021 CHI Conference on Human Factors in Computing Systems* (pp. 1–6). Association for Computing Machinery (ACM)

Koley, S., Bhunia, A. K., Sain, A., Chowdhury, P. N., Xiang, T., & Song, Y. Z. (2023). Picture that sketch: Photorealistic image generation from abstract sketches. In *Proceedings of the ieee/cvf Conference on Computer Vision and Pattern Recognition* (pp. 6850–6861). IEEE. https://doi.org/10.1109/CVPR52729.2023.00662

Kudless, A. (2023). Hierarchies of bias in artificial intelligence architecture: Collective, computational, and cognitive. *International Journal of Architectural Computing*, *21*(2), 256–279. https://doi.org/10.1177/14780771231170272

Kurniawan, A., Saragih, V. C., Wiranti, A., Rahmadi, D., & Poerbo, H. W. (2024, September). Revealing the future of urban design: Exploring human and artificial intelligence collaboration in planning and design for the sustainable built environment (Case Study: IKN Nusantara). In *IOP Conference Series: Earth and Environmental Science* (Vol. 1394, No. 1, p. 012032). IOP Publishing.

Li, L., Sun, H., Li, S., Dong, P., Luo, W., Xue, W., Liu, Q., & Guo, Y. (2025). Auto-GAS: Automated proxy discovery for training-free generative architecture search. In *European Conference on Computer Vision* (pp. 38–55). Springer.

Li, M., Zhang, J., Hong, Y., Xie, X., Meng, Z., & Gu, S. (2024). Advanced product personalization in the blockchain-enabled metaverse: A diffusion model for automatic style generation. *IEEE Internet of Things Journal*, *12*(8), 10304–10315. https://doi.org/10.1109/JIOT.2024.3511667

Liu, R., Khakzar, A., Gu, J., Chen, Q., Torr, P., & Pizzati, F. (2025). Latent guard: A safety framework for text-to-image generation. In *European Conference on Computer Vision* (pp. 93–109). Springer.

Lv, Z. (2023). Generative artificial intelligence in the metaverse era. *Cognitive Robotics*, *3*, 208–217.

Maganga, M. (2023). *The GAI-IS tool: The limits of the algorithm and human biases*. ArchDaily. Retrieved April 9, 2023, from https://www.archdaily.com/992012/the-ai-image-generator-thelimits-of-the-algorithm-and-human-biases

Malakan, Z. M., Hassan, G. M., & Mian, A. (2022, December). Vision transformer based model for describing a set of images as a story. In *Australasian Joint Conference on Artificial Intelligence* (pp. 15–28). Springer International Publishing.

Manovich, L. (2024). *Diffusions in architecture: Artificial intelligence and image generators*. John Wiley & Sons.

Martinez, J. M. C., Neves, L. A., Longo, L. H. D. C., Rozendo, G. B., Roberto, G. F., Tosta, T. A. A., …& do Nascimento, M. Z. (2024). Exploring deepDream and XAI representations for classifying hhistological images. *SN Computer Science*, *5*(4), 362.

Masterpiece Studio. (2025). *Masterpiece X: AI-powered 3D model generation*. Masterpiece Studio. Retrieved [2, 1, 2025], from https://www.masterpiecex.com/

Meeran, A. (2021). *AI and Architecture: Image-based Machine Learning for early-stage design conceptualization* (Doctoral dissertation, Singapore University of Technology and Design).

Nalbant, K. G., Aydın, S., & Uyanık, Ş. (2023). Generative adversarial network and digital art interactions with Metaverse marketing. *Trakya Üniversitesi Sosyal Bilimler Dergisi, 25*(2), 375–396.

Nasır, C. A., Yüksel, E. Ö., & Mayuk, S. G. (2024). Field study on determining the impact of design decisions on energy efficiency at settlement scale. *Journal of Design for Resilience in Architecture and Planning*, *5*(3), 378–400.

Ng, D. T. K. (2022). What is the metaverse? Definitions, technologies and the community of inquiry. *Australasian Journal of Educational Technology*, *38*(4), 190–205.

Open AI. (2023). *DALL·E 3*. OpenAI. Retrieved from https://openai.com/dall-e-3

Oppenlaender, J. (2022, November). The creativity of text-to-image generation. In *Proceedings of the 25th International Academic Mindtrek Conference* (pp. 192–202).ACM. https://doi.org/10.1145/3569219.3569352

Oppenlaender, J., Silvennoinen, J., Paananen, V., & Visuri, A. (2023). Perceptions and realities of text-to-image generation. *arXiv preprint arXiv, 2306,* 08363.

Paananen, V., Oppenlaender, J., & Visuri, A. (2024). Using text-to-image generation for architectural design ideation. *International Journal of Architectural Computing, 22*(3), 458–474.

Park, K. (2024). A study on the color of AI-generated images for fashion design-focused on the use of Midjourney. *The Journal of the Convergence on Culture Technology*, *10*(2), 343–348.

Patashnik, O., Wu, Z., Shechtman, E., Cohen-Or, D., & Lischinski, D. (2021). Styleclip: Text-driven manipulation of stylegan imagery. In *Proceedings of the IEEE/CVF International Conference on Computer Vision* (pp. 2085–2094). IEEE.

O. Patashnik, Z. Wu, E. Shechtman, D. Cohen-Or, and D. Lischinski, "StyleClip: Text-driven manipulation of StyleGAN imagery," in *Proceedings of the IEEE/CVF International Conference on Computer Vision (ICCV)*, Virtual Event, Oct. 2021, pp. 2085–2094.

Ploennigs, J., & Berger, M. (2022). AI art in architecture. *arXiv preprint arXiv, 2212,* 09399.

Ratican, J., Hutson, J., & Wright, A. (2023). A proposed meta-reality immersive development pipeline: Generative ai models and extended reality (xr) content for the metaverse. *Journal of Intelligent Learning Systems and Applications*, *15*(1), 24–35. https://doi.org/10.4236/jilsa.2023.151002

Reddy, M. D. M., Basha, M. S. M., Hari, M. M. C., & Penchalaiah, M. N. (2021). Dall-e: Creating images from text. *UGC Care Group I Journal*, *8*(14), 71–75.

Riemer, K., & Peter, S. (2024). Conceptualizing generative AI as style engines: Application archetypes and implications. *International Journal of Information Management*, *79*, 102824.

Roose, K. (2022). An AI-generated picture won an art prize. Artists Aren't Happy. *The New York Times*, *2*, 2022.

Saparamadu, P. V. I. N., Sepasgozar, S., Guruge, R. N. D., Jayasena, H. S., Darejeh, A., Ebrahimzadeh, S. M., & Eranga, B. A. I. (2025). Optimising contract interpretations with large language models: A comparative evaluation of a vector database-powered Chatbot vs. ChatGPT. *Buildings*, *15*(7), 1144.

Sardar, P., Abbott, J. D., Kundu, A., Aronow, H. D., Granada, J. F., & Giri, J. (2019). Impact of artificial intelligence on interventional cardiology: from decision-making aid to advanced interventional procedure assistance. *Cardiovascular Interventions*, *12*(14), 1293–1303.

Sargazi Moghadam, T., Darejeh, A., Delaramifar, M., & Mashayekh, S. (2024). Toward an artificial intelligence-based decision framework for developing adaptive e-learning systems to impact learners' emotions. *Interactive Learning Environments*, *32*(7), 3665–3685.

Sauer, A., Karras, T., Laine, S., Geiger, A., & Aila, T. (2023). Stylegan-t: Unlocking the power of gans for fast large-scale text-to-image synthesis. In *International Conference on Machine Learning* (pp. 30105–30118). PMLR.

Sun, C., Zhou, Y., & Han, Y. (2022). Automatic generation of architecture facade for historical urban renovation using generative adversarial network. *Building and Environment, 212*, 108781.

Temsah, M. H., Alhuzaimi, A. N., Almansour, M., Aljamaan, F., Alhasan, K., Batarfi, M. A., …& Nazer, R. (2024). Art or artifact: evaluating the accuracy, appeal, and educational value of AI-generated imagery in DALL· E 3 for illustrating congenital heart diseases. *Journal of Medical Systems*, *48*(1), 54.

Toğaçar, M., Cömert, Z., & Ergen, B. (2021). Enhancing of dataset using DeepDream, fuzzy color image enhancement and hypercolumn techniques to detection of the Alzheimer's disease stages by deep learning model. *Neural Computing and Applications*, *33*(16), 9877–9889.

Verheijden, M. P., & Funk, M. (2023, April). Collaborative diffusion: Boosting designerly Co-creation with generative AI. In *Extended Abstracts of the 2023 CHI Conference on Human Factors in Computing Systems* (pp. 1–8). ACM.

Walmsley, K. (2022). Cutting-edge practical research on generative design, IoT and digital twins. In *Innovation in Construction: A Practical Guide to Transforming the Construction Industry* (pp. 117–145). Springer International Publishing.

Wang, Y., Xu, J., & Sun, Y. (2022, June). End-to-end transformer based model for image captioning. In *Proceedings of the AAAI Conference on Artificial Intelligence* (Vol. 36, No. 3, pp. 2585–2594). AAAI Press.

Wen, R., & Li, X. (2025). *3D effect image solution of landscape design model based on generative adversarial networks. Computer-Aided Design & Applications*, *22*(S1), 1–15.

Witteveen, S., & Andrews, M. (2022). Investigating Prompt Engineering in Diffusion Models. *arXiv preprint arXiv, 2211,* 15462.

Wu, A. N., Stouffs, R., & Biljecki, F. (2022). Generative adversarial networks in the built environment: A comprehensive review of the application of GANs across data types and scales. *Building and Environment, 223*, Article 109477. https://doi.org/10.1016/j.buildenv.2022.109477

Yan, H., Zhang, H., Liu, L., Zhou, D., Xu, X., Zhang, Z., & Yan, S. (2022). *Toward intelligent design: An ai-based fashion designer using generative adversarial networks aided by sketch and rendering generators*. IEEE Transactions on Multimedia.

Yang, C., Shen, Y., & Zhou, B. (2021). Semantic hierarchy emerges in deep generative representations for scene synthesis. *International Journal of Computer Vision, 129*, 1451–1466.

Ye, X., Du, J., & Ye, Y. (2022). MasterplanGAN: Facilitating the smart rendering of urban master plans via generative adversarial networks. *Environment and Planning B: Urban Analytics and City Science*, *49*(3), 794–814.

Yildirim, E. (2022a). AI's vision of future cites. *Urbanizm: An International Scientific Journal of Urban Planning & Sustainable Development*, *27*(7):72–83. https://doi.org/10.58225/urbanizm.2022-27-60-71

Yildirim, E. (2022b). Text-to-image generation AI in architecture. In H. H. Kozlu (Ed.), *Art And Architecture: Theory, Practice and Experience* (pp. 97–120). Livre de Lyon.

Ying, H., Sacks, R., & Degani, A. (2023). Synthetic image data generation using BIM and computer graphics for building scene understanding. *Automation in Construction, 154*, 105016.

Zhou, H., & Xiang, S. (2024). Applicability evaluation and reflection on artificial intelligence-based "image to image" generation of landscape architecture masterplans. *Landscape Architecture Frontiers, 12*(2), 58–67.
Zhu, Y., Fukuda, T., & Yabuki, N. (2024). A mixed reality design system for interior renovation: Inpainting with 360-degree live streaming and generative adversarial networks after removal. *Technologies*, *12*(1), 9.

12 Leveraging Brain–Computer Interfaces to Advance Interaction Within the Digital Landscape

Ali Darejeh, Colby Rehn, Dylan Ngo, Dyllanson So, Sawat Wongsaroj, and Samad M. E. Sepasgozar

1 Introduction

A brain–computer interface (BCI), also referred to as a brain-machine interface (BMI), is a system that enables direct communication between the human brain and external devices by interpreting brain signals to execute commands. Initially developed as an assistive tool for individuals with motor impairments, BCI eliminates the need for physical movement, providing a hands-free, intuitive control method. This technology has been instrumental in helping individuals with injuries, paralysis, or neurological conditions communicate, control prosthetics, and undergo rehabilitation. Early applications of BCI included assisting individuals with speech impediments and offering innovative solutions in the medical field (Nicolas-Alonso & Gomez-Gil, 2012; Värbu, Muhammad, & Muhammad, 2022).

BCIs function by analyzing patterns in brain activity, typically recorded as electroencephalograms (EEG), to perform specific tasks (Darejeh & Mashayekh, 2023). Through training, these systems can interpret a user's imagination, thoughts, and feelings to control devices such as computers, speech synthesizers, assistive appliances, and neural prostheses. For example, the widely used P300 speller, a reliable BCI application, has enabled individuals with severe motor impairments to communicate by spelling words without the use of their limbs (Ryan et al., 2017). This highlights how BCIs provide transformative solutions for those with severe disabilities, potentially reducing caregiving costs as technology becomes more accessible and affordable.

In addition to enabling communication and control, BCI applications have expanded into rehabilitation for conditions such as stroke. Research has shown that BCIs can support motor recovery and offer an effective alternative to traditional treatment methods (Mane, Chouhan, & Guan, 2020). As the reliability of BCI systems improves, these applications continue to demonstrate success, paving the way for their adoption in various medical contexts.

Beyond healthcare, BCIs have transformative potential for human–computer interaction in digital environments. Studies have demonstrated the ability to move virtual avatars through imagined walking (Alchalabi, Faubert, & Labbé, 2021), communicate via inner speech recognition of silently spoken words (Nieto et al., 2022), and decode imagined handwriting gestures into real-time text (Willett et al., 2021). Emerging advancements even suggest the possibility of reconstructing images from brain signals

DOI: 10.1201/9781003557555-12

(Benchetrit, Banville, & King, 2023). These findings showcase the ability of BCI to create unique and engaging interactions in immersive environments such as the metaverse.

As BCI research advances, its applications are becoming more robust, reliable, and cost-effective, offering inclusive and innovative opportunities to enhance everyday lives and revolutionize how users interact with both physical and digital environments. One of the areas that can benefit from BCI technology is the metaverse—a virtual 3D space where users can interact with objects and others in real-time, typically accessed through technologies such as virtual reality (VR) and augmented reality (AR) (Bernal et al., 2022; Cheng, 2023; Darejeh, 2023). Metaverse recreates real-world social settings, enabling users to explore and communicate without leaving their physical environment (Darejeh et al., 2024a). Although companies like Apple, NVIDIA, and Meta are heavily investing in immersive devices to enhance digital interactions (Creed et al., 2024), current metaverse experiences remain constrained by physical input devices such as keyboards, VR controllers, and motion sensors (Darejeh et al., 2024b). These limitations make interaction in the metaverse nearly impossible for individuals with motor impairments or those who are non-verbal. BCIs have the potential to overcome these barriers by enabling interaction through thought alone.

However, the integration of BCI in the metaverse and other applications presents several challenges. As a nascent technology, BCI requires further exploration to address its limitations, understand its advantages, and identify the specific needs of its users. Issues such as reliability, scalability, and ethical concerns must be addressed to ensure the successful adoption of BCI in immersive environments.

This chapter explores the transformative potential of BCI technology by examining its applications, benefits, limitations, and the challenges associated with its integration into diverse contexts, such as interacting with user interfaces and facilitating communication within the metaverse environment. To provide a comprehensive perspective, it analyzes three empirical studies that highlight the practical implementations and outcomes of BCI systems. Ultimately, this chapter aims to expand the possibilities for human interaction and connection, unlocking new dimensions of engagement through the seamless integration of BCI technology into everyday life.

2 Literature Review

This section explores the current landscape of BCI technologies, beginning with their classification into invasive, semi-invasive, and non-invasive types. It outlines the underlying mechanisms of BCI systems and their expanding applications in healthcare, gaming, creative expression, AR/VR, the metaverse, and social media. The literature also highlights key challenges, including technical limitations, cognitive demands, and ethical concerns. These insights lay the groundwork for understanding the real-world applications presented in this chapter's empirical studies.

2.1 Classification and Functioning of BCIs

There are three primary ways to categorize BCIs: invasive, semi-invasive, and non-invasive. Invasive BCIs involve placing sensors directly into the brain; semi-invasive BCIs position electrodes on the brain's surface; and non-invasive

BCIs, the most practical and widely used form, involve placing electrodes on the user's scalp (Moumgiakmas & Papakostas, 2022).

Invasive BCI: This type of BCI requires surgical implantation of microelectrodes into the brain to record the electrical activity of neurons. These systems target deep brain structures and often utilize technologies such as microelectrode arrays (MEA), stereo-electroencephalography (sEEG) electrodes, and deep brain stimulation (DBS) electrodes. While invasive BCIs provide highly detailed and accurate neural data, their use is limited by the significant risks associated with surgical procedures. These risks include infection, brain tissue damage, and other potential complications, making their application in research and healthcare ethically and logistically complex (Xuanmin, 2021; Zhao et al., 2023).

Semi-Invasive BCI: This type of BCI involves the placement of sensors directly on the brain's cortical surface to capture activity from specific regions. Most of these systems employ electrocorticography (ECoG) electrodes, which are surgically implanted onto the brain's surface. Although these systems offer more reliable and clearer data compared to non-invasive BCIs, they still carry significant surgical risks. The requirement for surgical intervention limits their adoption to scenarios where highly precise neural data is essential. As with invasive BCIs, the benefits of semi-invasive systems must be weighed against the risks associated with their use (Zhu et al., 2024).

Non-Invasive BCI: This type of BCI is the most commonly used form of this technology due to its lower risks and ease of application. These systems involve placing electrodes on the scalp to record neural activity without requiring surgery. The most prevalent type of non-invasive BCI is electroencephalography (EEG)-based BCI, which captures electrical signals generated by the brain. Other forms of non-invasive BCI include magnetoencephalography (MEG), which records neural activity through the magnetic fields produced by the brain. While noninvasive BCIs offer lower resolution and accuracy compared to invasive or semi-invasive systems, their safety and accessibility make them ideal for widespread use in research and practical applications (Menon et al., 2009). Figure 12.1 illustrates the various types of BCI devices.

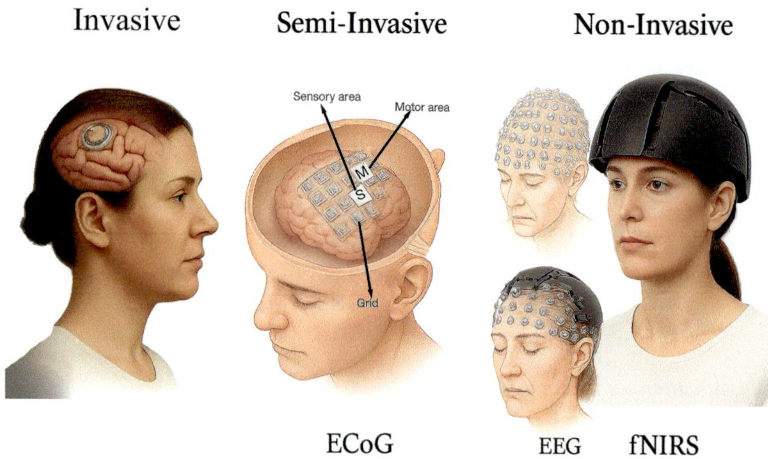

Figure 12.1 Different types of BCIs.

Despite the existence of multiple types of BCIs—categorized as invasive, semi-invasive, and non-invasive—the fundamental concept of BCI technology operates similarly across all types. At its core, BCI technology focuses on capturing and interpreting the electrical activity of the brain, which occurs as neurons fire in response to specific actions, thoughts, or behaviors. This neural activity is recorded through specialized sensors, with the captured data serving as the foundation for further analysis and processing (Naser & Bhattacharya, 2023).

The recorded data encompasses several key metrics, including the frequency of brain waves, the number of active neurons, and the amplitude of these neural signals. These brain waves are typically classified into five distinct categories: gamma, beta, alpha, delta, and theta waves. Each category corresponds to different states of cognitive and physiological activity. Gamma waves are associated with high-level cognitive functioning and information processing, while beta waves often indicate active thinking and focused mental states. Alpha waves, on the other hand, are linked to relaxation, delta waves to deep sleep, and theta waves to states of meditation or light sleep.

By analyzing the behavior and patterns of these waves, BCIs can uncover a wealth of information about the user's mental state and cognitive processes. For example, variations in brain wave activity can reveal changes in attention levels, stress, and even specific emotional states. This ability to detect human emotions and cognitive states extends the utility of BCIs beyond simple command execution, opening possibilities in fields such as mental health monitoring, cognitive enhancement, and personalized human-computer interaction (Kumar & Singh, 2021).

2.2 *Diverse Applications of BCI*

Although originally designed for medical use, BCIs are now applied across diverse domains—from healthcare and neurorehabilitation to immersive entertainment, creative expression, and even autonomous vehicles. The following sections examine how BCI technologies are currently being used across various industries and highlight their expanding potential.

2.2.1 *Medical Applications of BCI*

BCIs offer transformative possibilities in the healthcare sector, addressing physical disabilities, neurological conditions, and neurorehabilitation needs. Initially, BCIs were developed as assistive devices to replace or restore lost functions for individuals with neuromuscular disorders, such as cerebral palsy, stroke, or spinal cord injuries (Shih et al., 2012). These systems bypass physical limitations by creating a neural link between the brain and external devices, enabling communication and motor function restoration in severely impaired individuals, including those with amyotrophic lateral sclerosis (ALS) or locked-in syndrome (LIS). This technology has proven instrumental in enhancing the quality of life and reducing caregiving costs for individuals with severe motor impairments (Nicolas-Alonso & Gomez-Gil, 2012). Figure 12.2 illustrates how BCI technologies enable patients to control bionic hands.

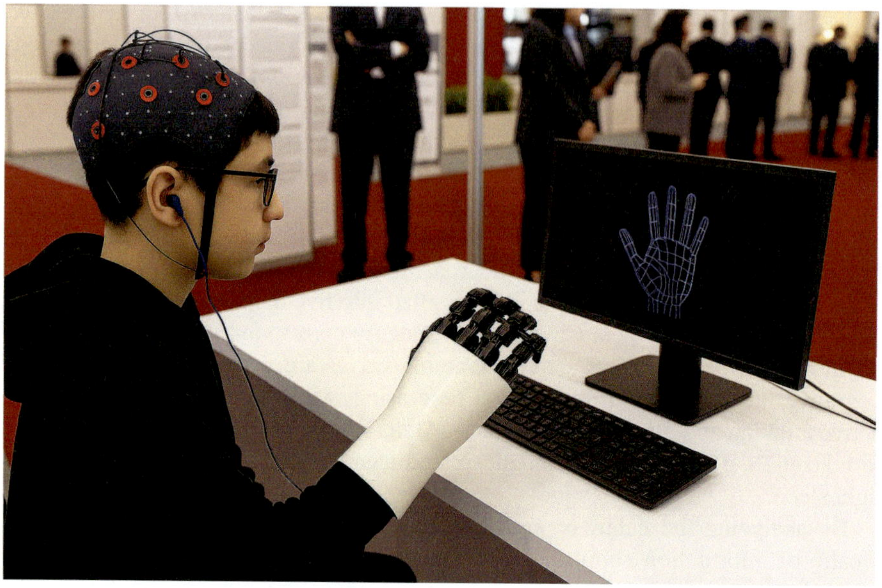

Figure 12.2 A patient controlling a bionic hand using a BCI.

According to Abdulkader et al. (2015), medical applications of BCIs can be categorized into three primary domains: a) prevention, b) detection and diagnosis, and c) rehabilitation and restoration. From restoring basic communication to enabling advanced therapeutic interventions, BCIs represent a significant leap forward in healthcare innovation. They also lay the groundwork for more complex systems—such as brain-controlled prosthetics, robotic arms, and wheelchairs—offering new possibilities for individuals with severe physical limitations (Rao & Scherer, 2010; Thompson et al., 2018).

In the area of prevention, BCIs are increasingly used to enhance brain functions such as memory, attention, and problem-solving—particularly among elderly individuals at risk of age-related cognitive decline. Studies have demonstrated that when used alongside cognitive training tasks, BCI systems can improve memory and cognitive performance in older adults (Chaudhary, Birbaumer & Ramos-Murguialday, 2016). By promoting sustained neural activity and mental engagement, BCIs provide a proactive means of preserving cognitive health. Emotion recognition capabilities further support preventative care by identifying emotional states in patients with limited communication abilities, such as those with disorders of consciousness (DOC) or elderly individuals, allowing caregivers to respond to psychological needs early and effectively.

BCIs also play a growing role in detection and diagnosis. Through continuous monitoring of brain activity, they offer the potential to identify neurological conditions such as Alzheimer's disease, epilepsy, and Parkinson's disease—often before clinical symptoms become evident—enabling earlier and more effective intervention

(Casson, 2019; Cinel, Valeriani & Poli, 2019). In the case of DOC patients, BCIs have been integrated with tools like the Coma Recovery Scale-Revised to assess awareness levels and guide appropriate clinical care (Xiao et al., 2018). Moreover, BCIs are increasingly applied in the assessment of mental health conditions, including Attention-Deficit Hyperactivity Disorder (ADHD) and autism spectrum disorders, helping clinicians detect cognitive and behavioral patterns that inform targeted treatment strategies (Amaral et al., 2018; Guger et al., 2020).

The most established applications of BCIs lie in rehabilitation and restoration, particularly in restoring communication and motor function (Yang, 2023). One of the most well-known examples is the P300 speller, a system that enables users to spell words by focusing on characters within a 6×6 grid. As rows and columns flash randomly, the system detects a P300 event-related potential when the user's target character appears, identifying the desired letter without any physical movement. This technology has been transformative for individuals with ALS, including the late physicist Stephen Hawking, who used a similar system to communicate (Lin et al., 2018; Ryan et al., 2017).

In neurorehabilitation, BCIs support recovery in patients with stroke or spinal cord injuries by training users to generate specific brain signals, often through motor imagery. This targeted brain training has been shown to significantly reduce impairment and promote the rebuilding of neural pathways through consistent use (Biasiucci et al., 2018; Mane, Chouhan, & Guan, 2020). Complementing these advances are brain-controlled exoskeletons—wearable robotic devices that amplify physical strength and endurance. These systems offer new possibilities for mobility rehabilitation in individuals with severe physical impairments. Although still in early stages of development, ongoing innovations in lightweight robotics and portable neural interfaces are making brain-controlled exoskeletons increasingly viable for clinical and eventual commercial use (Colucci et al., 2022).

2.2.2 Applications of BCI in Gaming

BCIs have the potential to create immersive and adaptive gaming experiences by responding directly to a player's emotions, intentions, and cognitive states. This enables users to perform tasks such as controlling game characters or navigating virtual environments using only their neural activity (Jayaraj et al., 2025). Moreover, BCI-enabled games can adapt in real time, adjusting content based on the player's mood or engagement levels to deliver a highly personalized and responsive experience (Keutayeva et al., 2025).

Research and development in BCI gaming has advanced considerably, particularly in the context of public applications using non-invasive EEG devices that are more accessible and user-friendly (Vasiljevic & de Miranda, 2020). Early BCI games featured simplified control mechanisms, but growing interest in neural interaction has led to the creation of more engaging and intuitive gameplay systems (Thomasson, 2023). A prominent example is Twitch streamer Perri Karyal, who trained extensively to convert motor imagery into in-game actions and successfully played Elden Ring using only her brainwaves via a hands-free controller (Nolan, 2023).

Beyond adapting traditional games for BCI control, researchers have also developed games specifically to explore and evaluate BCI paradigms. Notable examples include a pinball game controlled through non-invasive BCI (Tangermann et al., 2008), a Tetris-based BCI game (Pires et al., 2011), and a spaceship game designed to address challenges in BCI setup and assessment (Coyle et al., 2011). Additionally, commercial BCI devices such as those developed by NeuroSky have contributed to wider accessibility. These devices allow users to interact with video games and VR environments using mental commands, making BCI-based gaming experiences available to broader audiences (NeuroSky, n.d.).

The continued integration of BCIs into gaming not only enriches entertainment but also highlights the potential for these technologies to achieve broader public adoption through engaging and intuitive user experiences.

2.2.3 Applications of BCI in Mixed Reality and Metaverse Environments

The integration of BCIs into AR, VR, and broader Metaverse environments represents a transformative step toward more intuitive, immersive, and inclusive forms of digital interaction. By enabling direct communication between the brain and digital systems, BCIs bypass traditional input devices such as controllers, keyboards, or voice commands, allowing users to interact with virtual environments through mental activity alone (Zhu et al., 2024).

In AR and VR settings, BCIs facilitate hands-free control over digital content, enabling users to manipulate objects, navigate environments, and express themselves through neural signals. These interactions can include imagined speech, motor imagery, mental handwriting, and even facial expression-based commands (Abbasi-Asl et al., 2019; Gholizadeh Hamlabadi et al., 2024). For instance, EEG-based BCI systems have been successfully employed to perform complex 3D object transformations—such as rotation and scaling—with greater ease and accuracy than comparable 2D interface controls, highlighting VR's suitability as a medium for BCI-based tasks (Abbasi-Asl et al., 2019).

In Metaverse environments, these capabilities are further amplified. One key application is thought-based avatar control, allowing users to navigate and interact without physical movement. Beyond control, BCIs support the creation of intelligent and emotionally responsive avatars—so-called "Human Digital Twins"—that reflect the user's mental and affective state. These digital extensions of the self can enhance social interaction, act as learning companions, or function as personalized assistants in collaborative and educational settings (Hieu et al., 2023).

A particularly compelling use of BCIs in the Metaverse lies in creative design and spatial construction. In virtual platforms where users can build dream homes or landscapes, BCI technology offers a radically new design paradigm. Rather than relying on complex menus or manual modelling tools, users can conceptualize spatial forms and translate those mental images directly into 3D environments. For example, by focusing on certain visual or motor imagery patterns, users can manipulate virtual structures—such as adding walls, reshaping terrain, or adjusting lighting—purely through thought. This opens up new possibilities not only for people with physical disabilities but also for anyone seeking a more intuitive and expressive mode of digital creation. In doing so, BCI-enabled environments empower users to bring their imagination to life in ways that traditional tools cannot achieve.

This form of neural interaction promotes digital inclusion by allowing individuals with motor impairments to fully participate in virtual spaces (Gholizadeh Hamlabadi et al., 2024; Zhu et al., 2024). Imaginative inputs, such as mentally visualizing drawing or storytelling sequences, can be captured and transformed into interactive outputs, supporting applications in collaborative design, education, and immersive entertainment (Hieu et al., 2023). In VR gaming contexts, BCIs have already demonstrated real-time control and heightened user engagement, such as in a VR cricket simulation where users could bowl using only brain signals, resulting in significantly increased immersion (Jayaraj et al., 2025).

Beyond interaction, BCIs provide a unified sensing interface that captures a broad range of user states—including cognitive workload, emotional valence, fatigue, and intention—through a single sensor modality (Zhu et al., 2024). This rich, multimodal data stream is especially powerful in metaverse and VR environments, where adaptive and responsive interaction is key to immersion. Unlike traditional sensors like eye trackers or heart rate monitors, BCIs can detect subtle cognitive and emotional fluctuations, enabling hyper-personalized and dynamic experiences tailored to the user's mental state (Gholizadeh Hamlabadi et al., 2024; Zhu et al., 2024).

2.2.4 BCI-Based Interaction with Desktop and Web Interfaces

The integration of BCIs into desktop and web applications presents promising opportunities for developing accessible, non-muscular interaction systems—particularly benefiting individuals with neuromuscular disabilities. These systems allow users to interact with software applications, control computer peripherals, and browse the web without relying on traditional input devices such as keyboards or mice (Hintermüller et al., 2013). By bypassing conventional physical interfaces, BCIs enable users to perform tasks like typing, navigating menus, or managing home automation systems using only mental effort (Aydın et al., 2016).

BCI-enabled desktop applications are typically built upon well-established paradigms—such as P300 (as previously discussed), Steady-State Visual Evoked Potentials (SSVEP), and motor imagery—which form the basis for interpreting user intent and translating neural signals into discrete system commands. For instance, P300-based speller systems allow users to select letters or interface options by focusing on specific stimuli, facilitating communication and document creation tasks (Hintermüller et al., 2013).

In the context of web interaction, VEP-based systems have been developed to create real-time brain-controlled web browsers. These systems rely on SSVEP responses generated by users' visual attention to flickering interface elements, enabling full-featured web navigation such as link selection, scrolling, and form entry through neural signals alone (Yin et al., 2009).

Beyond individual browsing, web-based BCI platforms have also been designed to support remote interaction and control over internet-connected services. Notably, systems like BCI2000Web and WebFM facilitate browser-based BCI applications by enabling real-time neural data exchange between user devices and web interfaces through WebSocket communication protocols (Milsap et al., 2019). This architecture supports the development of flexible, scalable, and portable BCI applications that run directly within standard web browsers, eliminating the need for standalone software installations and improving interoperability.

Practical implementations of these systems are already evident in smart home environments, where users can control appliances, lighting, and media players through BCI-enabled web interfaces connected to embedded servers (Aydın et al., 2016). By transmitting only processed BCI outputs rather than raw EEG signals, these systems minimize latency, reduce power consumption, and enhance reliability—key considerations for real-world deployment.

In addition to interaction and control, BCIs have also been investigated as biometric authentication tools for desktop and web platforms. Given the uniqueness of EEG signal patterns, BCI-based authentication offers a secure and non-replicable alternative to traditional password systems. Such systems hold potential for enabling continuous or multi-factor authentication in web and cloud environments, particularly in applications involving sensitive data such as online banking, healthcare access, and remote education (Orenda et al., 2017).

Despite these advances, BCI-based desktop and web applications still face technical challenges. Integrating BCI systems with conventional web technologies requires specialized interfaces that can ensure real-time responsiveness, low-latency signal transmission, and compatibility with existing user interface frameworks (Hintermüller et al., 2013; Milsap et al., 2019). Addressing these challenges will be essential for the widespread adoption of BCI in mainstream desktop and web environments.

2.2.5 Applications of BCI in Art

BCI technology can also be applied to the field of art, enabling both experienced artists and those with no prior artistic background to create unique pieces of art. Although this creative application is less explored compared to other areas, as evidenced by its limited representation in scientific literature (Prpa & Pasquier, 2019), numerous examples exist in non-scientific contexts, such as YouTube videos showcasing artistic creations using BCIs.

Unlike traditional BCI applications that prioritize robustness and efficiency, art-focused BCIs aim to enhance the user experience (Nijholt et al., 2022). This shift allows researchers to explore the intersection of technology and creativity rather than solely focusing on system performance (Vanutelli et al., 2023). Through training, users can harness BCI technology to express their graphic or musical creativity, as demonstrated by initiatives like brain painting (Muenssinger et al., 2010). Such advancements pave the way for artistic expression through unconventional means, offering new opportunities to create art without relying on traditional tools.

2.2.6 Applications of BCI in Autonomous-Driving Vehicles

With the convergence of neuroscience, AI, and automotive technology, BCIs are emerging as a significant innovation in autonomous-driving vehicles. For healthy users, the integration of BCI technology allows some vehicle functions to be replaced or augmented with BCI control. For instance, Mercedes-Benz showcased a concept car that enabled users to perform actions like selecting navigation destinations, adjusting ambient interior lighting, and changing radio stations using thought control alone (Mercedes-Benz Group, 2021; Rosso, 2021).

A study conducted by the Free University of Berlin further explored this concept by using EEG patterns to map mental commands to vehicle functions, such as steering and speed adjustment (Göhring et al., 2013). The results demonstrated the feasibility of enabling basic car control through BCI, though challenges such as accuracy, reaction times, and the need for frequent training were identified. This integration has the potential to provide greater control over vehicle operations, paving the way for more intuitive and accessible interactions between users and autonomous vehicles.

2.3 Challenges and Limitations of BCI Integration

Despite their promise, BCIs present a range of significant challenges and limitations that must be carefully considered. Ensuring the successful and ethical integration of BCI technology requires continuous attention to these issues and proactive strategies for mitigation. The following sections outline key concerns that must be addressed to foster trust, usability, and broader adoption of BCIs across real-world applications.

2.3.1 Hardware Limitations of BCI Devices

While BCI technology has advanced considerably in recent years, both noninvasive and invasive systems continue to face significant technical challenges related to signal quality, reliability, user comfort, and practical deployment.

Non-invasive BCIs, the most widely used due to their accessibility, rely on electrodes placed on the scalp to capture EEG signals. These systems often utilize multiple electrodes—ranging from a few to several dozen—to enhance signal resolution and provide richer data for neural classification. However, increasing the number of electrodes can compromise user comfort, particularly during prolonged sessions. As such, developers must navigate a trade-off between maximizing data accuracy and ensuring user wearability and convenience.

A further challenge lies in the ongoing debate between the use of wet and dry electrodes. Wet electrodes typically provide more accurate and stable signals but are impractical for extended or mobile applications due to their complex preparation and maintenance. In contrast, dry electrodes are quicker to apply and more user-friendly but often deliver less consistent signal quality, and empirical findings regarding their overall effectiveness remain inconclusive (Padfield et al., 2019).

Non-invasive systems also face persistent issues with signal interference. EEG recordings can be easily distorted by environmental electrical noise, variable scalp impedance, and inconsistent electrode contact, all of which negatively affect classification accuracy (Saha et al., 2021). These issues are especially pronounced in users with long or thick hair, which complicates achieving stable and uniform contact between electrodes and the scalp.

Reliable signal acquisition becomes even more complex in multichannel systems, where ensuring consistent scalp contact across all electrodes is technically demanding—particularly in mobile or dynamic environments. In addition, user-related factors such as mental fatigue, distractions, and cognitive overload can further degrade EEG signal quality and reduce system performance (Cohen et al., 2014; Pfurtscheller et al., 2006).

Practical deployment is further hindered by the bulkiness and cost of advanced BCI systems. High-density EEG setups, while capable of delivering more precise data, are often cumbersome and expensive, limiting their portability and accessibility for everyday users (Maiseli et al., 2023; Saha et al., 2021). This imbalance between performance and affordability remains a key barrier to broader adoption beyond clinical and academic environments.

In contrast, invasive BCIs deliver higher precision by recording neural signals directly from within the brain. However, they require surgical implantation, which carries significant risks, including infection, inflammation, and potential damage to neural tissue (Waldert, 2016). These risks restrict their use to controlled medical or research settings. In addition, invasive systems face challenges such as neural plasticity and micromovements of implanted electrodes, which can cause signal drift and necessitate frequent recalibration (Willett et al., 2021). This recalibration process is time-consuming and further limits the practicality of invasive BCIs for daily use.

Addressing these limitations will require progress in several key areas. Advances in electrode design, particularly the development of high-fidelity, user-friendly multichannel systems, will be critical to improving both accuracy and comfort. Additionally, reducing production costs and increasing manufacturing scalability are necessary steps to make BCI technologies more affordable and widely accessible. Without these advancements, BCIs are likely to remain confined to niche applications—particularly among individuals with severe motor impairments—while their broader potential in education, entertainment, and general human–computer interaction will remain underexploited.

2.3.2 Software Limitations of BCI Devices

BCI technology remains relatively nascent in the commercial landscape, resulting in a limited availability of software resources—particularly Application Programming Interfaces (APIs) compatible with major game engines, which are essential for developing BCI-driven video games and metaverse applications. Developers also face significant challenges in designing high-performance BCI controllers due to the lack of comprehensive and reliable training datasets. This scarcity limits the creation of robust systems capable of delivering consistent and accurate performance across diverse applications (Noury et al., 2016; Saha et al., 2021).

Compounding these issues is the absence of pre-built templates, modular frameworks, and educational tools that could serve as foundational resources for developers. The lack of standardized APIs and middleware further complicates BCI integration with engines like Unity and Unreal, often requiring custom-built plugins or reliance on experimental third-party tools. To mitigate these challenges, open-source platforms such as OpenViBE and BCI2000 have begun providing modular frameworks and signal processing pipelines that support real-time BCI application development (Renard et al., 2010; Schalk et al., 2004).

In parallel, publicly available EEG datasets—such as those hosted on Kaggle—offer valuable resources for training machine learning models for BCI signal classification (Roy et al., 2019). However, to enable broader adoption and accelerate innovation, stronger collaboration between academia, industry, and the open-source community is essential. This includes the development of shared repositories of labeled brain signal data, standardized communication protocols, and reusable software libraries to promote compatibility across systems (Lotte et al., 2018).

Addressing these barriers requires not only technical solutions but also a concerted effort to educate and support the development community. Accessible, open-source tools and platforms are critical for lowering the entry barrier, encouraging collaboration, and fostering a shared knowledge base. Such initiatives would not only shorten the learning curve for newcomers but also contribute to the collective advancement and maturity of BCI-integrated software systems.

2.3.3 Lack of Standardization to Train and Define BCI Classifiers

A BCI classifier is responsible for identifying brain signals or EEG patterns to activate computer functionalities. However, there is no universal classifier adopted across all BCIs, requiring each study to develop and train its classifiers independently. This lack of standardization creates inconsistencies, as studies are often limited by the number of participants available for training. Additionally, some individuals may find it more challenging to produce brain signal patterns that can be reliably detected using existing classifiers and features (Amini Gougeh & Falk, 2023).

Despite the absence of standardization, there is limited understanding of which EEG patterns are the easiest to replicate. While imagery of limb movements is a common approach, numerous other methods, such as imagining pictures, writing letters, or keyboard typing, remain underexplored. Unlike virtual controllers, which are constrained by physical movements, the brain's ability to imagine concepts seems boundless, offering opportunities to experiment with diverse EEG patterns. However, identifying the most effective and consistent patterns for BCI remains a significant challenge.

One major limitation in training classifiers on EEG patterns is determining whether the produced EEG pattern accurately matches the user's intended imagination. In some cases, classifiers may inadvertently train on unrelated brain activity, leading to unreliable functionality. If the classifier is trained on normal or unintended brain activity, users might involuntarily activate the system. Mitigating this issue requires more reliable and consistent EEG patterns, yet there is currently no method to validate whether a dataset aligns with the intended imagery. Processing EEG data is inherently challenging, as classification approaches often suffer from poor accuracy (Padfield et al., 2019). Moreover, the feature selection process necessary to remove noisy data is complex, and EEG data itself is unstable due to the limited understanding of brain signals.

Another limitation with EEG classification is the potential overlap of similar EEG patterns, which can mistakenly trigger unintended commands. This misclassification leads to errors and user frustration, especially in scenarios requiring precise control. To mitigate these issues, it is recommended to limit the number of EEG patterns in a classifier, reducing the likelihood of misinputs. However, this limitation also restricts the ability of classifiers to map multiple commands effectively. This is particularly problematic in environments like the metaverse, where immersive interactions often demand a broad range of commands.

2.3.4 Learning Curve Associated with BCI Control

Learning to use BCIs presents challenges for the general population and is especially demanding for individuals with neuromuscular disorders. A fundamental difficulty lies in the mental tasks required to activate BCI functionalities. Users must perform tasks such as imagining movements, images, or concepts, which demand significant concentration and cognitive effort. Unlike physical controllers like a mouse or VR controllers, which rely on innate limb movements and are intuitive to use, BCIs require long-term training to restructure the brain for effective use (Padfield et al., 2019). Approximately 15–30% of the population find BCIs challenging to utilize due to distractions, intrusive thoughts, environmental noise, or moving objects, which can easily disrupt concentration and break control (Amini Gougeh & Falk, 2023). This highlights the inherent difficulty in teaching and describing mental tasks compared to physical actions, further exacerbating the learning curve for BCIs.

For individuals with neuromuscular disorders, these challenges are even more pronounced. Studies have shown that the target population, such as those with LIS or ALS, often experiences limited success in controlling BCI systems. In trials involving an auditory EEG-based interface designed to restore yes-no communication, only 2 out of 7 severely motor-impaired patients were able to control the system online, while 3 out of 7 succeeded with offline signal processing (Séguin et al., 2024). In contrast, 17 out of 18 healthy participants were able to use the system effectively. This disparity raises questions about the underlying factors that limit BCI usability among those who stand to benefit the most. Séguin et al. (2024)

propose several hypotheses to explain the reduced efficacy of BCIs in patients with severe motor impairments. One explanation is that mechanical ventilation, erratic muscle activity, and electrical interference from hospital equipment can degrade the quality of EEG signals, making it more difficult to extract meaningful data. Additionally, motor impairments are often accompanied by cognitive deficits, which can further hinder the individual's ability to perform and sustain the mental tasks required for effective BCI control.

These findings underscore the need for more inclusive research and robust solutions tailored to the unique needs of individuals with severe disabilities. Current studies often focus on healthy participants or are confined to small-scale, closely monitored trials (Séguin et al., 2024; Shih et al., 2012), limiting their applicability to real-world settings. Addressing these challenges requires advancements in signal processing, noise reduction, and adaptive systems that can accommodate the cognitive and physical limitations of the target population.

2.3.5 *Impact of User States and Environment on EEG Signal Quality*

A person's physical and emotional state has a profound impact on BCI performance. Maintaining overall well-being can enhance EEG signal quality, while negative emotions, stress, and fatigue tend to degrade it (Padfield et al., 2019). Even brief sessions of mental fatigue—such as during extended BCI use—have been shown to reduce signal strength within an hour (Cohen et al., 2014). Environmental factors such as posture, attention, and distraction further contribute to signal noise, complicating interpretation and classification accuracy.

This is particularly important when BCI is integrated into immersive environments like the metaverse, where the dynamic and emotionally engaging nature of the experience can both support and impair signal quality. While positive stimuli may enhance focus and improve performance, overwhelming sensory input or frustration can lead to cognitive overload, which disrupts stable EEG pattern generation (Padfield et al., 2019; Zhu et al., 2024). As studies have noted, even two seconds of lost concentration can substantially reduce classification accuracy (Padfield et al., 2019).

Moreover, BCI technology requires sustained mental effort, which exceeds the cognitive demand of traditional peripherals such as keyboards or mice. Immersive platforms like the metaverse often involve flashing visuals, moving objects, and changing audio—all of which can compete for cognitive resources and hinder users from maintaining the clear mental imagery required for effective BCI control (Gholizadeh Hamlabadi et al., 2024). This creates a tension between immersive engagement and cognitive stability.

From a technical standpoint, classifier training is also affected by variability in user mental states. Classifiers trained on EEG data from healthy individuals often fail to generalize to users with neurological conditions, underscoring the need for personalized or adaptive classifiers (Amini Gougeh & Falk, 2023). Additionally, headset design contributes to performance: many systems require firm electrode-to-scalp contact for signal fidelity, but this can cause discomfort over time and distract users from focusing on the required mental tasks (Padfield et al., 2019).

These findings emphasize the importance of addressing both technical and human-centered limitations. Improving electrode comfort, reducing cognitive load through interface design, and developing emotionally adaptive systems are essential steps for reliable BCI integration—especially in complex, immersive environments like the metaverse (Zhu et al., 2024).

2.3.6 Societal Impact Issues

The integration of BCIs into daily life can significantly impact users' psychological well-being and social dynamics. Users of BCIs may struggle with altered perceptions of self, particularly when the technology enables capabilities that extend beyond natural human limits. Early adopters of BCIs may face societal bias, especially in cultures less accepting of technological augmentation. This bias can manifest as mistrust or stigmatization, particularly toward disabled individuals who rely heavily on this technology (Ienca & Andorno, 2017).

Over-reliance on BCIs could potentially lead to diminished natural abilities, such as memory or problem-solving, as users become increasingly dependent on technological assistance. For instance, individuals using memory-augmentation BCIs might inadvertently weaken their natural memory capabilities as they grow accustomed to outsourcing these functions to technology. This phenomenon, similar to "cognitive offloading" commonly observed with modern technology, could extend to problem-solving and decision-making, raising concerns about long-term dependency (Grinschgl, Papenmeier & Meyerhoff, 2021).

2.3.7 High Cost of BCI Systems

The high cost of BCI systems poses a major barrier to adoption, limiting accessibility for most consumers. Advanced technologies—such as high-end wireless EEG headsets and invasive neural implants—remain prohibitively expensive due to the costly materials, technical complexity, and specialized expertise required for their development and deployment (EMOTIV, n.d.; Guger, Allison, & Edlinger, 2014). Invasive BCIs, which provide greater precision and functionality, are even more expensive due to the need for surgical implantation, ongoing maintenance, and specialized medical supervision.

This financial barrier creates a significant disparity in access to BCI technology. Reliable and advanced systems are typically only available to affluent individuals, well-funded research institutions, or private organizations with substantial resources. Meanwhile, economically disadvantaged populations, who might benefit the most from BCI applications—such as individuals with severe disabilities—often cannot afford these systems. Even emerging consumer-grade BCI devices, marketed as affordable alternatives, still remain out of reach for many, particularly in low-income regions.

The high cost of BCIs also impacts healthcare providers, limiting their ability to integrate these systems into routine medical care. For instance, hospitals or rehabilitation centers may prioritize other technologies that offer broader applications at a lower cost, further delaying the widespread adoption of BCI technology. This cost-driven inequality restricts the potential benefits of BCIs, leaving many unable to access life-changing advancements such as brain-controlled prosthetics, communication tools, and rehabilitation systems.

2.3.8 Ethical Concerns and Data Privacy

BCIs raise profound ethical concerns related to autonomy, privacy, and societal impact, particularly given their ability to collect highly sensitive neural data that may reveal private thoughts, emotions, or medical conditions. The unauthorized access to such data introduces serious privacy risks, including the potential for "brain hacking" or exploitation by malicious actors. These vulnerabilities underscore the urgent need for robust regulatory frameworks and technological safeguards to protect users' cognitive autonomy and prevent the misuse of neural data (Gordon & Seth, 2024; Ienca & Haselager, 2016).

A major concern is the potential misuse of EEG data, especially in emerging applications that integrate BCIs with immersive platforms like the metaverse. EEG signals—essentially recordings of brain activity—could theoretically be decoded to infer emotional states, thoughts, or even intentions. Classifiers trained to identify specific patterns could be repurposed to extract private information or predict user behavior (Padfield et al., 2019). This opens the door to ethically troubling scenarios such as targeted advertising, psychological profiling, or discriminatory practices—for example, adjusting service rates based on detected illnesses or mental health conditions.

Without stringent security protocols, EEG data may be accessed or manipulated by unauthorized parties, compromising user privacy and mental well-being. Malicious actors could create neural profiles that expose users' vulnerabilities, leading to intrusive and unethical interventions. These risks highlight the critical need for secure system architectures that prevent data breaches and uphold user trust.

BCIs also exist within a complex regulatory landscape, spanning medical devices, consumer electronics, and research tools. Informed consent becomes crucial in this context, ensuring users understand not only the immediate benefits—such as enhanced mobility or communication—but also the longer-term risks, including surveillance, data misuse, and cognitive intrusion. Transparency about these risks is ethically essential to support informed decision-making and maintain user trust in both the technology and its providers.

Another key challenge lies in determining accountability in the event of BCI malfunction or misuse. For instance, if a BCI-enabled prosthetic malfunctions and causes harm, it remains unclear whether liability falls on the user, the developer, or the manufacturer. This ambiguity calls for the development of clear legal and ethical frameworks that define accountability, supported by international collaboration to ensure consistent protection for users (Ienca & Haselager, 2016).

To address these concerns, the following measures are critical for the ethical advancement of BCI technologies:

a Implementation of robust security protocols to prevent unauthorized access to neural data and mitigate risks of manipulation.
b Transparent communication with users regarding the potential risks and benefits, including long-term implications of data collection and processing.
c Development of international ethical and legal frameworks that define standards for data protection, informed consent, and accountability.
d Continuous updating of ethical safeguards in parallel with technological evolution to address emerging challenges and vulnerabilities.

Without these safeguards, the transformative promise of BCIs risks being undermined by ethical controversies and public mistrust. Ensuring the protection of user privacy, autonomy, and psychological integrity is vital for the responsible and equitable integration of BCI technologies into society.

Building on the insights from the literature review, which examined the classifications, applications, and limitations of BCI technologies, the following section presents three empirical studies. These studies were conducted to explore the practical implementation of non-invasive BCIs in communication, artistic creation, and gaming environments.

3 Empirical Studies

This section explores our three studies, highlighting different facets of BCI applications. The first investigates the usability of two non-invasive BCI-operated communication systems, focusing on their effectiveness in enabling real-life interactions for individuals with motor impairments. The second study explores the creative potential of BCIs in the field of art through a virtual brain-controlled painting application, examining how such systems enable individuals with physical disabilities to engage in artistic expression. Lastly, the third study evaluates BCI integration into gaming, comparing it with traditional keyboard controls to assess its impact on user experience and cognitive performance.

All participants across the studies were recruited in accordance with a human research ethics protocol approved by the University of New South Wales (UNSW) and sanctioned by its institutional review board.

3.1 Empirical Study One: Comparing Chat Interfaces for BCI-Based Communication

This study aims to evaluate the effectiveness of two novel, non-invasive BCI-based communication methods designed for interaction in VR. The first method requires users to manually select a contextual category before choosing a specific response, while the second method automatically presents context-specific categories based on the VR scenario. By comparing these two interfaces, the study aims to identify which design provides a more suitable and user-friendly solution for BCI-mediated communication within the UNSW Metaverse platform, with the broader objective of enhancing accessibility for individuals with speech and motor impairments.

3.1.1 Application Development for Study One

The two communication methods were developed using the Unity game engine and the EMOTIV Insight five-channel EEG headset (Figure 12.3). The first interface requires users to manually select a category before choosing a response, whereas the second interface automatically presents a context-specific category based on the scenario. Both systems were tested in a VR environment using the HTC VIVE headset to evaluate their usability and performance in immersive settings.

Figure 12.3 EMOTIV Insight five-channel EEG headset.

Interface Featuring Manual Selection of Contextual Categories: This system is designed to offer users greater flexibility by allowing them to navigate across three main categories—General, Emotion, and Assistive—using a circular interface. The wheel visually divides these contexts into equal segments, enabling users to select the appropriate category when a question is asked (Figure 12.4). Once a category is selected, the interface displays a new wheel containing options

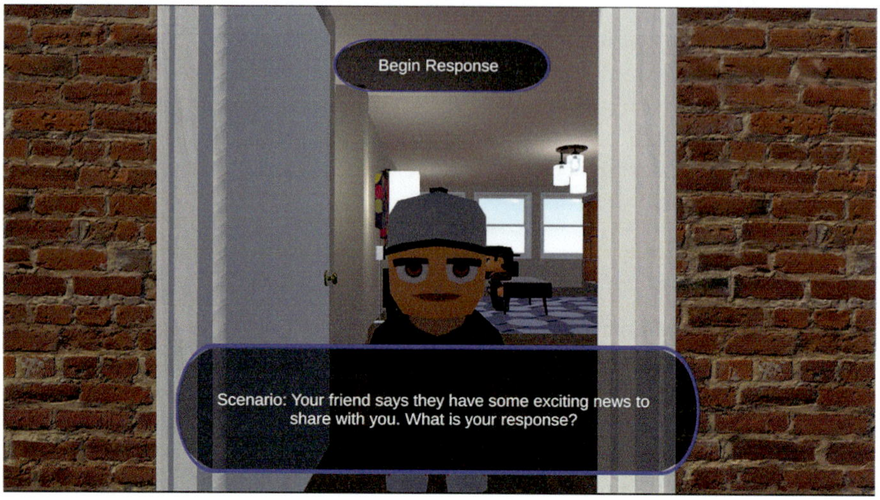

Figure 12.4 Image of the context-selection layer of our wheel system.

specific to that context. For example, to answer a yes/no question, users navigate to the General category and then choose between "Yes" and "No" (Figure 12.5).

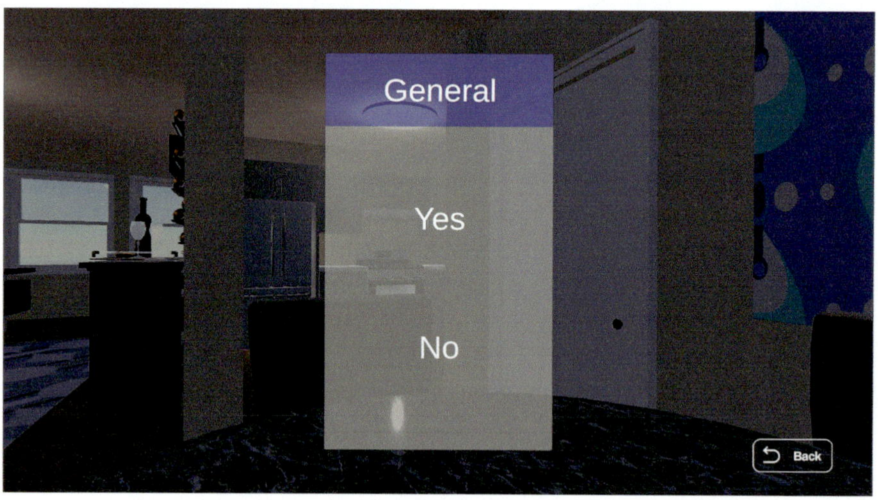

Figure 12.5 Image of the general context wheel options.

Similarly, the Emotion category might include options like "Happy" and "Sad" (Figure 12.6), while the Assistive category presents responses tailored to specific support needs (Figure 12.7).

Figure 12.6 Image of the emotions' wheel context.

Figure 12.7 Image of the assistive wheel context.

Interface Featuring Automatic Selection of Contextual Categories: In this system, when the user is asked a question, they do not need to manually select a category—such as General, Assistive, or Emotion—as required in the circular version. Instead, users simply click the "Begin Response" button (Figure 12.8), after which the system automatically presents a context-specific table containing two predefined options drawn from the relevant category. This eliminates the need to use brain signals to first navigate to a category before selecting an answer. For example, if the question allows a "Yes" or "No" response, the system automatically loads the General category and displays "Yes" and "No" as selectable options (Figure 12.9). Similarly, the Emotion and Assistive categories are automatically triggered based on the scenario, presenting context-appropriate options without requiring manual category selection (Figures 12.10 and 12.11).

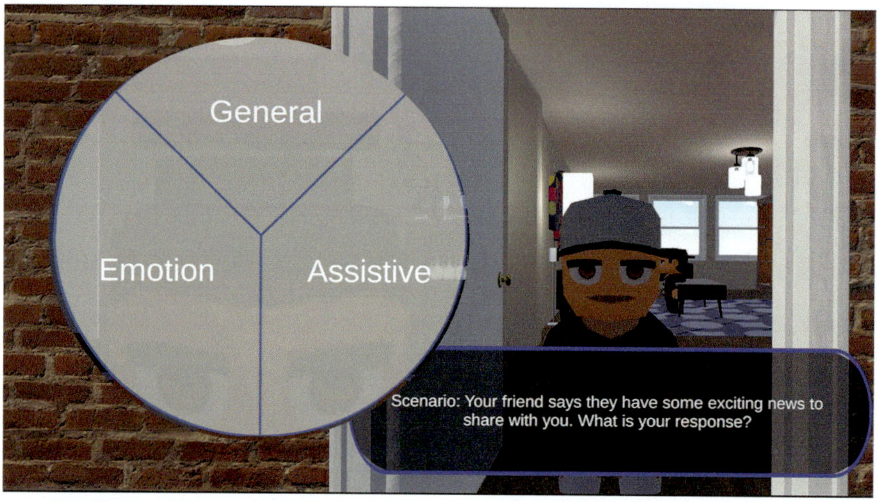

Figure 12.8 UI example of a scenario within the experiment.

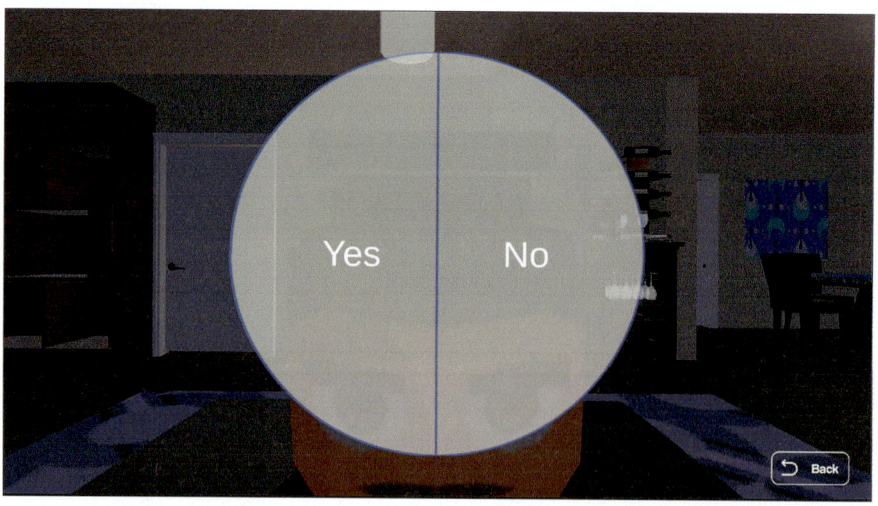

Figure 12.9 The general pre-defined context.

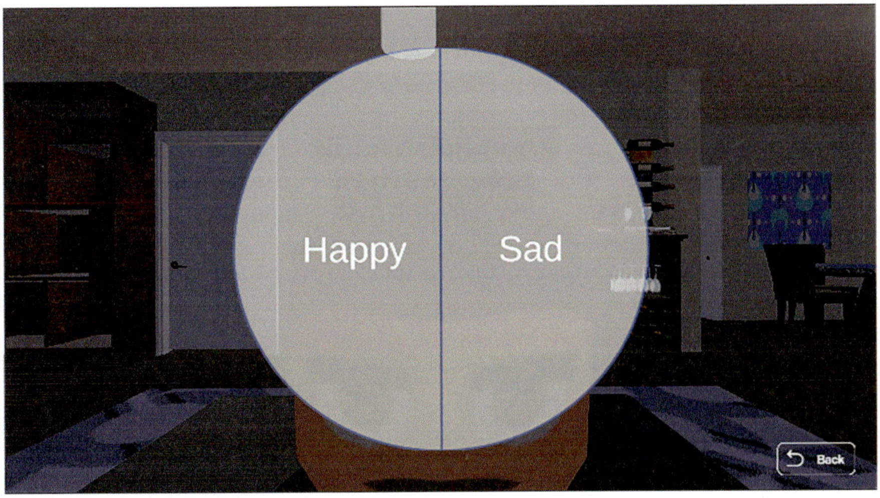

Figure 12.10 The emotions pre-defined context.

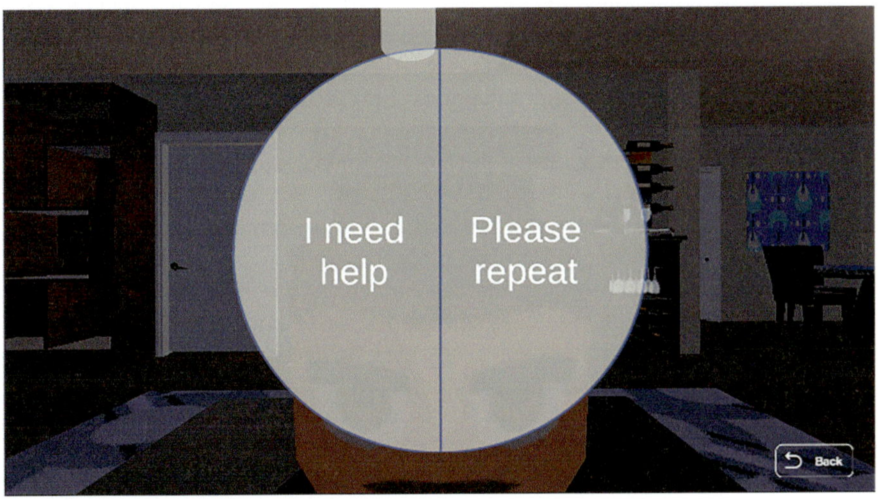

Figure 12.11 The assistive pre-defined context.

3.1.2 Methodology for Conducting Study One

The study recruited five participants aged between 18 and 24, all with minimal prior experience with BCIs, to evaluate the usability and accessibility of the systems from the perspective of novice users. Each participant completed an individualized session, allowing for focused observation and detailed data collection throughout the experiment.

The experimental procedure consisted of three phases. In the pre-experiment phase, participants signed consent forms, completed a screening test for English proficiency, and filled out a pre-questionnaire to capture baseline information on their technical skills and initial perceptions.

During the main experiment, participants received training on both BCI communication methods before engaging in a series of experimental scenarios using each system. All participants interacted with both systems: some began with the Interface Featuring Automatic Selection of Contextual Categories, followed by the Interface Featuring Manual Selection of Contextual Categories, while others used the systems in the reverse order. Data on response time, error rate, and skipped stages were collected through Unity's integrated tracking system.

In the post-experiment phase, participants completed a post-questionnaire and participated in a brief interview to share their feedback regarding the usability, effectiveness, and preferences for each communication method.

3.1.3 *Results from Study One*

The results revealed significant differences in performance and user preferences between the two communication systems. Quantitative data indicated that the interface featuring manual selection of contextual categories demonstrated significantly faster response times ($M = 18.84$, $SD = 8.50$) compared to the Interface featuring automatic selection of contextual categories ($M = 40.40$, $SD = 11.00$), $t(4) = -3.47$, $p < .01$. Although both systems exhibited comparable error rates, the manual selection interface slightly outperformed the automatic one, with average errors of ($M = 0.85$, $SD = 0.18$) versus ($M = 1.15$, $SD = 0.22$), $t(4) = -2.36$, $p < .05$. Additionally, only 3.33% of scenarios were skipped across both systems, suggesting that each provided adequate response options for most situations.

Qualitative feedback, however, indicated a general preference for the automatic selection interface, which participants found simpler and more practical. Users appreciated its clear instructions and ease of use, whereas the manual interface was critiqued for its limited visual engagement and lack of intuitiveness.

3.1.4 *Insights from Study One*

Both systems demonstrated strong potential for enabling hands-free communication. However, each system offered distinct advantages and trade-offs based on usability and performance metrics.

Quantitative results showed that the manual selection interface yielded significantly faster response times and slightly lower error rates compared to its automatic counterpart. This outcome was somewhat unexpected, but it can likely be attributed to the system's streamlined, user-controlled design. By allowing users to actively select the relevant category before executing a response, the interface aligns with established user interface design principles that emphasize user control and autonomy. This responsiveness makes it particularly well-suited for scenarios requiring user-directed decision-making, such as dynamic conversational exchanges in immersive environments.

In contrast, the automatic selection interface—though slower in performance—received more favorable qualitative feedback. Participants found it simpler and more intuitive, largely due to its context-aware structure that bypassed manual category selection. By automatically generating appropriate response options based on the scenario, it reduced cognitive load and offered a more guided and accessible user experience. This was especially beneficial for novice users or those with limited mental stamina.

These contrasting findings suggest that the suitability of each system depends on the intended use case. The manual selection interface excels in contexts where speed, control, and flexibility are critical, while the automatic selection interface is more effective in environments where ease of use and reduced mental effort are prioritized.

Additionally, the study identified broader challenges in BCI-mediated communication. Participants reported difficulties in maintaining focus and achieving consistent EEG signal quality—factors that impacted the reliability of both systems. These challenges highlight the need for continued improvements in hardware design and user training protocols to enhance signal stability and overall system robustness.

Finally, while the study reinforces the value of BCI technologies in promoting accessibility and inclusivity, the generalizability of these findings is limited by the small sample size and specific test conditions. Future research should expand to include more diverse participant groups and varied interaction scenarios to develop a more comprehensive understanding of these systems' real-world applications.

3.2 Empirical Study Two: Painting Using BCI

This study aimed to investigate the potential of BCI technology as an assistive tool for enabling creative expression through art. The research focused on the development and evaluation of a virtual painting application controlled entirely by users' mental activity, addressing the needs of individuals with limited or no motor abilities.

The virtual painting application was designed to interpret specific mental commands—such as selecting tools, drawing geometric shapes, and erasing mistakes—enabling users to create digital art without physical interaction. The study evaluated the system's usability, accuracy, and user satisfaction, while also identifying its limitations and areas for future enhancement. More broadly, this project highlights the potential of BCI systems to extend beyond traditional assistive functions, opening new possibilities for creative and therapeutic applications.

3.2.1 Application Development for Study Two

The application was developed using the Unity game engine for use with the EMOTIV Insight five-channel EEG headset. Designed to replicate a basic digital painting tool (see Figure 12.12), the system enabled users to perform a range of creative functions through mental commands. Users could select a brush tool for drawing lines, choose geometric shapes such as squares, triangles, or circles, and access an eraser tool to make corrections. Each of these actions was triggered by visualizing specific motor imagery (MI) tasks, allowing for hands-free interaction with the interface.

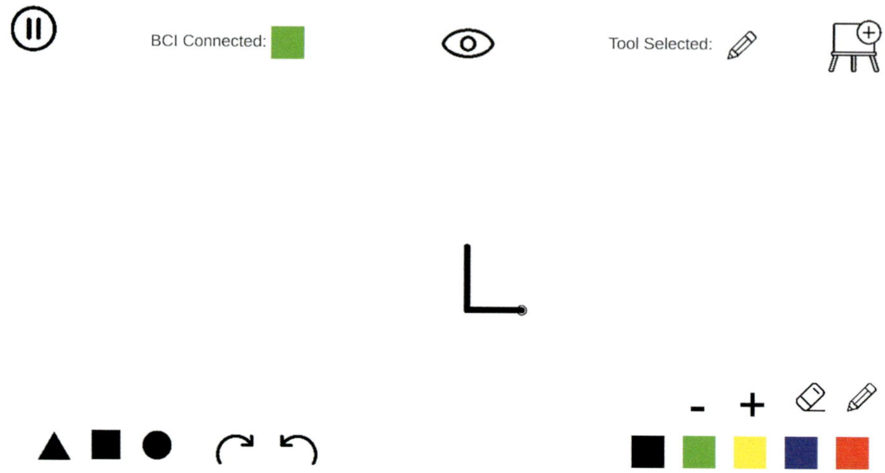

Figure 12.12 The interface of the paint application.

Participants interacted with the BCI painting system by training it to recognize two distinct mental commands: a neutral cognitive state, used to activate the drawing function with the currently selected tool, and left-hand MI, used to switch between and select tools such as the brush, shape selector, and eraser. The neutral state involved maintaining a relaxed mental focus, functioning as the default command to initiate drawing actions. In contrast, the left-hand MI task required users to mentally simulate the movement of their left hand, prompting the system to cycle through and select the available tools.

3.2.2 Methodology for Conducting Study Two

Five participants aged between 18 and 24 were recruited to evaluate the painting application. All participants had no prior experience with BCIs or digital painting, ensuring that their feedback reflected unbiased, first-time user interactions.

The experimental design involved three drawing tasks of increasing complexity: a triangle, a square, and a trapezium using the brush tool. For each task, participants used the painting application to visualize and execute the desired shapes through mental commands. This progression was intended to evaluate both the system's ability to interpret accurate commands and the users' growing proficiency with the interface.

Data collection involved both quantitative and qualitative measures to evaluate system performance and user interaction. Shape accuracy was assessed by instructing participants to draw specific geometric shapes, which were then rated on a scale from 1 to 10, with higher scores indicating a closer resemblance to the intended form. Tool-switching efficiency was measured by tracking the success rate of mental commands used to switch between tools, along with observations of any delays, errors, or difficulties encountered during tool activation.

Qualitative feedback was collected through participant interviews and questionnaires administered after the experiment. These instruments captured subjective perspectives on the system's usability, the overall user experience, and the perceived effectiveness of the mental command interface.

3.2.3 Results from Study Two

The results of the study revealed both promising aspects and notable challenges in the implementation and functionality of the painting application. Quantitative findings showed that participants achieved average shape ratings of 8.3 for squares, 7.5 for triangles, and 7.2 for trapeziums out of 10, indicating moderate success in translating mental commands into precise shapes. However, tool-switching efficiency received an average score of 6.73 out of 10, highlighting difficulties in maintaining consistent focus when transitioning between painting and other tools.

Qualitative insights further illustrated the user experience. Participants described the interaction as novel and engaging, expressing appreciation for the opportunity to control the system through their mental activity. At the same time, frustration was reported due to the inconsistency of tool-switching and occasional lack of responsiveness, which negatively affected overall satisfaction. Nonetheless, many users conveyed a sense of accomplishment upon successfully completing the tasks, pointing to the system's potential to foster creativity and promote a sense of independence.

3.2.4 Insights from Study Two

The painting application offered users a unique opportunity to engage in artistic expression without the need for physical input devices. For both the able-bodied participants in this study and individuals with limited mobility, this represents a significant advancement in accessibility and inclusivity—enabling participation in creative activities that were previously out of reach for some. While regular users primarily experienced the system as an enjoyable and novel form of interaction, for individuals with physical disabilities, it introduced a meaningful new capability: the ability to draw and paint independently.

However, several challenges and limitations emerged during the study. One notable issue was the learning curve associated with using the BCI system. Participants found it challenging to maintain consistent focus on the mental commands, particularly when switching between tools. Training the system to recognize neural signals required substantial mental effort and was time-intensive, highlighting the need for more efficient training processes to improve accessibility and reduce cognitive strain.

Another limitation was the inconsistent accuracy of the application. The system occasionally misinterpreted commands, leading to unintended actions. This inconsistency was attributed to variability in EEG signals and the limited dataset used for training the classifier. Such inaccuracies, though expected at this stage of technological development, underscore the importance of refining signal processing algorithms and expanding training datasets to enhance system reliability.

Participants also noted areas for improvement in immersion and feedback. While the application provided real-time visual feedback, the interface could be more intuitive and responsive to enhance the user experience. A more immersive and user-friendly interface design could significantly improve engagement and satisfaction, particularly for users who may already face challenges with focus and mental fatigue.

Additionally, generalization issues with the classifier posed a concern. The model, trained on a small and homogeneous sample, may not perform as effectively for a broader, more diverse user base, particularly individuals with neurological impairments. This limitation underscores the need for further research involving diverse participants to ensure that BCI systems can accommodate a wide range of user needs and capabilities.

3.3 Empirical Study Three: Playing Video Games Using BCI

This study investigates the integration of BCI technology into video gaming, focusing on its impact on cognitive performance and user experience. It examines the feasibility of using BCI for playing 2D games and compares its performance with traditional keyboard controls. Key aspects include the cognitive load of BCI systems, their impact on task efficiency and game performance, and their influence on player engagement and satisfaction.

3.3.1 Application Development for Study Three

A game was developed using the Unity game engine as a 2D interactive environment that required players to navigate through levels filled with various obstacles and objects (see Figures 12.13A and 12.13B). The gameplay involved collecting items such as apples while avoiding obstacles by jumping over them—requiring precise timing and control. In this game, jumping was the sole task assigned to users.

Figure 12.13A Screenshot of one of the game levels (beginning of the level).

Figure 12.13B Screenshot of one of the game levels (end of the level).

The design prioritized simplicity and consistency, maintaining a clear focus on the control mechanisms being tested rather than incorporating complex visuals or narrative elements. The levels were meticulously structured to provide identical layouts for both the keyboard control method and the BCI method, allowing players to complete tasks using either keyboard inputs or BCI-supported actions. Visual feedback and sound effects were integrated to enhance player engagement, offering immediate reinforcement for successful actions and highlighting errors or delays to improve the user experience.

For the BCI control condition, the experiment employed the EMOTIV Insight five-channel EEG device. The game consisted of two levels designed to compare traditional and BCI-based control methods. In Level 1, participants used standard keyboard input, jumping by pressing the spacebar. In Level 2, participants performed the same task using BCI-supported controls.

The BCI control scheme was based on two mental commands: a neutral command, triggered by a relaxed cognitive state and used as a baseline, and a jump command, activated through motor imagery of jumping. To ensure reliable performance, the BCI system was individually calibrated for each participant to accurately map their neural signals to the intended in-game actions. Figure 12.14 illustrates the user interaction with the game using the BCI device.

Figure 12.14 Playing the game using a BCI device.

3.3.2 Methodology for Conducting Study Three

Eleven participants (8 males and 3 females), aged between 18 and 59, were recruited from UNSW. Each participant provided informed consent and completed a pre-survey to document their demographic information, prior experience with BCIs, and familiarity with video games.

The experiment consisted of three main phases. In the training phase, participants were introduced to BCI principles and completed practice sessions to become familiar with the required commands. During the gameplay phase, participants engaged in two levels of a game, each utilizing a different control method. Finally, in the post-experiment phase, participants completed a questionnaire and participated in interviews to provide qualitative feedback on their experience.

Data collection included both quantitative and qualitative measures. Quantitative metrics encompassed response times, task completion rates, and error counts, while qualitative feedback was obtained through structured interviews aimed at capturing participants' subjective experiences and perceptions of the different control methods.

3.3.3 Results from Study Three

The game performance analysis revealed significant differences between traditional keyboard controls and BCI-supported controls. For response time, participants completed Level 1 using keyboard controls $M = 52.64$, $SD = 6.8$, while completion time for Level 2 using BCI controls increased substantially $M = 194.45$, $SD = 9.1$—a 270% rise. This difference was statistically significant $t(10) = -41.40$, $p < .001$, indicating a clear performance gap between the two control methods. In terms of task completion rate, participants collected an average of 13.91 $SD = 5.9$ apples out of 15 using keyboard controls, compared to 11.82 $SD = 8.2$ apples with BCI controls. However, this difference was not statistically significant $t(10) = 0.69$, $p = .505$, suggesting that while BCI controls reduced efficiency, they did not substantially impact overall task success.

User experience feedback revealed that while participants found keyboard controls easier and less effortful, they considered BCI controls more engaging and immersive. Despite the added difficulty, many expressed enthusiasm about the unique nature of using BCI, noting that with additional practice, BCI controls could become more intuitive and enjoyable.

3.3.4 Insights from Study Three

While keyboard controls clearly outperform BCI systems in terms of speed, accuracy, and intuitiveness, BCI controls offer a distinct advantage by enabling immersive and engaging interactions through thought alone. However, the increased response times and reduced task completion rates associated with BCI systems reflect current technological limitations—particularly in signal processing speed, cognitive demand, and the variability of neural signal interpretation. These challenges place BCIs at a disadvantage compared to the near-instantaneous responsiveness of traditional input methods.

The heightened mental effort required to generate reliable commands, combined with inconsistent translation of brain signals into actionable inputs, underscores the need for advancements in interface design and signal processing algorithms to reduce latency and improve accuracy. Despite these limitations, qualitative feedback from participants revealed a strong appreciation for the novelty and immersive potential of BCI interaction. Users found the experience engaging and expressed optimism about its future applicability, especially in scenarios where immersion and experiential engagement are valued over raw performance—such as exploratory or narrative-driven gaming environments.

This feedback also suggests that user adaptation plays a critical role in the effectiveness of BCI controls. With additional practice and well-designed training protocols, users may overcome initial challenges and achieve greater control precision. This aligns with prior research emphasizing the importance of structured learning and adaptive systems that respond to individual user behavior in improving BCI usability over time.

4 Discussion

The exploration of BCI systems across communication, creative expression, and gaming domains in this chapter reinforces the dual narrative highlighted in the literature: while BCIs offer unprecedented possibilities for human–computer interaction, they remain limited by significant technical and cognitive constraints. The empirical studies critically translate the theoretical promise of BCI technology into practical applications, offering nuanced insights into usability, inclusivity, and the pathways toward broader adoption.

A recurring theme across all three studies is the tension between novelty and reliability. As highlighted in the literature, BCIs enable intuitive, hands-free interaction through thought alone (Gholizadeh Hamlabadi et al., 2024; Zhu et al., 2024), bypassing conventional motor interfaces. However, real-world application—especially as demonstrated in the communication and painting studies—reveals inconsistencies in EEG signal interpretation and system responsiveness. While some participants achieved high levels of control, others struggled to maintain cognitive focus, reflecting previously reported challenges related to mental fatigue and user variability (Amini Gougeh & Falk, 2023; Padfield et al., 2019).

One critical insight is the domain-dependent suitability of BCI systems. In communication-based tasks within VR and metaverse environments (Study 1), efficiency and reliability were essential; the manual category selection interface outperformed the automatic system in terms of response time, though it fell short in user satisfaction. This trade-off reflects a broader design dilemma noted in the literature—whether to prioritize system performance or user intuitiveness (Saha et al., 2021). Conversely, in the painting application (Study 2), users were more forgiving of technical limitations, as the focus shifted from speed to expressive potential. Here, the immersive nature of BCI-enabled art echoed Vanutelli et al. (2023), who argue that BCI's strength lies not just in utility but in its ability to expand modes of self-expression, especially for individuals with physical disabilities.

In gaming (Study 3), the complexity of translating neural commands into real-time control became especially apparent. The significant gap in performance between keyboard and BCI controls, while expected, raises deeper questions about cognitive load and task suitability. Unlike creative tasks where imprecision may be tolerable, time-sensitive environments such as games expose the weaknesses of current BCI systems, including slow signal processing and false activations (Noury et al., 2016). Still, participants' positive feedback on the novelty and immersion of the BCI experience suggests that such systems could excel in exploratory or narrative-driven game genres where interaction style, rather than speed, is the focus (De Castro-Cros et al., 2020).

Another key dimension across studies was user learning and adaptation. In line with literature emphasizing steep BCI learning curves (Séguin et al., 2024), participants reported initial difficulty but demonstrated improved control with brief training. This indicates the potential value of adaptive interfaces and personalized classifiers, which can accommodate cognitive differences and adjust based on user behavior over time (Amini Gougeh & Falk, 2023). Such systems would be particularly beneficial in metaverse or educational settings, where prolonged interaction is expected and user variability is high (Darejeh et al., 2024a).

Importantly, the findings also raise equity and ethical concerns. As outlined in the literature, high system costs, limited dataset generalizability, and the potential for data misuse in immersive platforms (Gordon & Seth, 2024; Ienca & Haselager, 2016) continue to constrain BCI deployment. While the first study's simplified communication models, the second study's artistic platform, and the third study's accessible method of video game interaction for individuals with disabilities suggest promising low-barrier use cases, broader implementation must be accompanied by robust consent protocols, inclusive design practices, and strong privacy safeguards—particularly in settings involving vulnerable populations.

Collectively, this chapter's findings call for a shift in how BCI success is measured. Rather than benchmarking solely against traditional input devices, future research should incorporate metrics such as user empowerment, emotional engagement, and creative autonomy. Additionally, the applications should be evaluated with a larger sample size, as the current findings are limited by a relatively small participant group.

5 Conclusion

BCIs represent a paradigm shift in human–computer interaction, offering applications across accessibility, creative expression, gaming, and immersive environments. Their integration into digital platforms—particularly within the metaverse—holds significant potential to dismantle participation barriers and enable novel, inclusive modes of interaction. This chapter examined the practical application of non-invasive BCI technology through three case studies, each highlighting their capacity to enhance communication, creativity, and interactive engagement.

However, the case studies also underscore the dual nature of BCI technology: while its potential is transformative, it is tempered by notable technical, cognitive, and ethical challenges. Issues such as signal instability, steep learning curves, and concerns over neural data privacy must be addressed before BCIs can be reliably integrated into mainstream digital environments. Despite these limitations, the demonstrated use cases—from communication modules to art applications and interactive gaming—illustrate how BCI-enabled systems can support more accessible and engaging user experiences.

Realizing the full potential of BCIs will require targeted advancements in signal processing, adaptive interface design, and user training protocols. Prioritizing user-centric design, incorporating adaptive feedback, and fostering interdisciplinary collaboration across neuroscience, engineering, and ethics will be critical to mitigating risks and enhancing usability.

Ultimately, the broader implications of BCI technology extend far beyond isolated use cases. As the technology matures, BCIs have the capacity to evolve from experimental tools into widely accessible platforms that redefine how we communicate, create, and interact in digital environments—bringing us closer to a more inclusive and interconnected world.

References

Abbasi-Asl, R., Keshavarzi, M., & Chan, D. Y. (2019). *Brain-computer interface in virtual reality*. In *9th International IEEE EMBS Conference on Neural Engineering* (pp. 1220–1224). IEEE. https://doi.org/10.1109/NER.2019.8717158

Abdulkader, S. N., Atia, A., & Mostafa, M.-S. M. (2015). Brain computer interfacing: Applications and challenges. *Egyptian Informatics Journal*, *16*(2), 213–230. https://doi.org/10.1016/j.eij.2015.06.002

Alchalabi, B., Faubert, J., & Labbé, D. R. (2021). A multi-modal modified feedback self-paced BCI to control the gait of an avatar. *Journal of Neural Engineering*, *18*(5), 56005-. https://doi.org/10.1088/1741-2552/abee51

Amaral, C., Mouga, S., Simões, M., Pereira, H. C., Bernardino, I., Quental, H., Playle, R., McNamara, R., Oliveira, G., & Castelo-Branco, M. (2018). A feasibility clinical trial to improve social attention in Autistic Spectrum Disorder (ASD) using a brain computer interface. *Frontiers in Neuroscience*, 12. Frontiers Media SA. https://doi.org/10.3389/fnins.2018.00477

Amini Gougeh, R., & Falk, T. H. (2023). Enhancing motor imagery detection efficacy using multisensory virtual reality priming. *Frontiers in Neuroergonomics*, *4*, 1080200–1080200. https://doi.org/10.3389/fnrgo.2023.1080200

Aydın, E. A., Bay, Ö. F., & Güler, İ. (2016). Implementation of an embedded web server application for wireless control of brain computer interface based home environments. *Journal of Medical Systems, 40*(27). https://doi.org/10.1007/s10916-015-0386-0

Benchetrit, Y., Banville, H., & King, J.-R. (2023). *Brain decoding: Toward real-time reconstruction of visual perception*. arXiv. Preprint. https://doi.org/10.48550/arxiv.2310.19812

Bernal, S. L., Pérez, M. Q., Beltrán, E. T. M., Pérez, G. M., & Celdrán, A. H. (2022). *When brain-computer interfaces meet the metaverse: Landscape, demonstrator, trends, challenges, and concerns*. arXiv. Preprint.

Biasiucci, A., Leeb, R., Iturrate, I., Perdikis, S., Al-Khodairy, A., Corbet, T., Schnider, A., Schmidlin, T., Zhang, H., Bassolino, M., Viceic, D., Vuadens, P., Guggisberg, A. G., & Millán, J. d. R. (2018). Brain-actuated functional electrical stimulation elicits lasting arm motor recovery after stroke. *Nature Communications*, *9*(1). https://doi.org/10.1038/s41467-018-04673-z

Casson, A. J. (2019). Wearable EEG and beyond. *Biomedical Engineering Letters*, *9*(1), 53–71. doi:https://doi.org/10.1007/s13534-018-00093-6.

Chaudhary, U., Birbaumer, N., & Ramos-Murguialday, A. (2016). Brain–computer interfaces for communication and rehabilitation. *Nature Reviews Neurology, 12*(9), 513–525. doi:https://doi.org/10.1038/nrneurol.2016.113.

Cheng, S. (2023). Metaverse. In S. Cheng (Ed.), *Metaverse: Concept, Content and Context* (pp. 1–23). Springer Nature Switzerland. https://doi.org/10.1007/978-3-031-24359-2_1

Cinel, C., Valeriani, D., & Poli, R. (2019). Neurotechnologies for human cognitive augmentation: current state of the art and future prospects. *Frontiers in Human Neuroscience, 13, 13*. https://doi.org/10.3389/fnhum.2019.00013

Cinel, C., Valeriani, D., & Poli, R. (2019). Neurotechnologies for human cognitive augmentation: current state of the art and future prospects. *Frontiers in Human Neuroscience, 13*, Article 13. https://doi.org/10.3389/fnhum.2019.00013

Cohen, O., Koppel, M., Malach, R., & Friedman, D. (2014). Controlling an avatar by thought using real-time fMRI. *Journal of Neural Engineering, 11*(3), 035006–035006. https://doi.org/10.1088/1741-2560/11/3/035006

Colucci, A., Vermehren, M., Cavallo, A., Angerhöfer, C., Peekhaus, N., Zollo, L., Kim, W. S., Paik, N. J., & Soekadar, S. R. (2022). Brain–computer interface-controlled exoskeletons in clinical neurorehabilitation: Ready or not? *Neurorehabilitation and Neural Repair, 36*(12), 747–756. doi:https://doi.org/10.1177/15459683221138751.

Coyle, D., Garcia, J., Satti, A. R., & McGinnity, T. M. (2011). EEG-based continuous control of a game using a 3 channel motor imagery BCI: BCI game. *2011 IEEE Symposium on Computational Intelligence, Cognitive Algorithms, Mind, and Brain (CCMB),* 1–7. https://doi.org/10.1109/CCMB.2011.5952128

Creed, C., Al-Kalbani, M., Theil, A., Sarcar, S., & Williams, I. (2024). Inclusive AR/VR: accessibility barriers for immersive technologies. *Universal Access in the Information Society*, *23*(1), 59–73.

Darejeh, A. (2023). Empowering education through eerp: A customizable educational vr escape room platform. In *2023 IEEE International Symposium on Mixed and Augmented Reality Adjunct (ISMAR-Adjunct)* (pp. 764–766). IEEE.

Darejeh, A., & Mashayekh, S. (2023). Mind-object control as a new form of interaction in metaverse for users with motor disability. *Journal of Biomedical Physics & Engineering*, 13(1), 257.

Darejeh, A., Huynh, L., Ji, X., Marcus, N., & Polly, P. (2024a). Assessing the efficacy of virtual reality-based laboratories versus real-life settings for PCR procedures, within medical science laboratories. In *2024 IEEE International Symposium on Mixed and Augmented Reality Adjunct (ISMAR-Adjunct)* (pp. 489–492). IEEE.

Darejeh, A., Mondal, M., Marcus, N., & Vassar, A. (2024b). Exploring the cognitive load effects of diverse virtual reality interaction methods in interactive educational platforms. In *2024 IEEE International Symposium on Mixed and Augmented Reality Adjunct (ISMAR-Adjunct)* (pp. 435–438). IEEE.

De Castro-Cros, M., Sebastian-Romagosa, M., Rodríguez-Serrano, J., Opisso, E., Ochoa, M., Ortner, R., Guger, C., & Tost, D. (2020). Effects of gamification in BCI functional rehabilitation. *Frontiers in Neuroscience*, *14*, 882–882. https://doi.org/10.3389/fnins.2020.00882

EMOTIV (n.d.). *Brain data measuring hardware and software solutions*. https://www.emotiv.com/

Gholizadeh Hamlabadi, K., Laamarti, F., & El Saddik, A. (2024). Meta-review on brain-computer interface (BCI) in the Metaverse. *ACM Transactions on Multimedia Computing, Communications, and Applications, 20*(12), Article 395. https://doi.org/10.1145/3696109

Göhring, D., Latotzky, D., Wang, M., & Rojas, R. (2013). Semi-autonomous car control using brain computer interfaces. In S. Lee, H. Cho, K.- J. Yoon, & J. Lee (Eds.), *Intelligent autonomous systems 12* (Vol. 194, pp. 393–408). Springer Berlin Heidelberg. https://doi.org/10.1007/978-3-642-33932-5_37

Gordon, E. C., & Seth, A. K. (2024). Ethical considerations for the use of brain–computer interfaces for cognitive enhancement. *PLoS Biology, 22*(10), e3002899–e3002899. doi:https://doi.org/10.1371/journal.pbio.3002899.

Grinschgl, S., Papenmeier, F., & Meyerhoff, H. S. (2021). Consequences of cognitive offloading: Boosting performance but diminishing memory. *Quarterly Journal of Experimental Psychology, 74*(9), 1477–1496. https://doi.org/10.1177/17470218211008060.

Guger, C., Allison, B. Z., & Edlinger, G. (2014). Emerging BCI opportunities from a market perspective. In G. Grübler & E. Hildt (Eds.), *Brain-computer-interfaces in their ethical, social and cultural contexts* (pp. 85–98). Springer Netherlands. https://doi.org/10.1007/978-94-017-8996-7_7

Guger, C., Prabhakaran, V., Spataro, R., Krusienski, D. J., & Hebb, A. O. (2020). Editorial: Breakthrough BCI applications in medicine. *Frontiers in Neuroscience, 14*, 598247. https://doi.org/10.3389/fnins.2020.598247

Hieu, N. Q., Hoang, D. T., Nguyen, D. N., & Dutkiewicz, E. (2023). Toward BCI-enabled Metaverse: A joint learning and resource allocation approach. *IEEE Global Communications Conference (GLOBECOM)*. (pp. 4088–4093). IEEE. https://doi.org/10.1109/GLOBECOM54140.2023.10437600

Hintermüller, C., Kapeller, C., Edlinger, G., & Guger, C. (2013). BCI integration: Application interfaces. In *Brain-Computer Interface Systems – Recent Progress and Future Prospects* (pp. 21–36). IntechOpen. https://doi.org/10.5772/55806

Ienca, M., & Andorno, R. (2017). Towards new human rights in the age of neuroscience and neurotechnology. *Life Sciences, Society and Policy, 13*, 1–27.

Ienca, M., & Haselager, P. (2016). Hacking the brain: brain–computer interfacing technology and the ethics of neurosecurity. *Ethics and Information Technology, 18*(2), 117–129. doi:https://doi.org/10.1007/s10676-016-9398-9.

Jayaraj, L., Wood, J., Reifer, A., & Braund, E. (2025, January). Experience in a VR cricket game using Brain Computer Interaction (BCI). In *2025 IEEE International Conference on Artificial Intelligence and eXtended and Virtual Reality (AIxVR)* (pp. 248–251). IEEE. https://doi.org/10.1109/AIxVR63409.2025.00047

Keutayeva, A., Nwachukwu, C. J., Alaran, M., Otarbay, Z., & Abibullaev, B. (2025). *Neurotechnology in gaming: A systematic review of visual evoked potential-based brain-computer interfaces*. IEEE Access.

Kumar, S., & Singh, S. K. (2021). Brain computer interaction (BCI): A way to interact with brain waves. *Insights2Techinfo, 1*, 1–12.

Lin, Z., Zhang, C., Zeng, Y., Tong, L., & Yan, B. (2018). A novel P300 BCI speller based on the Triple RSVP paradigm. *Scientific Reports, 8*(1), 3350.

Lotte, F., Larrue, F., & Mühl, C. (2018). Flaws in current human training protocols for spontaneous brain–computer interfaces: Lessons learned from instructional design. *Frontiers in Human Neuroscience, 12*, 34.

Maiseli, B., Abdalla, A. T., Massawe, L. V., Mbise, M., Mkocha, K., Nassor, N. A., Ismail, M., Michael, J., & Kimambo, S. (2023). Brain–computer interface: Trend, challenges, and threats. *Brain Informatics, 10*(1), 20. https://doi.org/10.1186/s40708-023-00199-3

Mane, R., Chouhan, T., & Guan, C. (2020). BCI for stroke rehabilitation: Motor and beyond. *Journal of Neural Engineering, 17*(4), 041001.

Menon, C., De Negueruela, C., Millán, J. D. R., Tonet, O., Carpi, F., Broschart, M., … & De Rossi, D. (2009). Prospects of brain–machine interfaces for space system control. *Acta Astronautica, 64*(4), 448–456.

Mercedes-Benz Group. (2021, September 6). Mercedes-Benz VISION AVTR. Mercedes-Benz Group. https://group.mercedes-benz.com/innovation/product-innovation/design/vision-avtr-bci.html

Milsap, G., Collard, M., Coogan, C., & Crone, N. E. (2019). BCI2000Web and WebFM: Browser-based tools for brain computer interfaces and functional brain mapping. *Frontiers in Neuroscience, 12*, 1030. https://doi.org/10.3389/fnins.2018.01030

Moumgiakmas, S. S., & Papakostas, G. A. (2022). Robustly effective approaches on motor imagery-based brain computer interfaces. *Computers, 11*(5), 61.

Muenssinger, J. I., Halder, S., Kleih, S. C., Furdea, A., Raco, V., Hoesle, A., & Kubler, A. (2010). Brain painting: First evaluation of a new brain–computer interface application with ALS-patients and healthy volunteers. *Frontiers in Neuroscience, 4*, 182. https://doi.org/10.3389/fnins.2010.00182

Naser, M. Y., & Bhattacharya, S. (2023). Towards practical BCI-driven wheelchairs: A systematic review study. *IEEE Transactions on Neural Systems and Rehabilitation Engineering, 31*, 1030–1044.

NeuroSky. (n.d.). *Games*. NeuroSky Store. https://store.neurosky.com/#games

Nicolas-Alonso, L. F., & Gomez-Gil, J. (2012). Brain computer interfaces, a review. *Sensors, 12*(2), 1211–1279.

Nieto, N., Peterson, V., Rufiner, H. L., Kamienkowski, J. E., & Spies, R. (2022). Thinking out loud, an open-access EEG-based BCI dataset for inner speech recognition. *Scientific Data, 9*(1), 52–17. https://doi.org/10.1038/s41597-022-01147-2

Nijholt, A., Contreras-Vidal, J. L., Jeunet, C., & Väljamäe, A. (2022). Editorial: Brain-computer interfaces for non-clinical (Home, Sports, Art, Entertainment, Education, Well-Being) applications. *Frontiers in Computer Science, 4, 860619*. https://doi.org/10.3389/fcomp.2022.860619

Nolan, M. (2023, May 24). *This gamer turned EEG tech into a game controller*. IEEE Spectrum. https://spectrum.ieee.org/elden-ring-hands-free-controller

Noury, N., Hipp, J. F., & Siegel, M. (2016). Physiological processes non-linearly affect electrophysiological recordings during concurrent EEG–fMRI. *NeuroImage, 140*, 125–137.

Orenda, M. P., Garg, L., & Garg, G. (2017). Exploring the feasibility to authenticate users of web and cloud services using a brain-computer interface (BCI). In *ICIAP 2017 International Workshops* (pp. 353–363). Springer. https://doi.org/10.1007/978-3-319-70742-6_33

Padfield, N., Zabalza, J., Zhao, H., Masero, V., & Ren, J. (2019). EEG-Based Brain-Computer Interfaces Using Motor-Imagery: Techniques and Challenges. *Sensors (Basel, Switzerland), 19*(6), 1423-. https://doi.org/10.3390/s19061423

Pfurtscheller, G., Leeb, R., Keinrath, C., Friedman, D., Neuper, C., Guger, C., & Slater, M. (2006). Walking from thought. *Brain Research, 1071*(1), 145–152. https://doi.org/10.1016/j.brainres.2005.11.083

Pires, G., Torres, M., Casaleiro, N., Nunes, U., & Castelo-Branco, M. (2011). Playing Tetris with non-invasive BCI. *2011 IEEE 1st International Conference on Serious Games and Applications for Health (SeGAH),* 1–6. https://doi.org/10.1109/SeGAH.2011.6165454

Prpa, M., & Pasquier, P. (2019). Brain-computer interfaces in contemporary art: A state of the art and taxonomy. In A. Nijholt (Ed.), Brain art: Brain-computer interfaces for artistic expression (pp. 65–115). Springer International Publishing. https://doi.org/10.1007/978-3-030-14323-7_3

Rao, R. P. N., & Scherer, R. (2010). Brain-computer interfacing [In the Spotlight]. *IEEE Signal Processing Magazine*, *27*(4), 152–150. IEEE Signal Processing Magazine. https://doi.org/10.1109/MSP.2010.936774

Renard, Y., Lotte, F., Gibert, G., Congedo, M., Maby, E., Delannoy, V., Bertrand, O., & Lécuyer, A. (2010). OpenViBE: An open-source software platform to design, test, and use brain–computer interfaces in real and virtual environments. *Presence: Teleoperators and Virtual Environments*, *19*(1), 35–53. https://doi.org/10.1162/pres.19.1.35

Rosso, C. (2021). Autos to Integrate AI-based Brain-Computer Interfaces (BCIs)| Psychology Today. https://www.psychologytoday.com/us/blog/the-future-brain/202109/autos-integrate-ai-based-brain-computer-interfaces-bcis-0

Roy, Y., Banville, H., Albuquerque, I., Gramfort, A., Falk, T. H., & Faubert, J. (2019). Deep learning-based electroencephalography analysis: A systematic review. *Journal of Neural Engineering*, *16*(5), 051001. https://doi.org/10.1088/1741-2552/ab260c

Ryan, D. B., Townsend, G., Gates, N. A., Colwell, K., & Sellers, E. W. (2017). Evaluating brain-computer interface performance using color in the P300 checkerboard speller. *Clinical Neurophysiology*, *128*(10), 2050–2057.

Saha, S., Mamun, K. A., Ahmed, K., Mostafa, R., Naik, G. R., Darvishi, S., Khandoker, A. H., & Baumert, M. (2021). Progress in brain computer interface: Challenges and opportunities. *Frontiers in Systems Neuroscience*, *15*, 578875. https://doi.org/10.3389/fnsys.2021.578875

Schalk, G., McFarland, D. J., Hinterberger, T., Birbaumer, N., & Wolpaw, J. R. (2004). BCI2000: A general-purpose brain–computer interface (BCI) system. *IEEE Transactions on Biomedical Engineering*, *51*(6), 1034–1043. https://doi.org/10.1109/TBME.2004.827072

Séguin, P., Maby, E., Fouillen, M., Otman, A., Luauté, J., Giraux, P., Morlet, D., & Mattout, J. (2024). The challenge of controlling an auditory BCI in the case of severe motor disability. *Journal of NeuroEngineering and Rehabilitation*, *21*, 9. https://doi.org/10.1186/s12984-023-01289-3

Shih, J. J., Krusienski, D. J., & Wolpaw, J. R. (2012). Brain-computer interfaces in medicine. *Mayo Clinic Proceedings*, *87*(3), 268–279. https://doi.org/10.1016/j.mayocp.2011.12.008

Tangermann, M., Krauledat, M., Grzeska, K., Sagebaum, M., Blankertz, B., Vidaurre, C., & Müller, K.-R. (2008). Playing Pinball with non-invasive BCI. *ResearchGate*, *21*, 1641–1648. https://www.researchgate.net/publication/221618400_Playing_Pinball_with_non-invasive_BCI

Thomasson, N. (2023). An introduction to BCI and its use in video games: A review. *Journal of Student Research*, *12*(1), 1–13. https://doi.org/10.47611/jsrhs.v12i1.3924

Thompson, A. K., Carruth, H., Haywood, R., Hill, N. J., Sarnacki, W. A., McCane, L. M., Wolpaw, J. R., & McFarland, D. J. (2018). Effects of sensorimotor rhythm modulation on the human flexor carpi radialis H-reflex. *Frontiers in Neuroscience*, *12*, 505. https://doi.org/10.3389/fnins.2018.00505

Vanutelli, M. E., Salvadore, M., & Lucchiari, C. (2023). BCI applications to creativity: Review and future directions, from little-c to C2. *Brain Sciences*, *13*(4), 665. https://doi.org/10.3390/brainsci13040665

Värbu, K., Muhammad, N., & Muhammad, Y. (2022). Past, present, and future of EEG-based BCI applications. *Sensors*, *22*(9), 3331.

Vasiljevic, G. A. M., & de Miranda, L. C. (2020). Brain–computer interface games based on consumer-grade EEG devices: A systematic literature review. *International Journal of Human–Computer Interaction, 36*(2), 105–142. https://doi.org/10.1080/10447318.2019.1612213

Waldert, S. (2016). Invasive vs. Non-invasive neuronal signals for brain-machine interfaces: Will one prevail? *Frontiers in Neuroscience, [online], 10*, 295. https://doi.org/10.3389/fnins.2016.00295

Willett, F. R., Avansino, D. T., Hochberg, L. R., Henderson, J. M., & Shenoy, K. V. (2021). High-performance brain-to-text communication via handwriting. *Nature (London), 593*(7858), 249–254. https://doi.org/10.1038/s41586-021-03506-2

Xiao, J., Pan, J., He, Y., Xie, Q., Yu, T., Huang, H., …& Li, Y. (2018). Visual fixation assessment in patients with disorders of consciousness based on brain-computer interface. *Neuroscience Bulletin, 34*, 679–690.

Xuanmin, L. (2021). Chinese start-up Neuracle tests emerging 'micro-invasive' brain-computer interface in pigs. *Global Times*. https://www.globaltimes.cn/page/202108/1230191.shtml

Yang, K. (2023). Brain-computer interfaces for rehabilitation. *Medium*. https://medium.com/@kellyang2006/brain-computer-interfaces-for-rehabilitation-80f3a78d9655

Yin, J., Jiang, D., & Hu, J. (2009). Design and application of brain-computer interface web browser based on VEP. *2009 International Conference on Future BioMedical Information Engineering*, 77–80. https://doi.org/10.1109/FBIE.2009.63

Zhao, Z. P., Nie, C., Jiang, C. T., Cao, S. H., Tian, K. X., Yu, S., & Gu, J. W. (2023). Modulating brain activity with invasive brain–computer interface: A narrative review. *Brain sciences, 13*(1), 134.

Zhu, H. Y., Hieu, N. Q., Hoang, D. T., Nguyen, D. N., & Lin, C. T. (2024). A human-centric Metaverse enabled by brain-computer interface: A survey. *IEEE Communications Surveys & Tutorials, 26*(3), 2120–2149. https://doi.org/10.1109/COMST.2024.3387124

13 A Roadmap for Intelligent Metaverse-Based Learning Environments

From Lessons Learned to Practical Frameworks

Ali Darejeh and Samad M. E. Sepasgozar

1 Introduction

Over the course of this book, we have witnessed how immersive technologies are increasingly transforming educational practices across a wide range of disciplines. From foundational discussions on the concept of the metaverse to detailed empirical studies in engineering, medicine, and creative arts, it has become clear that these tools are no longer speculative—they are actively shaping how students engage with content, peers, and instructors in today's learning environments.

Across previous chapters, we explored how virtual reality (VR), generative artificial intelligence (GenAI), and brain–computer interfaces (BCIs) are converging to form powerful, interactive ecosystems that redefine teaching and learning. We examined how VR enhances procedural learning in STEM fields, how GenAI-guided tutors personalize education in creative disciplines, how virtual museums create new pathways for cultural engagement, and how BCIs empower learners with physical impairments. We also addressed the ethical, pedagogical, and technical challenges of deploying these innovations, particularly in real-world academic contexts.

These investigations have not only demonstrated what is currently possible but also highlighted critical gaps and opportunities for future development. This final chapter turns toward those future possibilities. Rather than offering a retrospective summary, it poses a forward-looking question: How can we intentionally design immersive learning environments that are intelligent, inclusive, scalable, and responsive to emerging educational needs?

To begin answering this question, we need more than speculation—we require structured approaches that can guide future development. In response, we present two integrated design frameworks, developed through our interdisciplinary research and experimentation at UNSW and grounded in the empirical studies detailed throughout this book. These frameworks synthesize empirical insights and practical lessons into actionable guidance for designing, implementing, and evaluating immersive learning platforms.

The first, the Design Framework for Educational Metaverse, focuses on the foundational principles required to build meaningful, scalable virtual learning

DOI: 10.1201/9781003557555-13

spaces. It outlines ten essential components—ranging from realism and usability to cognitive load management, feedback, and privacy preservation—that together support effective learning in immersive contexts.

The second, the Design Framework for AI and BCI-Enhanced Educational Metaverse, takes a step further by incorporating adaptive and neuroresponsive technologies. GenAI enables real-time personalization of content and feedback, while BCIs introduce new possibilities for sensing cognitive effort and emotional state, making immersive environments more responsive and inclusive. Together, these technologies underpin what we define as intelligent immersive learning environments.

These frameworks serve multiple purposes:

- As practical tools for educators and developers designing immersive educational experiences;
- As strategic resources for institutions adopting or scaling emerging technologies;
- And as a foundation for future research into ethical, adaptive, and accessible digital learning systems.

In the sections that follow, we introduce each framework in detail, supported by design strategies, implementation considerations, and real-world applications. Ultimately, this chapter seeks to bridge research and practice—offering a roadmap to ensure that as immersive technologies evolve, they do so in ways that are pedagogically sound, ethically grounded, and learner-centered.

2 Design Framework for Educational Metaverse

This section presents our first proposed framework, the Design Framework for Educational Metaverse, which outlines the fundamental design principles essential for creating immersive, interactive, and effective learning experiences within metaverse environments. Unlike the second framework, which explores the integration of GenAI and BCI to develop next-generation metaverse environments, this framework focuses on the core design principles essential for building an interactive and efficient learning space. It establishes the foundational pillars of metaverse-based education, offering structured guidance without incorporating AI or neuroadaptive technologies.

2.1 Components of the Framework

As illustrated in Figure 13.1, the framework is divided into ten key components, each addressing a critical aspect of designing an effective metaverse learning environment. These components include (1) *realism and accuracy*, which ensure that the virtual environment closely mirrors real-world objects and scenarios; (2) *interactivity*, which promotes active learner engagement; (3) *interaction methods*, focusing on how users communicate and navigate within the environment;

(4) *adaptability*, allowing the system to adjust to individual learning needs; (5) *Cognitive Load*, ensuring that learners can process information efficiently without being overwhelmed; (6) *Feedback*, which provides timely and meaningful responses to learner actions; (7) *Collaboration*, supporting shared and cooperative learning experiences; (8) *Accessibility*, ensuring inclusive design for users of all abilities; (9) *Infrastructure*, referring to the technical foundation required for smooth operation; and (10) *Privacy Preservation*, which safeguards user data and promotes a secure learning space. The following sections provide a detailed breakdown of each component, supplemented with examples to enhance clarity and understanding.

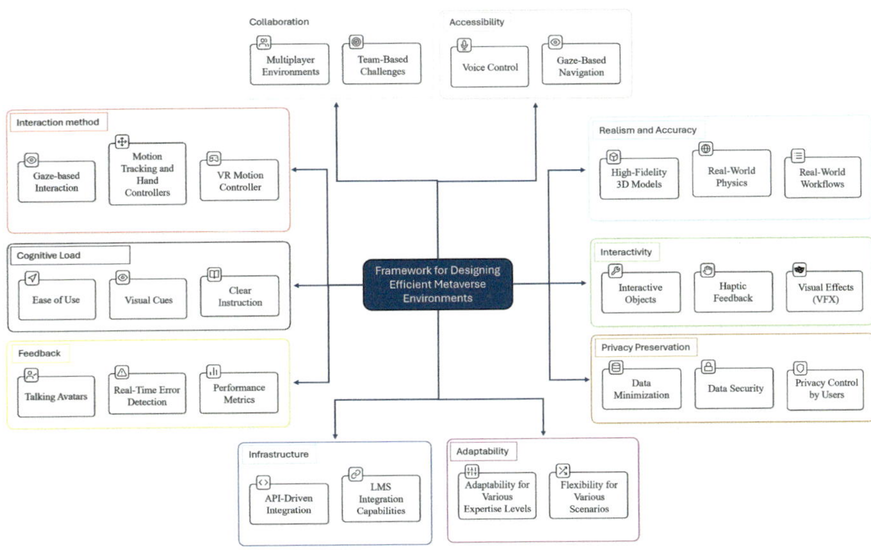

Figure 13.1 Framework for designing efficient educational metaverse environments.

2.1.1 Realism and Accuracy

To support effective knowledge transfer and skill acquisition, metaverse-based learning environments must closely mirror real-world conditions. This includes the use of high-fidelity 3D models that accurately reflect physical forms, realistic simulation of material behaviors and physical laws, and procedural tasks that replicate authentic workflows. Together, these elements enhance immersion, deepen conceptual understanding, and ensure learners develop transferable skills aligned with real-world expectations.

High-fidelity 3D Models: The fidelity of 3D models is a critical factor in creating immersive and pedagogically effective virtual environments. High-fidelity models incorporate accurate textures, realistic shading, and precise geometric representations that closely replicate the form and behavior of real-world objects

(Darejeh et al., 2024a; Grande et al., 2024; Kasapakis, Dzardanova, & Agelada, 2023; Lowell & Tagare, 2023). In educational settings, especially those involving complex systems like human anatomy or mechanical structures, such realism supports deeper cognitive engagement and enhances conceptual understanding.

For example, in a metaverse-based virtual anatomy lab, students interact with anatomically accurate 3D models of human organs that not only respond to dissection gestures but are also modeled to match the exact size, shape, and structure of real-world organs. This level of detail enables learners to develop spatial awareness and procedural knowledge comparable to that gained from working with real cadavers—without the associated ethical, financial, or logistical constraints. In contrast, if abstract or simplified models are used—such as cube-like representations of hearts or kidneys—students may struggle to transfer their learning to real-world scenarios, thereby limiting the effectiveness of the simulation.

Real-World Physics: To ensure authenticity and support effective knowledge transfer, objects within a virtual environment must behave in accordance with the physical laws that govern their real-world counterparts (Bogusevschi, Muntean, & Muntean, 2020). This includes accurate simulation of forces, material properties, and environmental responses. Physics-based interactions—such as gravity, friction, or fluid dynamics—should be realistically modeled to provide learners with meaningful, hands-on experience (Georgiou, Tsivitanidou, & Ioannou, 2021).

For example, in a metaverse-based physics laboratory, when a student drops a metal ball into a container of liquid, the system should simulate the effects of density, buoyancy, and viscosity with high fidelity. The ball's rate of descent, resistance through the fluid, and eventual resting position must reflect the actual physics involved, enabling students to observe and manipulate variables as they would in a real lab setting. This type of precise, rule-based interaction reinforces theoretical understanding and helps students build accurate mental models of scientific phenomena. When virtual environments reflect real-world constraints, they move beyond visual resemblance to offer experiential learning grounded in authentic physical behavior.

Real-World Workflows: To develop procedural expertise, immersive learning environments should guide students through structured, sequential tasks that mirror industry-standard workflows. This not only supports the acquisition of technical skills but also reinforces the importance of accuracy, consistency, and regulatory compliance in professional practice. Each step should be logically ordered, with system feedback reinforcing correct execution and highlighting errors when deviations occur (Bernardo & Duarte, 2022; Ipsita et al., 2025).

For example, in a metaverse-based pharmaceutical research lab, students are tasked with mixing drug compounds following a specific protocol. If they overlook a critical safety step—such as sterilizing equipment or verifying ingredient concentrations—the system immediately flags the error and prevents progression until the mistake is addressed. This ensures that learners internalize proper procedures and understand the consequences of non-compliance. By simulating real-world checks and balances, this approach prepares students to operate confidently and responsibly in high-stakes environments where procedural accuracy is essential.

2.1.2 Interactivity

To move beyond passive learning, metaverse environments must foster active engagement through interactive tools, real-time feedback, and tactile experiences. By enabling learners to manipulate virtual objects, observe visual effects (VFX), and receive haptic feedback, these environments enhance understanding, support skill development, and improve knowledge retention—effectively bridging the gap between theory and practical application.

Interactive Objects: Hands-on interaction is a crucial component of effective learning in immersive environments, particularly for developing spatial awareness, procedural knowledge, and problem-solving skills. In metaverse-based environments, virtual objects and tools should go beyond serving as mere decorative or aesthetic elements. Instead, they must be fully interactive, allowing users to engage with them in diverse and purposeful ways. To support meaningful practice, these tools should accurately replicate the behavior and functionality of their real-world counterparts (Chen et al., 2020; Khorasani et al., 2023).

For example, in a metaverse-based mechanical engineering workshop, students use virtual wrenches to assemble components of an automobile engine. The simulation requires learners to position, rotate, and tighten parts with realistic precision, closely mirroring real-world mechanical procedures. When using the virtual wrench, students are expected to grasp it via a VR controller and perform natural hand movements to align and fasten components—replicating the motor skills needed in actual workshop settings. This risk-free environment allows students to experiment, make and correct mistakes, and repeat tasks as needed—ultimately reducing the likelihood of errors in real-world applications and reinforcing core engineering competencies.

Visual Effects: VFX enhance the sensory dimension of immersive learning environments by adding dynamic, context-sensitive animations and visual transitions that simulate real-world phenomena. These effects—such as explosions, environmental shaking, light flares, smoke, or particle emissions—extend the interactivity and realism of virtual experiences and increase emotional and cognitive engagement. When integrated meaningfully, VFX can capture learner attention, reinforce conceptual understanding, and transform abstract ideas into memorable, embodied experiences (Dinur, 2021; Pardeshi & Karbhari, 2019). In immersive educational settings, VFX can be particularly effective for teaching subjects that involve dynamic systems, invisible forces, or large-scale natural events (Cidota et al., 2016).

For example, in a metaverse-based earth science lesson, students studying tectonic movements experience a simulated earthquake where the environment begins to tremble, shelves collapse, and ground fissures appear in real time. The inclusion of visual tremors and synchronized sound effects not only evokes a visceral reaction but also helps students comprehend the suddenness and destructive nature of seismic activity.

Similarly, in a chemistry lab simulation, combining VFX with chemical interactions—such as glowing reactions, simulated smoke, or safe virtual explosions—can deepen learner engagement while conveying concepts like volatility, energy release, and reaction dynamics. These effects do not merely serve entertainment purposes; they function as cognitive anchors, helping learners associate visual stimuli with scientific principles and facilitating long-term retention.

Haptic Feedback: Tactile interaction plays a crucial role in developing fine motor skills and procedural accuracy, particularly in disciplines that require precision and physical sensitivity. Haptic feedback technologies enhance interactivity and virtual training by simulating sensations such as pressure, texture, and resistance, effectively bridging the gap between visual immersion and physical realism (Gibbs, Gillies, & Pan, 2022; Varney, May, & Richert, 2024).

For example, in a VR-based dental training lab, students perform virtual drilling procedures on teeth using haptic-enabled controllers. The system replicates the distinct tactile resistance of enamel and dentin, allowing learners to experience variations in material density and adjust their hand pressure accordingly. This realistic feedback trains students to perform delicate, high-stakes procedures with the appropriate level of force and control, ultimately improving their readiness for real-world clinical practice.

2.1.3 Interaction Method

Effective metaverse environments should support diverse and intuitive interaction methods that enhance user immersion and accessibility. Techniques such as gaze-based selection, motion tracking, and tool-based VR controllers allow learners to navigate, manipulate objects, and perform complex tasks naturally. These interaction modes not only improve engagement but also help develop motor skills and accommodate various user needs, including those with physical impairments.

VR Motion Controller: In immersive environments, VR motion controllers are the most common interface for interacting with virtual elements, enabling users to precisely grab, manipulate, and navigate objects. These controllers can be mapped to a wide range of virtual tools, allowing users to perform context-specific tasks with a high degree of realism and control. When a task requires the use of a specific instrument—such as a hammer, scalpel, or screwdriver—the system dynamically transforms the virtual controller into a realistic representation of that tool (Alizadehsalehi, Hadavi, & Huang, 2019; Darejeh et al., 2024b). When combined with haptic feedback, this setup simulates sensations such as weight, resistance, and impact, further enhancing the realism and immersiveness of the experience.

For example, in a virtual construction training module, a student can swing a VR controller that visually and functionally replicates a hammer, using it to drive nails into virtual surfaces. The system provides force-based haptic feedback, simulating the resistance and impact experienced in real-world hammering tasks. This intuitive interaction enables learners to engage naturally with their environment, improving task accuracy and facilitating the transfer of skills to real-life scenarios.

Gaze-based Interaction: Gaze-based interaction enables users to navigate and select objects using their eye movements, reducing reliance on manual controllers and enhancing accessibility. This hands-free interaction method streamlines user experience and allows for more intuitive engagement, especially in contexts that demand minimal physical effort or where fine motor control may be limited (Kevin, Pai, & Kunze, 2018; Nilsson, Gustafsson, & Carleberg, 2009).

For example, in a metaverse-based chemistry lab, students can select and manipulate virtual test tubes simply by directing their gaze toward them. This approach not only accelerates interaction by eliminating additional input steps but also provides an inclusive solution for learners with mobility impairments, ensuring equitable access to immersive educational experiences.

Motion Tracking and Hand Controllers: Tasks that require fine motor skills demand high-precision input and accurate tracking of hand and finger movements. Motion-sensitive VR headsets equipped with motion/hand-tracking sensors allow users to interact with virtual objects in ways that closely replicate real-world gestures and techniques (Buń, Husár, & Kaščak, 2022; Hameed, Perkis, & Möller, 2021).

For example, in a metaverse-based orthopedic surgery training scenario, students use their own hands—captured by motion-tracking sensors—to align and fixate fractured bones. This immersive setup accurately reflects the tactile and spatial demands of surgical procedures. By repeatedly practicing in this virtual environment, learners can develop muscle memory, improve manual dexterity, and gain procedural confidence—significantly enhancing their readiness for clinical practice while minimizing real-world risk.

2.1.4 Adaptability

Adaptive metaverse environments tailor learning experiences to individual users by adjusting content based on their expertise, pace, and progress. Through customizable difficulty levels and exposure to diverse real-world scenarios, these systems promote personalized learning while building critical thinking and decision-making skills in dynamic, high-stakes environments.

Adaptability for Various Expertise Levels: Effective immersive learning environments should support differentiated instruction by adapting task complexity to match the learner's proficiency level. This ensures that beginners are guided through foundational concepts while advanced learners are appropriately challenged (Marougkas et al., 2021).

For example, in a metaverse-based software engineering workshop, novice students receive step-by-step assistance during coding tasks, including visual prompts and instructional overlays. In contrast, more experienced users are presented with complex debugging scenarios that require independent problem-solving and minimal guidance. This tiered approach maintains engagement across skill levels, promotes learner autonomy, and supports mastery-based progression tailored to individual needs.

Flexibility for Various Scenarios: Immersive learning environments should offer a diverse range of real-world scenarios to help learners develop critical thinking, adaptability, and situational decision-making skills. By encountering varied and unpredictable challenges, students can practice applying theoretical knowledge in dynamic, high-stakes contexts (Lee et al., 2023; Vats & Joshi, 2023).

For example, in a metaverse-based emergency medical response training, learners engage with multiple trauma cases—ranging from vehicle collisions to industrial accidents—each requiring different clinical protocols and prioritization strategies. This exposure to a wide spectrum of realistic situations prepares learners to make rapid, informed decisions under pressure, builds resilience, and increases their confidence in managing real-life emergencies.

2.1.5 Cognitive Load

To support effective learning, metaverse environments must reduce unnecessary mental effort by offering intuitive interfaces, clear instructions, and visual cues. By simplifying navigation and guiding user actions, these environments help learners concentrate on the content rather than struggling with interface mechanics—leading to improved focus, comprehension, and overall learning outcomes.

Ease of Use: User interfaces in metaverse-based learning environments should be designed to be intuitive, structured, and accessible, minimizing unnecessary complexity that could distract from the learning experience. Clear layouts, familiar interaction patterns, and simplified controls enable learners to navigate virtual spaces efficiently and focus on educational tasks (Ahn et al., 2017; Cárdenas-Sainz et al., 2022; Hulusic et al., 2023).

For example, in a metaverse-based business simulation, students manage a virtual startup using intuitive visual tools—such as dragging icons to recruit team members, adjusting sliders to allocate budgets, or tapping dashboards to monitor performance metrics. By eliminating complex menu systems and avoiding cluttered dashboards that display multiple options simultaneously, the platform reduces cognitive load and enables learners to focus on higher-order decision-making. This usability-focused design ensures that technological complexity does not become a barrier to learning.

Visual Cues: Well-designed visual cues play a crucial role in guiding user attention, reducing ambiguity, and enhancing overall comprehension in immersive learning environments. Elements such as arrows, color highlights, pulsing effects, and animations can direct focus to relevant objects, actions, or next steps, enabling learners to process information more efficiently (Hwang & Shin, 2018; Kasapakis, Dzardanova, & Agelada, 2023).

For example, in a metaverse-based electrical circuit design lab, the environment provides proactive visual cues that guide learners through the assembly process. When a student reaches the next step, the required component—such as a resistor, LED, or wire—glows softly or is highlighted with an animated outline, indicating that it should be selected and placed. Additionally, the virtual representation of the correct tool, like a wire stripper or soldering iron, appears next to the workspace or is visually overlaid on the controller. This approach not only supports procedural learning but also reduces cognitive load by minimizing the need to reference external instructions, making the learning process smoother and more engaging—especially for beginners.

Clear Instruction: Providing clear, step-by-step guidance in a 3D environment is essential to prevent users from becoming disoriented or cognitively overwhelmed. In immersive spaces, learners may struggle to understand how to navigate the environment, which objects are interactive, or what actions they are expected to take. Ambiguity in navigation or interaction increases cognitive load, diverting attention from the actual learning objectives to unnecessary problem-solving about interface usage (Ros et al., 2020; Wijewickrema et al., 2016).

For example, in a metaverse-based biology lab, students are required to complete a virtual dissection. If the platform fails to indicate which tools can be used, how to activate them, or which organs are interactive versus static, students may spend significant time exploring the interface instead of focusing on biological content. However, by integrating layered instructional aids—such as tooltips, step-by-step onboarding animations, or floating indicators—the system can guide users through the task in a structured and intuitive manner. These elements reduce mental effort, allowing learners to engage more deeply with the subject matter rather than the interface mechanics.

2.1.6 Feedback

Effective metaverse learning environments rely on timely, data-driven feedback to guide skill development and improve outcomes. Real-time error detection, talking avatars, and performance metrics help learners monitor progress, correct mistakes, and stay engaged. When automated and personalized, these feedback tools transform immersive experiences into structured and responsive learning journeys.

Real-Time Error Detection: Real-time error detection is crucial for reinforcing correct techniques and preventing the consolidation of procedural mistakes (Davaris et al., 2019; Eaves et al., 2011). In immersive environments, these systems continuously monitor user actions and provide immediate corrective feedback—delivered through simple text messages, audio cues, or talking avatars (which will be discussed in the next section).

For example, in a virtual surgical simulation, if a student sutures a wound incorrectly—such as applying uneven tension or misaligning tissue—the system detects the error, highlights the affected area, and delivers step-by-step corrective guidance. This just-in-time feedback accelerates skill acquisition, reduces the risk of internalizing incorrect methods, and supports the development of procedural fluency critical for high-precision tasks.

Talking Avatars: Talking avatars are embedded agents designed to provide task-specific feedback either during or immediately after user interactions. There are two primary models of talking avatars. The first model—which is the focus of the current framework—relies on predefined actions and delivers fixed feedback based on common errors or interacts with users by asking scripted questions. In contrast, the second, more advanced model leverages AI to generate personalized, adaptive feedback without relying on predefined content (Darejeh, Marcus, & Sweller, 2022; Laine et al., 2022; Nofal et al., 2025; Woodworth, Lipari, & Borst, 2019). The second model will be explored in the next framework.

For example, in a metaverse-based job interview preparation course, a talking avatar plays the role of an interviewer and asks students questions about their background and work experience. However, since this avatar lacks AI integration, it cannot adjust its follow-up questions based on the student's responses. Similarly, in a mechanical engineering lab, if students are assembling a virtual car engine and place a component incorrectly, the talking avatar provides an alert and demonstrates the correct part and its placement.

Performance Metrics: Effective immersive learning environments should incorporate performance tracking mechanisms that capture quantitative indicators of learner progress and present them to users through appropriate formats, such as structured text or voice-based feedback. Metrics such as task completion time, accuracy, error rates, and behavioral patterns provide objective data that can be used to assess proficiency and guide improvement (Wang, Pedaste, & Huang, 2024).

For example, in a metaverse-based flight simulation, pilot trainees receive real-time assessments on key performance factors, including reaction speed, maneuvre precision, and fuel efficiency. These metrics enable learners to reflect on their performance, identify areas for improvement, and iteratively refine their skills. By grounding feedback in measurable outcomes, performance metrics support evidence-based learning and help ensure alignment with real-world competency standards.

2.1.7 Collaboration

Collaboration is a critical component of effective learning in STEM and professional training contexts, where successful problem-solving often relies on communication and teamwork. Metaverse-based environments should facilitate multiplayer experiences—such as group simulations and team-based challenges—to foster skills like coordination, negotiation, and collective decision-making, thereby preparing learners for real-world collaborative scenarios.

Multiplayer Environments: To support collaborative experiences, immersive learning platforms should include multiplayer functionality that enables shared problem-solving and co-creation of knowledge. Allowing multiple users to engage within the same virtual environment promotes interaction, perspective-sharing, and the development of essential teamwork competencies (Darejeh, 2023; Du et al., 2020; Ke et al., 2021).

For example, in a metaverse-based classroom, groups of students gather in a shared virtual space and interact with one another through personalized avatars. They can respond to lecture questions by raising their avatar's hand, engage in real-time discussions, and dynamically form breakout groups for collaborative activities. These interactions closely mirror the social dynamics of a physical classroom, enhancing engagement, promoting active participation, and fostering a sense of community among remote learners.

Team-based Challenges: Metaverse environments should incorporate collaborative scenarios that require learners to work together toward shared objectives, fostering interdependence, role distribution, and strategic coordination (Doumanis et al., 2019; Holly, Resch, & Pirker, 2023). These challenges simulate the complexity of real-world tasks where success depends on effective communication and teamwork.

For example, in a metaverse-based surgical room simulation, a group of medical students collaboratively performs a complex procedure that would not typically be carried out individually in the real world, with each participant assuming a specialized role—such as lead surgeon, anesthesiologist, or surgical assistant. Success depends on the team's ability to coordinate actions in real time, respond to dynamic complications, and make joint decisions under pressure. This type of collaborative learning not only strengthens procedural knowledge but also cultivates essential soft skills such as leadership, communication, and trust, preparing learners for high-stakes, team-oriented professional environments.

2.1.8 Accessibility

Inclusive design is essential to ensure that metaverse-based learning environments are usable by individuals with diverse physical and sensory abilities. By integrating features such as voice control and gaze-based navigation, these platforms can provide hands-free interaction, reduce barriers to engagement, and support equitable access to virtual education for all learners.

Voice Control: Speech recognition technology enables hands-free interaction in immersive environments, allowing users to navigate interfaces, control virtual tools, and issue commands through natural language. This functionality is particularly useful for users with mobility impairments or those engaged in tasks that require the use of both hands (Fang & Wang, 2024; Polap, 2018).

For example, in a metaverse-based chemistry lab simulation, students can verbally instruct the system to mix chemicals, adjust equipment parameters, or initiate experiments. By minimizing the need for motor input, voice control streamlines interaction and supports a more intuitive and inclusive user experience.

Gaze-based Navigation: As previously discussed, gaze-based navigation enables users to navigate virtual environments, interact with objects, and trigger system responses simply by directing their gaze. This hands-free method of interaction is particularly valuable for regular users performing simple tasks—such as selecting a menu item without a controller—as well as for individuals with severe motor and speech impairments who are unable to issue voice commands (Kevin, Pai, & Kunze, 2018; Nilsson, Gustafsson, & Carleberg, 2009).

For example, in a metaverse-based accessibility-focused museum, visitors can move between exhibits by looking at designated visual markers, activating transitions and interactions without the need for controllers or physical movement. This approach not only supports users with physical disabilities but also contributes to a more seamless and inclusive immersive experience for all users.

2.1.9 Infrastructure

A robust technical infrastructure is essential for the reliable operation, scalability, and institutional integration of metaverse-based educational platforms. Key components include application programming interfaces (API)-driven integration for embedding real-time tools and services and compatibility with learning management systems (LMS) to ensure seamless deployment, data synchronization, and streamlined user access. Together, these elements support a consistent and scalable learning experience within institutional ecosystems.

API-Driven Integration: Educational metaverse platforms should support open, well-documented APIs that enable seamless integration with third-party tools, institutional systems, and external services. This interoperability allows developers and educators to extend platform functionality, embed customized features, and maintain alignment with evolving technological ecosystems (Fırat, 2023; Mowla & Kolekar, 2020).

For example, integrating real-time communication APIs—such as Zoom or Microsoft Teams SDKs—can enable embedded video conferencing within a virtual classroom. Natural language processing APIs like OpenAI's GPT can support intelligent tutors for tutoring or FAQs, while data visualization APIs such as Google Charts or Tableau can power dynamic analytics dashboards that track student engagement, performance, and participation. Additionally, authentication APIs like OAuth2 can streamline secure single sign-on (SSO) with institutional credentials. These integrations ensure that the platform remains flexible, scalable across diverse educational settings, and responsive to emerging innovations in digital learning infrastructure.

LMS Integration Capabilities: Compatibility with LMS is essential for embedding metaverse experiences within existing educational infrastructures (Chakravarthy & Raman, 2014; Fırat, 2023). Seamless LMS integration allows immersive modules—such as virtual labs, simulations, or collaborative environments—to be launched directly from platforms like Moodle or Canvas.

For example, a virtual lab simulation can be accessed through the LMS interface, with learner participation, task completion, and assessment data automatically recorded and synchronized. This level of integration streamlines adoption, simplifies administrative processes, and ensures that immersive learning environments function as coherent extensions of the broader digital learning ecosystem.

2.1.10 Privacy Preservation

As metaverse platforms gather increasing volumes of behavioral and biometric data, safeguarding user privacy must be a foundational design priority. Effective privacy preservation involves minimizing data collection, empowering users with control over their personal information, and ensuring secure data transmission and storage. By embedding these practices into the platform's architecture, immersive learning environments can maintain user trust and comply with ethical and legal standards.

Data Minimization: Metaverse learning platforms should adhere to the principle of data minimization by collecting only the information necessary to support educational objectives. Avoiding the accumulation of excessive or sensitive personal data helps mitigate privacy risks and aligns with ethical data governance practices (Fiaz et al., 2024).

For example, instead of continuously recording a learner's every movement or voice interaction within the virtual environment, the system can focus on capturing key learning milestones—such as task completion or time spent on essential activities. This targeted approach ensures that educators receive meaningful insights into student performance while safeguarding user privacy and maintaining regulatory compliance.

Privacy Control by Users: Learners should retain full control over their personal data within metaverse-based educational platforms. Providing transparent, user-friendly privacy dashboards empowers individuals to manage their data proactively and make informed decisions about how it is used (Fiaz et al., 2024). In addition to offering settings that allow users to enable or disable the sharing of specific data types, the platform should require explicit user consent before accessing sensitive hardware-based inputs such as the microphone, eye tracker, or motion-sensing cameras (Zhao et al., 2023).

For example, a student using a virtual lab learning platform can access a personal privacy dashboard to opt out of data sharing with third-party systems, such as LMS, or choose to anonymize their records for added protection. Providing such controls not only reinforces user autonomy but also builds trust and aligns with best practices in ethical data management and digital consent.

Data Security: Ensuring the security of user data is critical in metaverse-based educational platforms, where sensitive personal, behavioral, and academic information may be collected and stored. All data—whether stored locally or transmitted across networks—should be encrypted using industry-standard protocols such as Transport Layer Security (TLS) and Advanced Encryption Standard (AES) to protect against unauthorized access and data breaches (Chen et al., 2022; Parlar, 2023).

In addition to encryption, platforms must implement robust access control measures—such as multi-factor authentication (MFA) and role-based access permissions—to ensure that only authorized individuals can access or modify sensitive information. For example, when users first enter an educational metaverse platform, they are required to authenticate their identity by entering a username and password, followed by a verification code from an MFA app on their mobile device. Additionally, for performing sensitive actions—such as changing their avatar, which visually represents them in the virtual environment, or adding a credit card for in-platform purchases—the system prompts users to re-enter an MFA code, even if they are already logged in. This layered security approach enhances protection against unauthorized access and reinforces trust in the platform's data governance practices.

3 Design Framework for AI and BCI-Enhanced Educational Metaverse

The convergence of BCI and GenAI presents a transformative opportunity for the evolution of the metaverse. By integrating these technologies, we can create intelligent, adaptive, and deeply immersive virtual environments that respond dynamically to user intent and cognition. The framework proposed in this section builds upon the earlier Design Framework for Educational Metaverse by focusing specifically on the integration of BCI and GenAI within educational metaverse environments. It offers a comprehensive and structured approach to designing and implementing immersive learning spaces that effectively leverage these two emerging technologies.

3.1 Components of the Framework

As illustrated in Figure 13.2, the framework is organized into four key sections, each addressing a distinct aspect of integrating GenAI and BCI in metaverse-based learning environments. The first section, *AI-Powered Metaverse*, explores how GenAI supports dynamic content creation, personalized learning pathways, and real-time instructional adaptation. The second section, *Neuroadaptive Metaverse*, examines the potential of BCI to enable neuroadaptive learning by monitoring and responding to learners' cognitive states. The third section, *Collaborative AI and BCI-Enhanced Metaverse*, discusses the synergy between AI and BCI in creating interactive, immersive, and cooperative educational experiences. Finally, the fourth section, *Ethics and Privacy*, highlights the ethical challenges and best practices associated with the use of these technologies, emphasizing the importance of transparency, privacy, and user autonomy. The following sections provide a detailed explanation of each component, supported by practical examples to aid understanding.

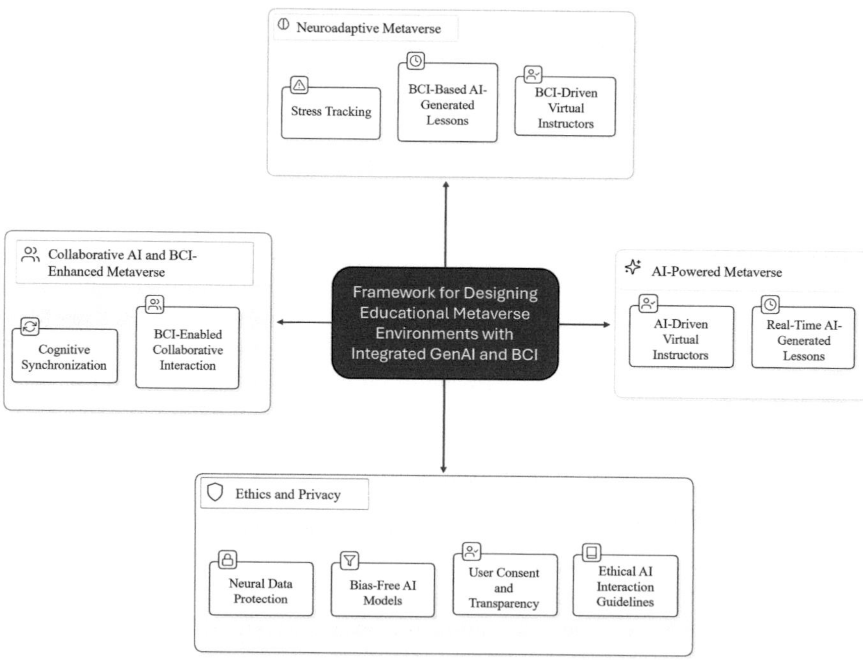

Figure 13.2 Framework for designing educational metaverse environments, integrating BCI and GenAI.

3.1.1 AI-Powered Metaverse

AI-powered learning environments in the metaverse harness the capabilities of GenAI to deliver fully interactive, adaptive, and immersive educational experiences. By integrating AI-driven virtual instructors and real-time lesson generation within spatial 3D environments, these platforms dynamically respond to learner behavior, adjusting content and assessments to match individual needs and performance.

AI-Driven Virtual Instructors: AI-driven virtual instructors function as intelligent, interactive tutors within the metaverse, providing adaptive feedback and guidance in real time. These instructors can take the form of AI-generated avatars that guide students through complex concepts, demonstrate procedures, and respond dynamically to individual progress. Unlike traditional virtual tutors, these virtual instructors leverage spatial computing to create hands-on, immersive experiences where students actively participate in learning rather than passively consuming information (Halim et al., 2024; Laine et al., 2022; Woodworth, Lipari, & Borst, 2019).

For example, in a virtual medical laboratory, a GenAI-powered instructor teaches students human anatomy through interactive lessons. During instruction, the AI instructor poses context-specific questions to reinforce learning—for instance, asking, "Can you identify the organ responsible for filtering blood?" If a student provides a partially correct answer—such as mistakenly naming the liver instead of the kidneys—the instructor responds with clarifying feedback: "You're close—the liver plays a key role in detoxification, but the kidneys are primarily responsible for filtering blood."

Another example of how a GenAI-based instructor can be used is in a virtual chemistry lab, where students interactively perform experiments. If a student attempts to combine two incompatible chemicals—for instance, mixing hydrochloric acid with sodium hydroxide in incorrect proportions—the AI instructor immediately intervenes. It might say, "Be careful—those substances can neutralize each other, but only under controlled conditions. Let's review the proper ratios for a safe reaction." The GenAI tutor can then dynamically generate a visual simulation of the correct procedure, provide a brief explanation of the underlying chemical principles, and offer a short quiz to reinforce the concept.

By continuously analyzing each student's performance and engagement, the instructor adjusts the pace, depth, and instructional approach in real time. It may offer additional practice, simplified explanations, or visual comparisons tailored to the learner's needs.

Real-Time AI-Generated Lessons: GenAI enables the dynamic creation of instructional content in real time within metaverse environments, allowing lessons to adapt continuously based on learners' engagement levels, performance data, and interaction patterns. This approach ensures that students are presented with appropriately challenging material, maintaining engagement while promoting deep learning (Osadchyi et al., 2020; Shirazi et al., 2024). By embedding adaptive learning strategies, GenAI-generated lessons can respond to learner needs on the fly—adjusting difficulty, pacing, or content type to enhance motivation and optimize cognitive load.

For example, in a metaverse-based computer science course, as students progress through an introductory programming module, the AI monitors their performance—such as the accuracy of their responses in short quizzes and the time spent on each screen when learning specific concepts—and identifies recurring difficulties in understanding conditional statements. In response, it generates a personalized lesson sequence that includes visual demonstrations, simplified code examples, and interactive coding tasks tailored to clarify conditional statements. For students who have already mastered the content, the AI bypasses basic exercises and instead generates more complex problem-solving challenges —such as designing scenarios involving nested conditionals or managing control flow in a multi-function program. This real-time adaptation transforms instruction into a dynamic, responsive experience that supports mastery-based learning across diverse learner profiles.

3.1.2 Neuroadaptive Metaverse

Neuroadaptive learning environments represent an advanced evolution of AI-enhanced metaverse platforms, integrating BCIs and AI to adapt instruction in real time based on learners' cognitive and emotional states. Unlike traditional adaptive systems that rely on behavioral inputs, these environments interpret neural signals to adjust pacing, complexity, and engagement strategies—maximizing learning efficiency while preventing overload and fatigue (Darejeh & Mashayekh, 2023). The following components illustrate how BCI creates an adaptive learning environment based on neural signals:

Stress Tracking: Stress tracking utilizes real-time electroencephalogram (EEG)-based monitoring to assess a learner's mental fatigue, stress levels, and engagement levels. By tracking brain activity through BCI sensors, the AI system dynamically modifies the learning environment, adjusting task difficulty and recommending personalized interventions such as breaks, alternative explanations, or interactive examples to enhance retention (Hernández-Mustieles et al., 2024; Premchand et al., 2024; Sargazi Moghadam et al., 2024).

For instance, in a virtual neuroscience lab, medical students perform a 3D brain surgery simulation in the metaverse while wearing a BCI-enabled VR headset. If the system detects high cognitive strain (e.g., increased mental fatigue and reduced engagement), the AI adapts by slowing down the procedure, offering additional visual cues, or integrating interactive tutorials to reinforce critical concepts. If stress levels rise significantly, the system pauses the session and introduces relaxation techniques, ensuring that students maintain an optimal mental state for effective learning.

BCI-Based AI-Generated Lessons: AI-powered lessons within neuroadaptive environments dynamically adjust instructional pacing, problem difficulty, and interaction complexity in real time based on neural feedback from BCI sensors. By interpreting brain activity patterns, the AI system ensures that learners progress at their optimal cognitive speed, reducing frustration and enhancing knowledge mastery (Baradari et al., 2025; Beauchemin et al., 2024; Darejeh et al., 2024c; Mladenović, Mattout, & Lotte, 2018).

For instance, in a metaverse-based physics class focused on quantum mechanics and wave functions, the AI monitors students' cognitive load through EEG signals and dynamically adjusts the complexity of its explanations accordingly. If a student shows signs of cognitive overload, the AI simplifies the lesson content and provides step-by-step breakdowns. Conversely, if high engagement and low fatigue are detected, the AI accelerates the lesson pace, introducing more challenging concepts and complex problem-solving scenarios. This real-time adjustment ensures that students remain engaged while avoiding unnecessary frustration or stagnation.

BCI-Driven Virtual Instructors: BCI-driven virtual instructors are GenAI-enhanced digital tutors that interpret neural and emotional responses to personalize learning interactions. By analyzing stress levels, focus, and emotional states through EEG, these instructors adjust their engagement strategies, modifying tone, pacing, and feedback methods to align with the learner's cognitive needs.

As an example, in a metaverse-based psychology course, students participate in simulated therapy sessions where they interact with an AI-driven virtual therapist. If BCI data indicates emotional distress, the AI adapts its responses by adopting a calmer tone and offering reassurance to help prevent emotional overload. If high engagement and motivation are detected, the AI challenges the student with deeper analytical questions, encouraging critical thinking and problem-solving. This neuroadaptive approach personalizes the learning experience, ensuring that each learner receives emotionally and cognitively optimized instruction.

3.1.3 Collaborative AI and BCI-Enhanced Metaverse

The integration of AI and BCI in metaverse-based learning environments significantly enhances collaborative education by enabling real-time adaptation to both individual and group cognitive states, while also fostering engagement through BCI-enabled collaborative interactions.

Cognitive Synchronization: By monitoring brain activity and behavioral engagement through BCI sensors, the system detects if a student is overloaded, under-stimulated, or disengaged (Beauchemin et al., 2024). AI then dynamically modifies group interactions to redistribute cognitive effort, ensuring that all participants contribute effectively and benefit equally from the learning experience. This approach is particularly valuable in focus group sessions, second-language speaking classes, collaborative problem-solving tasks, and other group-based learning activities where balanced participation and shared cognitive load are critical to success.

For example, in a metaverse-based biomedical engineering course, students work in teams to design a brain-controlled prosthetic limb. AI continuously monitors their cognitive states via BCI and detects if one student is dominating the discussion while others remain passive. To balance the workload, the system adjusts task distribution, prompts quieter students with AI-generated suggestions, and introduces real-time collaborative problem-solving exercises to keep all participants equally engaged. This ensures an optimal team dynamic where every student actively contributes to the project.

BCI-Enabled Collaborative Interaction: BCI-powered metaverse environments enable users to interact, share ideas, and make collaborative decisions using neural commands rather than voice input (Gholizadeh HamlAbadi, Laamarti, & El Saddik, 2024). This fosters a more inclusive and accessible learning experience, particularly for students with physical or speech impairments, by eliminating the need for manual input devices.

As an instance, in a metaverse-based space mission simulation, students form a team of astronauts tasked with repairing a damaged spacecraft. Using BCI-controlled interfaces, team members communicate and vote on the best course of action simply by focusing on specific options within the virtual control panel. The AI system interprets collective neural inputs and executes decisions in real time. This approach ensures a democratic and accessible team-based experience, allowing all members—regardless of physical ability—to contribute equally.

3.1.4 Ethics and Privacy

The integration of BCI and AI into metaverse-based learning environments introduces a range of complex ethical, privacy, and security challenges. Addressing these concerns requires a comprehensive framework focused on safeguarding neural data, promoting fairness, ensuring transparency, and maintaining learner well-being. The following principles outline key considerations for the responsible design and implementation of AI- and BCI-enhanced educational metaverse platforms.

Bias-Free AI Models: AI algorithms used in educational metaverse environments must be regularly audited for inclusivity, fairness, and representational balance to prevent algorithmic bias. AI-generated instructional content, assessments, and interactions should be tested with diverse datasets to ensure representation of culture, gender, and neurodiversity and to avoid reinforcing systemic biases that could disadvantage certain groups of learners (Cantini et al., 2024; Lohyna & Pico Oristrell, 2023).

As an example, in a metaverse-based AI tutoring system for history education, AI-driven narratives should be fact-checked against multiple cultural perspectives to avoid bias in storytelling. If an AI-generated history lesson disproportionately highlights certain geopolitical viewpoints while ignoring others, an ethics-driven AI auditing system should flag and correct these biases, ensuring that students receive well-rounded and inclusive educational content.

Neural Data Protection: The collection and processing of neural data through BCI devices raise serious privacy and cybersecurity risks. To prevent unauthorized access, misuse, or breaches of sensitive brainwave data, secure encryption protocols and zero-trust architecture should be implemented for both cloud-based and local storage solutions (Takabi, Bhalotiya, & Alohaly, 2016). Additionally, BCI-specific cybersecurity standards must be established to regulate data transmission and prevent hacking attempts that could manipulate cognitive input.

For instance, in a metaverse-based science course, students use BCI headsets to navigate virtual environments and operate laboratory tools through neural commands. The system encrypts real-time EEG signals using end-to-end encryption before transmitting them to AI-driven analysis tools. Furthermore, multi-layer authentication is required before any external entity can access stored neural data, ensuring that students' brainwave information remains protected from unauthorized access.

User Consent and Transparency: Clear and explicit policies regarding the collection, processing, and storage of neural data must be implemented, ensuring that learners maintain full autonomy over their data. Users should have the ability to opt-in or opt-out of BCI-driven personalization and control how their neural activity data is stored, shared, and used in AI-powered metaverse learning environments (Naufel & Klein, 2020; Van Stuijvenberg et al., 2024).

For example, in a metaverse-based medical simulation training, students wearing BCI-integrated VR headsets should receive a detailed consent agreement before their cognitive data is analyzed. The system should provide a dashboard allowing users to review, delete, or anonymize their brainwave data at any time. If AI-driven feedback is based on BCI data, the system should provide clear explanations on how decisions were made, ensuring transparency in AI-driven learning adaptations.

Ethical AI Interaction Guidelines: AI-powered tutors, mentors, and virtual assistants within the metaverse must operate under clearly defined ethical principles to ensure transparency, accountability, and fairness in educational interactions. AI-driven feedback should be explainable (Hashemi, Darejeh, & Cruz, 2024, 2025) and free from misleading or manipulative responses (Pretolesi et al., 2024). Additionally, AI should avoid overstepping the roles of human educators and function as a supportive rather than authoritative entity (Saparamadu et al., 2025).

For example, in a metaverse-based law school course, an AI-powered virtual professor provides in-depth analysis of legal cases. If the AI generates a legal interpretation, it should cite its sources and provide students with access to the original legal texts used in its analysis. If a student challenges the AI's explanation, the system should offer alternative perspectives rather than presenting a singular "correct" answer. This approach promotes critical thinking and ethical accountability in AI development within educational settings.

4 Discussion and Applications of the Proposed Frameworks

The frameworks proposed in this chapter provide a practical and forward-thinking foundation for designing and implementing metaverse-based educational platforms. Grounded in empirical insights from the preceding chapters, they address both the immediate needs of current immersive learning environments and the future potential of integrating advanced technologies such as GenAI and BCI. This discussion highlights the significance of these frameworks and examines their practical applications in various educational contexts.

The first framework, Design Framework for Educational Metaverse, which focuses on the foundational design of metaverse learning environments, emphasizes ten essential components: realism, interactivity, interaction methods, adaptability, cognitive load reduction, feedback systems, collaboration, accessibility, infrastructure, and privacy preservation. These elements are not isolated features; they are interdependent principles that, when applied collectively, create immersive, inclusive, and pedagogically effective learning spaces.

For instance, realism and interactivity are critical in disciplines that rely on hands-on training, such as engineering and medical science. In the VR-based labs presented in Chapters 2, 3, and 4, realistic simulations of a manufacturing robot, oscilloscope, and polymerase chain reaction (PCR) machine provided learners with authentic, risk-free environments to develop procedural knowledge. Applying these framework principles ensures that such experiences are not only visually accurate but also functionally realistic, promoting deeper understanding through interaction.

Adaptability and feedback systems, likewise, are essential in tailoring the learning experience to individual needs. As seen in the programming escape room discussed in Chapter 5 and the GenAI-powered cinematography tutor in Chapter 6, learners benefit from environments that can respond to their actions, progress, and performance in real time. By embedding these components into platform design, educators can create more responsive systems that foster self-paced, mastery-oriented learning.

Cognitive load reduction is another cornerstone of effective immersive design. Multiple chapters in this book reported cognitive strain as a barrier to learning in VR environments. The first framework encourages designers to streamline information presentation, integrate clear instructional cues, and avoid overwhelming users with unnecessary sensory input—practices supported by research in Chapters 2 and 4. Addressing this concern is especially vital when combining complex simulations with novel technologies, such as BCI or AI interfaces.

The addition of infrastructure further reinforces the framework's relevance to real-world educational deployment. Infrastructure ensures that immersive systems are not only functionally rich but also technically viable within institutional settings. By supporting API-driven integration and LMS compatibility, metaverse platforms can be seamlessly embedded into existing digital ecosystems, enabling institutions to adopt immersive technologies without disrupting existing workflows. Such integrations allow centralized grading, analytics, and management, which are crucial for scalability and administrative efficiency.

The second framework, Design Framework for AI and BCI-Enhanced Educational Metaverse, builds upon this foundation by incorporating GenAI and BCI into immersive educational platforms. Its four dimensions—GenAI applications, the role of BCI, collaborative learning environments, and ethical considerations—outline a roadmap for the next generation of smart, adaptive learning systems. While still emerging, these technologies offer transformative possibilities that can greatly enhance the personalization, inclusivity, and efficiency of virtual learning.

As demonstrated in Chapters 6, 10, and 11, GenAI has the capability to personalize feedback and generate adaptive learning pathways. By applying this framework, educators can design platforms where learners are not passive recipients of information but active co-creators of their educational experience—engaging with content that dynamically evolves in response to their interactions, progress, and learning goals.

Meanwhile, BCI opens the door to neuroadaptive learning systems—platforms that detect cognitive or emotional states and adjust learning experiences accordingly. Chapter 12 illustrated how BCI applications can enhance engagement and accessibility, particularly for users with motor impairments or those requiring alternative input methods. Incorporating BCI into platform design based on this framework means creating environments that are not only interactive but also attentive and sensitive to learner stress, fatigue, focus, and emotional engagement.

Furthermore, the collaborative learning dimension within both frameworks encourages the integration of multiplayer features, shared workspaces, and real-time feedback tools. In the context of the virtual labs and lecture halls described throughout this book, collaboration is key to mimicking real-world academic and professional environments. By supporting peer interaction and cooperative problem-solving, metaverse platforms can prepare students for interdisciplinary teamwork, a skill that is increasingly essential across various fields.

Crucially, ethical considerations and privacy are not peripheral to these two frameworks but are embedded as core design imperatives. Principles such as transparency, informed consent, algorithmic fairness, and digital accessibility must guide every phase of development. As behavioral and biometric data are increasingly collected through technologies such as VR, eye tracking, and BCI devices—and further interpreted by GenAI systems—it becomes imperative that platforms adopt a privacy-by-design approach. As discussed in Chapter 9, the risks associated with data overcollection and limited user control underscore the need for accountability in system design. Reflecting these concerns, the privacy-preservation component within both frameworks ensures that only essential data is collected, users retain control over their personal information, and strong security protocols are in place to prevent misuse.

Overall, the application of these frameworks empowers stakeholders—educators, instructional designers, developers, researchers, and policymakers—to move from experimentation to sustainable implementation. By translating empirical findings into structured design strategies, the frameworks bridge the gap between innovation and educational impact, offering a roadmap for what immersive learning environments can become and how they can be responsibly and effectively realized. As immersive technologies continue to evolve, these frameworks serve as adaptable tools that grow alongside them. Whether institutions are piloting virtual labs, scaling fully immersive degree programs, or integrating AI tutors into hybrid classrooms, the frameworks provide essential scaffolding to ensure that the future of learning is not only immersive and intelligent but also inclusive, ethical, and learner-centered.

5 Conclusion

Value of this Book and Frameworks: This book offers a timely and practical contribution to the evolving discourse on immersive technologies in education, with a central focus on the design and implementation of metaverse-based learning platforms. As the metaverse emerges as a unifying virtual ecosystem—integrating spatial computing, AI, and real-time interaction—there is a growing need for educational strategies that are not only innovative but also grounded in pedagogical integrity.

At its core, this book introduces two frameworks that connect immersive learning design with cognitive science and human-computer interaction, offering both conceptual insight and actionable guidance. Technologies such as VR, GenAI, and BCIs are positioned as integral components that enhance the functionality, adaptability, and personalization of educational experiences in the metaverse. These frameworks provide structured, evidence-informed approaches for creating scalable, inclusive, and ethically designed immersive learning environments. Rather than prescribing fixed blueprints, they offer flexible structures that support iterative development—enabling educators to move beyond isolated pilot projects toward the thoughtful, long-term integration of immersive technologies in education.

In addition, the frameworks offer forward-looking perspectives on how BCI and GenAI could be leveraged to create more adaptive and equitable learning experiences. They propose that metaverse-based content might dynamically respond to learners' neural data, enabling intelligent tutors to adjust instructional strategies in real time based on cognitive signals. The frameworks also explore AI-driven approaches to fostering balanced collaboration by monitoring neural engagement and managing group dynamics through a GenAI-powered facilitator.

Intended Audience of this Book: This book and its proposed frameworks are intended for a multidisciplinary audience spanning both academic and professional domains, offering valuable insights for educators, researchers, instructional designers, institutional leaders, and industry professionals.

For educators, instructional designers, and curriculum developers, the frameworks serve as practical tools to guide the design and evaluation of immersive modules that prioritize cognitive load, motivation, and user agency. These readers will find guidance on integrating immersive technologies into teaching and assessment in ways that align with pedagogical intent rather than merely showcasing technological novelty.

Academic managers and institutional leaders may draw strategic value from this book when shaping digital learning policies, investing in technology infrastructure, or leading innovation initiatives. The frameworks help align institutional decision-making with contemporary demands for personalization, adaptability, and inclusivity in digital learning.

Researchers in learning engineering, medical sciences, human–computer interaction, and educational psychology can use the frameworks as a foundation for empirical inquiry—particularly in evaluating their effectiveness, refining design principles, and addressing the cognitive and affective dimensions of immersive learning. This book also supports postgraduate students and early-career academics exploring applied technology-enhanced learning, offering conceptual depth alongside real-world case studies.

In industry contexts—particularly EdTech, e-learning, and professional training—this book offers design principles and evaluation criteria that respond to the evolving need for user-centered, adaptable learning environments that support the preparation of future engineers, professionals, and scientists across diverse industries.

Limitations and Future Directions: While this book offers practical examples and conceptual models, the frameworks have not yet undergone large-scale empirical validation. Their application remains exploratory and context-dependent. Further research is needed to test, refine, and prioritize the framework elements through large-scale studies conducted across diverse disciplines and learning environments. Additionally, this book primarily focuses on higher education, and further investigation is needed to assess applicability in K–12, vocational training, and corporate learning environments.

References

Ahn, J., Choi, S., Lee, M., & Kim, K. (2017). Investigating key user experience factors for virtual reality interactions. *Journal of the Ergonomics Society of Korea*, *36*(4), 267–280.

Alizadehsalehi, S., Hadavi, A., & Huang, J. C. (2019). Virtual reality for design and construction education environment. *AEI, 2019*, 193–203.

Baradari, D., Kosmyna, N., Petrov, O., Kaplun, R., & Maes, P. (2025). NeuroChat: A neuroadaptive AI Chatbot for customizing learning experiences. *arXiv preprint arXiv, 2503*, 07599.

Beauchemin, N., Charland, P., Karran, A., Boasen, J., Tadson, B., Sénécal, S., & Léger, P. M. (2024). Enhancing learning experiences: EEG-based passive BCI system adapts learning speed to cognitive load in real-time, with motivation as catalyst. *Frontiers in Human Neuroscience*, *18*, 1416683.

Bernardo, N., & Duarte, E. (2022). Immersive virtual reality in an industrial design education context: what the future looks like according to its educators. *Computer-Aided Design & Applications*, *19*(2), 238–255.

Bogusevschi, D., Muntean, C., & Muntean, G. M. (2020). Teaching and learning physics using 3D virtual learning environment: A case study of combined virtual reality and virtual laboratory in secondary school. *Journal of Computers in Mathematics and Science Teaching*, *39*(1), 5–18.

Buń, P., Husár, J., & Kaščak, J. (2022, March). Hand tracking in extended reality educational applications. In *International Scientific-Technical Conference Manufacturing* (pp. 317–325). Springer International Publishing.

Cantini, R., Cosenza, G., Orsino, A., & Talia, D. (2024, October). Are large language models really bias-free? Jailbreak prompts for assessing adversarial robustness to bias elicitation. In *International Conference on Discovery Science* (pp. 52–68). Springer Nature Switzerland.

Cárdenas-Sainz, B. A., Barrón-Estrada, M. L., Zatarain-Cabada, R., & Ríos-Félix, J. M. (2022). Integration and acceptance of natural user interfaces for interactive learning environments. *International Journal of Child-Computer Interaction*, *31*, 100381.

Chakravarthy, S. S., & Raman, A. C. (2014, April). Educational data mining on learning management systems using experience API. In *2014 Fourth International Conference on Communication Systems and Network Technologies* (pp. 424–427). IEEE.

Chen, J. C., Huang, Y., Lin, K. Y., Chang, Y. S., Lin, H. C., Lin, C. Y., & Hsiao, H. S. (2020). Developing a hands-on activity using virtual reality to help students learn by doing. *Journal of Computer Assisted Learning, 36*(1), 46–60.

Chen, Z., Wu, J., Gan, W., & Qi, Z. (2022, December). Metaverse security and privacy: An overview. In *2022 IEEE International Conference on Big Data (Big Data)* (pp. 2950–2959). IEEE.

Cidota, M. A., Clifford, R. M., Lukosch, S. G., & Billinghurst, M. (2016, September). Using visual effects to facilitate depth perception for spatial tasks in virtual and augmented reality. In *2016 IEEE International Symposium on Mixed and Augmented Reality (ISMAR-Adjunct)* (pp. 172–177). IEEE.

Darejeh, A. (2023). Empowering education through eerp: A customizable educational vr escape room platform. In *2023 IEEE International Symposium on Mixed and Augmented Reality Adjunct (ISMAR-Adjunct)* (pp. 764–766). IEEE.

Darejeh, A., & Mashayekh, S. (2023). Mind-object control as a new form of interaction in metaverse for users with motor disability. *Journal of Biomedical Physics & Engineering, 13*(1), 257.

Darejeh, A., Huynh, L., Ji, X., Marcus, N., & Polly, P. (2024a). Assessing the efficacy of virtual reality-based laboratories versus real-life settings for PCR procedures, within medical science laboratories. In *2024 IEEE International Symposium on Mixed and Augmented Reality Adjunct (ISMAR-Adjunct)* (pp. 489–492). IEEE.

Darejeh, A., Marcus, N., & Sweller, J. (2022). Increasing learner interactions with E-learning systems can either decrease or increase cognitive load depending on the nature of the interaction. *L'Année psychologique, 122*(3), 405–437.

Darejeh, A., Moghadam, T. S., Delaramifar, M., & Mashayekh, S. (2024c). A framework for AI-powered decision making in developing adaptive e-learning systems to impact learners' emotional responses. In *2024 11th International and the 17th National Conference on E-Learning and E-Teaching (ICeLeT)* (pp. 1–6). IEEE.

Darejeh, A., Mondal, M., Marcus, N., & Vassar, A. (2024b). Exploring the cognitive load effects of diverse virtual reality interaction methods in interactive educational platforms. In *2024 IEEE International Symposium on Mixed and Augmented Reality Adjunct (ISMAR-Adjunct)* (pp. 435–438). IEEE.

Davaris, M., Wijewickrema, S., Zhou, Y., Piromchai, P., Bailey, J., Kennedy, G., & O'Leary, S. (2019). The importance of automated real-time performance feedback in virtual reality temporal bone surgery training. In *Artificial Intelligence in Education: 20th International Conference, AIED 2019, Chicago, IL, USA, June 25–29, 2019, Proceedings, Part I* 20 (pp. 96–109). Springer International Publishing.

Dinur, E. (2021). *The complete guide to photorealism for visual effects, visualization and games*. Routledge.

Doumanis, I., Economou, D., Sim, G. R., & Porter, S. (2019). The impact of multimodal collaborative virtual environments on learning: A gamified online debate. *Computers & Education, 130*, 121–138.

Du, Y. C., Fan, S. C., & Yang, L. C. (2020). The impact of multi-person virtual reality competitive learning on anatomy education: a randomized controlled study. *BMC Medical Education, 20*, 1–10.

Eaves, D. L., Breslin, G., Van Schaik, P., Robinson, E., & Spears, I. R. (2011). The short-term effects of real-time virtual reality feedback on motor learning in dance. *Presence: Teleoperators and Virtual Environments, 20*(1), 62–77.

Fang, K., & Wang, J. (2024). Interactive design with gesture and voice recognition in virtual teaching environments. *IEEE Access*, *12*, 4213–4224.

Fiaz, F., Sajjad, S. M., Iqbal, Z., Yousaf, M., & Muhammad, Z. (2024). Metassi: A framework for personal data protection, enhanced cybersecurity and privacy in metaverse virtual reality platforms. *Future Internet*, *16*(5), 176.

Fırat, M. (2023). Integrating AI applications into learning management systems to enhance e-learning. *Instructional Technology and Lifelong Learning*, *4*(1), 1–14.

Georgiou, Y., Tsivitanidou, O., & Ioannou, A. (2021). Learning experience design with immersive virtual reality in physics education. *Educational Technology Research and Development*, *69*(6), 3051–3080.

Gholizadeh HamlAbadi, K., Laamarti, F., & El Saddik, A. (2024). Meta-review on Brain-Computer Interface (BCI) in the metaverse. *ACM Transactions on Multimedia Computing, Communications and Applications*, *20*(12), 1–42.

Gibbs, J. K., Gillies, M., & Pan, X. (2022). A comparison of the effects of haptic and visual feedback on presence in virtual reality. *International Journal of Human-Computer Studies*, *157*, 102717.

Grande, R., Albusac, J., Vallejo, D., Glez-Morcillo, C., & Castro-Schez, J. J. (2024). Performance evaluation and optimization of 3D models from low-cost 3D scanning technologies for virtual reality and metaverse E-commerce. *Applied Sciences*, *14*(14), 6037.

Halim, N. A. A., Azlan, M. H., Ismail, A. W., Fazli, F. E., Ahmad, M. A., & Aladin, M. Y. F. (2024, September). Edu-metaverse classroom with AI-driven virtual avatar assistant. In *2024 5th International Conference on Smart Electronics and Communication (ICOSEC)* (pp. 1548–1554). IEEE.

Hameed, A., Perkis, A., & Möller, S. (2021, June). Evaluating hand-tracking interaction for performing motor-tasks in vr learning environments. In *2021 13th International Conference on Quality of Multimedia Experience (QoMEX)* (pp. 219–224). IEEE.

Hashemi, M., Darejeh, A., & Cruz, F. (2024, October). A user-centric exploration of axiomatic explainable AI in participatory budgeting. In *Companion of the 2024 on ACM International Joint Conference on Pervasive and Ubiquitous Computing* (pp. 126–130).

Hashemi, M., Darejeh, A., & Cruz, F. (2025). Understanding user preferences in explainable artificial intelligence: A mapping function proposal. *ACM Transactions on Intelligent Systems and Technology*, *16*(4), 1–37. https://doi.org/10.1145/3733837

Hernández-Mustieles, M. A., Lima-Carmona, Y. E., Pacheco-Ramírez, M. A., Mendoza-Armenta, A. A., Romero-Gómez, J. E., Cruz-Gómez, C. F., … & Lozoya-Santos, J. D. J. (2024). Wearable biosensor technology in education: A systematic review. *Sensors*, *24*(8), 2437.

Holly, M., Resch, S., & Pirker, J. (2023). An asymmetric multiplayer learning environment for room-scale virtual reality and a handheld device. *Proceedings of the ACM on Human-Computer Interaction*, *7*(MHCI), 1–19.

Hulusic, V., Gusia, L., Luci, N., & Smith, M. (2023). Tangible user interfaces for enhancing user experience of virtual reality cultural heritage applications for utilization in educational environment. *ACM Journal on Computing and Cultural Heritage*, *16*(2), 1–24.

Hwang, Y., & Shin, D. (2018). Visual cues enhance user performance in virtual environments. *Social Behavior and Personality: An International Journal*, *46*(1), 11–24.

Ipsita, A., Kaki, R., Liu, Z., Patel, M., Duan, R., Deshpande, L., … & Ramani, K. (2025). Virtual reality in manufacturing education: A scoping review indicating state-of-the-art, benefits, and challenges across domains, levels, and entities. *arXiv preprint arXiv, 2503,* 18805.

Kasapakis, V., Dzardanova, E., & Agelada, A. (2023). Virtual reality in education: The impact of high-fidelity nonverbal cues on the learning experience. *Computers & Education: X Reality*, *2*, 100020.

Ke, F., Dai, Z., Pachman, M., & Yuan, X. (2021). Exploring multiuser virtual teaching simulation as an alternative learning environment for student instructors. *Instructional Science*, *49*, 831–854.

Kevin, S., Pai, Y. S., & Kunze, K. (2018, November). Virtual gaze: Exploring use of gaze as rich interaction method with virtual agent in interactive virtual reality content. In *Proceedings of the 24th ACM Symposium on Virtual Reality Software and Technology* (pp. 1–2).

Khorasani, S., Syiem, B. V., Nawaz, S., Knibbe, J., & Velloso, E. (2023). Hands-on or hands-off: Deciphering the impact of interactivity on embodied learning in VR. *Computers & Education: X Reality*, *3*, 100037.

Laine, J., Lindqvist, T., Korhonen, T., & Hakkarainen, K. (2022). Systematic review of intelligent tutoring systems for hard skills training in virtual reality environments. *International Journal of Technology in Education and Science*, *6*(2), 178–203.

Lee, Y. S., Rashidi, A., Talei, A., Beh, H. J., & Rashidi, S. (2023, February). A comparison study on the learning effectiveness of construction training scenarios in a virtual reality environment. In *Virtual Worlds* (Vol. 2, No. 1, pp. 36–52). MDPI.

Lohyna, D., & Pico Oristrell, A. (2023). *Towards bias-free AI-supported decision-making: Exploring organisational and technology characteristics for algorithmic bias mitigation* [Master's thesis, Lund University]. Lund University Publications. https://lup.lub.lu.se/student-papers/record/9120214

Lowell, V. L., & Tagare, D. (2023). Authentic learning and fidelity in virtual reality learning experiences for self-efficacy and transfer. *Computers & Education: X Reality*, *2*, 100017.

Marougkas, A., Troussas, C., Krouska, A., & Sgouropoulou, C. (2021). A framework for personalized fully immersive virtual reality learning environments with gamified design in education. In C. Frasson, K. Kabassi, & A. Voulodimos (Eds.), *Novelties in Intelligent Digital Systems* (pp. 95–104). IOS Press.

Mladenović, J., Mattout, J., & Lotte, F. (2018). A generic framework for adaptive EEG-based BCI training and operation. In C. S. Nam, A. Nijholt, & F. Lotte (Eds.), *Brain–Computer Interfaces Handbook: Technological and Theoretical Advances* (pp. 595–612). CRC Press.

Mowla, S., & Kolekar, S. (2020). Development and integration of E-learning services using rest APIs. *International Journal of Emerging Technologies in Learning (iJET)*, *15*(4), 53–72.

Naufel, S., & Klein, E. (2020). Brain–Computer interface (BCI) researcher perspectives on neural data ownership and privacy. *Journal of Neural Engineering*, *17*(1), 016039.

Nilsson, S., Gustafsson, T., & Carleberg, P. (2009). Hands free interaction with virtual information in a real environment: Eye gaze as an interaction tool in an augmented reality system. *PsychNology Journal*, *7*(2), 175–196.

Nofal, A. B., Ali, H., Hadi, M., Ahmad, A., Qayyum, A., Johri, A., … & Qadir, J. (2025). AI-enhanced interview simulation in the metaverse: Transforming professional skills training through VR and generative conversational AI. *Computers and Education: Artificial Intelligence*, *8*, 100347.

Osadchyi, V. V., Chemerys, H. Y., Osadcha, K. P., Kruhlyk, V. S., & Koniukhov, S. L. (2020, November). *Conceptual model of learning based on the combined capabilities of augmented and virtual reality technologies with adaptive learning systems*. CEUR Workshop Proceedings.

Pardeshi, A. S., & Karbhari, V. B. (2019). Recent Trends in VFX (Virtual Effects) and SFX (Special Effects). *International Journal of Engineering Research and Technology*, *8*(07), 882–884.

Parlar, T. (2023). Data privacy and security in the metaverse. In F. S. Esen, H. Tinmaz, & M. Singh (Eds.), *Metaverse: Technologies, Opportunities and Threats* (pp. 123–133). Springer Nature Singapore.

Polap, D. (2018, September). Voice control in mixed reality. In *2018 Federated Conference on Computer Science and Information Systems (FedCSIS)* (pp. 497–500). IEEE.

Premchand, B., Liang, L., Phua, K. S., Zhang, Z., Wang, C., Guo, L., ... & Ang, K. K. (2024). Wearable EEG-based brain–computer interface for stress monitoring. *NeuroScience*, *5*(4), 407–428.

Pretolesi, D., Zechner, O., Schrom-Feiertag, H., & Tscheligi, M. (2024, October). Can I trust You? exploring the impact of misleading AI suggestions on user's trust. In *2024 IEEE International Conference on Metrology for eXtended Reality, Artificial Intelligence and Neural Engineering (MetroXRAINE)* (pp. 1230–1235). IEEE.

Ros, M., Debien, B., Cyteval, C., Molinari, N., Gatto, F., & Lonjon, N. (2020). Applying an immersive tutorial in virtual reality to learning a new technique. *Neurochirurgie*, *66*(4), 212–218.

Saparamadu, P. V. I. N., Sepasgozar, S., Guruge, R. N. D., Jayasena, H. S., Darejeh, A., Ebrahimzadeh, S. M., & Eranga, B. A. I. (2025). Optimising contract interpretations with large language models: A comparative evaluation of a vector database-powered Chatbot vs. ChatGPT. *Buildings*, *15*(7), 1144.

Sargazi Moghadam, T., Darejeh, A., Delaramifar, M., & Mashayekh, S. (2024). Toward an artificial intelligence-based decision framework for developing adaptive e-learning systems to impact learners' emotions. *Interactive Learning Environments*, *32*(7), 3665–3685.

Shirazi, B. N., Safavi, A. A., Aftabi, E., & Salimi, G. (2024, February). The integration of virtual reality and artificial intelligence in Educational Paradigms. In *2024 11th International and the 17th National Conference on E-Learning and E-Teaching (ICeLeT)* (pp. 1–6). IEEE.

Takabi, H., Bhalotiya, A., & Alohaly, M. (2016, November). Brain computer interface (BCI) applications: Privacy threats and countermeasures. In *2016 IEEE 2nd International Conference on Collaboration and Internet Computing (CIC)* (pp. 102–111). IEEE.

Van Stuijvenberg, O. C., Samlal, D. P., Vansteensel, M. J., Broekman, M. L. D., & Jongsma, K. R. (2024). The ethical significance of user-control in AI-driven speech-BCIs: A narrative review. *Frontiers in Human Neuroscience*, *18*, 1420334.

Varney, P., May, D., & Richert, A. (2024). Exploring the impact of haptic feedback on learning in virtual and augmented reality environments: A (comprehensive) literature review. *INTED2024 Proceedings*, (Vol. 1, pp. 4385–4389). IATED Academy. https://doi.org/10.21125/inted.2024.1130

Vats, S., & Joshi, R. (2023, December). The impact of virtual reality in education: A comprehensive research study. In *International Working Conference on Transfer and Diffusion of IT* (pp. 126–136). Springer Nature Switzerland.

Wang, W. S., Pedaste, M., & Huang, Y. M. (2024, July). The impact of feedback mechanism in VR learning environment. In *International Conference on Innovative Technologies and Learning* (pp. 134–142). Springer Nature Switzerland.

Wijewickrema, S., Zhou, Y., Bailey, J., Kennedy, G., & O'Leary, S. (2016, November). Provision of automated step-by-step procedural guidance in virtual reality surgery simulation. In *Proceedings of the 22nd ACM Conference on Virtual Reality Software and Technology* (pp. 69–72).

Woodworth, J. W., Lipari, N. G., & Borst, C. W. (2019, March). Evaluating teacher avatar appearances in educational VR. In *2019 IEEE conference on virtual reality and 3D user interfaces (VR)* (pp. 1235–1236). IEEE.

Zhao, R., Zhang, Y., Zhu, Y., Lan, R., & Hua, Z. (2023). Metaverse: Security and privacy concerns. *Journal of Metaverse*, *3*(2), 93–99.

Index

For Product Safety Concerns and Information please contact our EU representative GPSR@taylorandfrancis.com Taylor & Francis Verlag GmbH, Kaufingerstraße 24, 80331 München, Germany